A Guide to Business Mathematics

The success of business today is dependent on the knowledge and expertise of its employees. The need for mathematics arises naturally in business such as in the work of the actuary in an insurance company, the financial mathematics required in the day-to-day work of the banker and the need to analyse data to extract useful information to enable the business to make the right decisions to be successful.

A Guide to Business Mathematics provides a valuable self-study guide to business practitioners, business students and the general reader to enable them to gain an appropriate insight into the mathematics used in business. This book offers an accessible introduction to essential mathematics for the business field.

A wide selection of topics is discussed with the mathematical material presented in a reader-friendly way. The business context motivates the presentation. The author uses modelling and applications to motivate the material, demonstrating how mathematics is used in the financial sector. In addition to the role of the actuary and the banker, the book covers operations research including game theory, trade discounts and the fundamentals of statistics and probability.

The book is also a guide to using metrics to manage and measure performance, and business economics. Foundations on algebra, number theory, sequences and series, matrix theory and calculus are included as is a complete chapter on using software.

Features

- Discusses simple interest and its application to promissory notes/ treasury bills.
- Discusses compound interest with applications to present and future values.
- Introduces the banking field including loans, annuities and the spot/forward FX market.
- Discusses trade discounts and markups/markdowns.
- Introduces the insurance field and the role of the actuary.

- Introduces the fields of data analytics and operations research.
- Discusses business metrics and problem solving.
- Introduces matrices and their applications.
- Discusses calculus and its applications.
- Discusses basic financial statements such as balance sheet, profit and loss and cash account.
- Reviews a selection of software to support business mathematics.

This broad-ranging text gives the reader a flavour of the applications of mathematics to the business field and stimulates further study in the subject. As such, it will be of great benefit to business students, while also capturing the interest of the more casual reader.

A Guide to Business Mathematics

Gerard O'Regan
University of Central Asia, Kyrgyzstan

CRC Press is an imprint of the
Taylor & Francis Group, an **informa** business
A CHAPMAN & HALL BOOK

First edition published 2023
by CRC Press
6000 Broken Sound Parkway NW, Suite 300, Boca Raton, FL 33487-2742

and by CRC Press
2 Park Square, Milton Park, Abingdon, Oxon, OX14 4RN

CRC Press is an imprint of Taylor & Francis Group, LLC

Library of Congress Cataloging-in-Publication Data
Names: O'Regan, Gerard, author.
Title: A guide to business mathematics / Gerard O'Regan.
Description: 1 Edition. | Boca Raton, FL : Chapman & Hall, CRC Press, 2022. | Includes bibliographical references and index.
Identifiers: LCCN 2022008895 (print) | LCCN 2022008896 (ebook) | ISBN 9781032311197 (hardback) | ISBN 9781032311166 (paperback) | ISBN 9781003308140 (ebook)
Subjects: LCSH: Business mathematics.
Classification: LCC HF5691 .O684 2022 (print) | LCC HF5691 (ebook) | DDC 650.01/51--dc23/eng/20220331
LC record available at https://lccn.loc.gov/2022008895
LC ebook record available at https://lccn.loc.gov/2022008896

ISBN: 978-1-032-31119-7 (hbk)
ISBN: 978-1-032-31116-6 (pbk)
ISBN: 978-1-003-30814-0 (ebk)

DOI: 10.1201/9781003308140

Typeset in Palatino
by MPS Limited, Dehradun

To
My friend
Maura Kenny Crowley

Contents

Preface

Overview

The objective of this book is to give the reader an insight into the application of mathematics to the business field and to act as a self-study guide for business students, business practitioners as well as the general reader. The goal is to show how mathematics is applied to business, rather than the study of mathematics for its own sake.

Organization and Features

The early chapters focus on the basic mathematical foundations with chapters on number theory, algebra, set theory, and sequences and series. The first chapter introduces number theory, and we discuss prime number theory including the greatest common divisor and least common multiple of two numbers. We discuss the various numbers systems such as the natural numbers, the integers and rational and real numbers. We show how fractions may be added or multiplied together, and we discuss ratios, proportions and percentages.

Chapter 2 introduces algebra, which uses letters to stand for variables or constants in mathematical expressions. Algebra is the study of such mathematical symbols and the rules for manipulating them, and it is a powerful tool for problem solving. We discuss simple and simultaneous equations and methods to solve them, including the method of elimination, the method of substitution and graphical techniques. We show how quadratic equations may be solved by factorization, completing the square, using the quadratic formula or graphical techniques. We present the laws of logarithms and indices, and discuss exponentials and natural logarithms.

Chapter 3 discusses sets, relations and functions. A set is a collection of well-defined objects and it may be finite or infinite. A relation $R(A, B)$ between two sets A and B indicates a relationship between members of the two sets, and it is a subset of the Cartesian product of the two sets. A function is a special type of relation such that for each element in A there is at most one element in the co-domain B. Functions may be partial or total, and injective, surjective or bijective.

Chapter 4 gives an introduction to sequences and series, and we discuss arithmetic sequences and series (where two successive terms differ by a constant d), and geometric sequences and series (where two successive terms differ by a ratio r to the previous element). We discuss the counting principle and the pigeon hole principle and permutations and combinations.

Chapter 5 discusses simple interest and applications, where simple interest is earned on the principal only, and the amount earned is determined from the principal invested, the rate of interest and the period of time that the principal is invested for. We discuss future and present values, where the future value is the amount that the principal will grow to at a given rate of simple interest over the period of time, whereas the present value of an amount to be received at a given date in the future is the principal that will grow to that amount at the given rate of simple interest over that period of time. We discuss short-term promissory notes that are a written promise by one party to pay a certain sum of money (with or without interest) on a particular date to another party. A promissory note may be interest bearing (where the rate of interest is stated in the note), or non-interest-bearing (where there is no rate of interest specified) and these are termed treasury bills.

Chapter 6 discusses compound interest, and while simple interest is calculated on the principal only, compound interest is more complicated in that interest is also applied to the interest earned in previous compounding periods. Compound interest is generally employed for long-term investments and loans, and we show how present and future values may be determined. The concept of the time value of money is that the earlier that cash is received the greater its value to the recipient. Similarly, the later that a cash payment is made, the lower its cost to the payer and the lower its value to the payee.

Chapter 7 discusses banking and financial services, and we discuss the roles of the various types of banks including the central bank, commercial banks and retail banks. We discuss how a mortgage or loan is paid back with an annuity. We discuss the foreign exchange market including spot and forward trading, and we discuss corporate bonds where a bond is a debt obligation of a corporation to an investor, and investors who buy corporate bonds are, in effect, lending money to the company that is issuing the bond. The investor receives regular interest payments during the lifetime of the bond, and the investor is repaid the capital in full on the maturity date of the bond.

Chapter 8 discusses trade discounts that are a reduction in the list price of a manufactured product, and the discount is generally stated as a percentage of the list price. Trade discounts are used by manufacturers, distributors and wholesalers as pricing tools for their products, and in communicating changes of prices to their customers. A cash discount may be given to encourage early or prompt payment of an invoice, and the rate of discount and the discount period for when the discount may be applied are specified on the invoice.

Chapter 9 discusses statistics and discusses sample spaces, sampling, the abuse of statistics, frequency distributions and various charts such as bar charts, histograms, pie charts and trend graphs. Various statistical measures such as the average of a sample including the arithmetic mean, mode and median are discussed, as well as the variance and standard deviation of a sample. We discuss correlation and regression, and hypothesis testing.

Chapter 10 discusses probability including a discussion on discrete and continuous random variables, and probability distributions such as the Binomial and Poisson distributions. We discuss the normal and unit normal distributions and confidence intervals. We discuss Bayesianism and how this helps in updating probabilities in the light of new information.

Chapter 11 discusses the fundamentals of the insurance field including the basic mathematics underlying motor and health insurance. The concept of a life annuity is discussed as well as the basic mathematics underlying pensions. We discuss the role of the actuary in the insurance field.

Chapter 12 discusses the important field of data science and data analytics, where data science is a multi-disciplinary field that extracts knowledge from data sets that consist of structured and unstructured data, and large data sets may be analysed to extract useful information.

Chapter 13 is concerned with metrics and problem solving, and this includes a discussion of the balanced scorecard which assists in identifying appropriate metrics for an organization. The Goal, Question, Metrics (GQM) approach is discussed, and this allows appropriate metrics related to the organization goals to be defined. A selection of sample metrics for an organization is presented, and problem-solving tools such as fishbone diagrams, Pareto charts and trend charts are discussed.

Chapter 14 discusses matrices including 2×2 and general $n \times m$ matrices. Various operations such as the addition and multiplication of matrices are considered, and the determinant and inverse of a matrix are discussed. The application of matrices to solving a set of linear equations using Gaussian elimination is considered, and we show how the inverse of a matrix is determined.

Chapter 15 discusses operations research, and we discuss linear programming, cost volume profit analysis (CVPA) and game theory. Linear programming is a mathematical model for determining the best possible outcome of a particular problem, where the problem is subject to various constraints that are expressed as a set of linear equations and linear inequalities. CVPA is used to analyse the relationship between the costs, volume, revenue and profitability of the products produced. Game theory is the study of mathematical models of strategic interaction among rational decision-makers.

Chapter 16 discusses basic financial statements including the balance sheet, the profit and loss account and cash management. The balance sheet is a snapshot of the net worth of a company and it is a summary of everything that is owned by the company less everything that it owes. The profit and loss account is the earnings statement of the company during the period.

Chapter 17 provides a short introduction to calculus, and provides an overview of limits, continuity, differentiation and integration. Chapter 18 presents applications of calculus, and includes a brief discussion of Fourier series, Laplace transforms and differential equations.

Chapter 19 provides a short introduction to economics, including a short discussion of macroeconomics and microeconomics. The important economic concepts of gross domestic product, inflation and employment are discussed, as well as the theoretical concepts of utility and elasticity.

Chapter 20 reviews a selection of software to support business mathematics, and includes a discussion of software such as Microsoft Excel, Python, Maple, Mathematica, Matlab, Minitab, R, Sage, SPSS, SQL and SAS.

Chapter 21 is the concluding chapter that summarizes the journey that we have travelled in this book.

Audience

This book aims to provide a useful self-study guide to business practitioners, business students and the general reader to enable them to gain an insight into the mathematics used in business. The audience includes business students who wish to obtain an overview of the mathematics used in the business field, and mathematicians who are curious as to how mathematics is applied in business.

Acknowledgements

I am deeply indebted to family and friends who supported my efforts in this endeavour. I would like to thank Maura Crowley for her friendship over many years, and I have been amazed at her strength in dealing with the loss of her husband (and my close friend) Kevin, as well as dealing with the loss of her close friend Lorraine, and her mother, and yet managing to find the strength to keep everything together while being a full-time mother with four young children (superwoman). I would especially like to thank her for her support during the early part of the pandemic on my return from Nepal, and this book is dedicated to her to express my deep appreciation.

I would also like to express my thanks to my editor, Robert Ross, who guided the publication of this book. Bob asked the tough questions, and his comments and suggestions helped to improve this book. I am, of course, responsible for all remaining errors and omissions, and although I was used to a different style of editor (this is the first time that Bob and I have worked together) I recognize the value of his contribution.

Gerard O'Regan
Cork, Ireland

About the Author

Dr. Gerard O'Regan is an Assistant Professor in Mathematics at the University of Central Asia in Kyrgyzstan. His research interests include software quality and software process improvement, mathematical approaches to software quality and the history of computing. He is the author of several books in the Mathematics and Computing fields.

1

Introduction to Number Theory

Arithmetic (or number theory) is the branch of mathematics that is concerned with the study of numbers and their properties. It includes the study of the integer numbers, and operations on them, such as addition, subtraction, multiplication and division.

Number theory studies various properties of integer numbers such as their parity and divisibility; their additive and multiplicative properties; whether a number is prime or composite; the prime factors of a number; the greatest common divisor of two numbers; the least common multiple of two numbers; and so on.

The natural numbers $\mathbb{N}$ consist of the numbers {1, 2, 3, …}. The integer numbers $\mathbb{Z}$ are a superset of the set of natural numbers, and they consist of {…, −2, −1, 0, 1, 2, …}. The rational numbers $\mathbb{Q}$ are a superset of the set of integer numbers, and they consist of all numbers of the form $\{{}^{p}/_{q}$ where p and q are integers and $q \neq 0\}$. The real numbers $\mathbb{R}$ are a superset of the set of rational numbers, and they are defined to be the set of converging sequences of rational numbers. They contain the rational and irrational numbers. The complex numbers $\mathbb{C}$ consist of all numbers of the form $\{a + bi$ where $a, b \in \mathbb{R}$ and $i = \sqrt{-1}\}$, and they are a superset of the set of real numbers.

Number theory has many applications including cryptography and coding theory in computing. For example, the RSA public key cryptographic system relies on its security due to the infeasibility of the integer factorization problem for large numbers.

There are several unsolved problems in number theory and especially in prime number theory. For example, Goldbach's[1] Conjecture states that every even integer greater than two is the sum of two primes, and this result has not been proved to date. Fermat's[2] Last Theorem (Fig. 1.1) states that there is no integer solution to $x^n + y^n = z^n$ for $n > 2$, and this result remained unproved for over 300 years until Andrew Wiles provided a proof in the mid-1990s. There is more detailed information on number theory in [Yan:98].

1.1 Basic Number Theory

The natural numbers $\mathbb{N}$ {1, 2, 3, …} are used for counting things starting from the number 1, and they form an ordered set $1 < 2 < 3 < \ldots$ and so on. The natural numbers are all positive (i.e., there are no negative natural numbers) and the

DOI: 10.1201/9781003308140-1

FIGURE 1.1
Pierre de Fermat.

number 0 is not usually included as a natural number. However, the set of natural numbers including 0 is denoted by $\mathbb{N}_0$, and it is the set {0, 1, 2, 3, …}.

The natural numbers are an ordered set and so given any pair of natural numbers (n, m) then either $n < m$, $n = m$ or $n > m$. There is no largest natural number as such since given any natural number we can immediately determine a larger natural number (e.g., its successor). Each natural number has a unique successor and every natural number larger than 1 has a unique predecessor.

The addition of two numbers yields a new number, and the subtraction of a smaller number from a larger number yields the difference between them. Multiplication is the mathematical operation of scaling one number by another: for example, $3 * 4 = 4 + 4 + 4 = 12$.

Peano's axiomatization of arithmetic is a formal axiomatization of the natural numbers, and they include axioms for the successor of a natural number and the axiom of induction. The number 0 is considered to be a natural number in the Peano system.

The natural numbers satisfy several nice algebraic properties such as closure under addition and multiplication (i.e., a natural number results from the addition or multiplication of two natural numbers); commutability of addition and multiplication: i.e., $a + b = b + a$ and $a \times b = b \times a$; addition and multiplication are associative: $a + (b + c) = (a + b) + c$ and $a \times (b \times c) = (a \times b) \times c$. Further, multiplication is distributive over addition: $a \times (b + c) = a \times b + a \times c$.

A square number is an integer that is the square of another integer. For example, the number 4 is a square number since $4 = 2^2$. Similarly, the number 9 and the number 16 are square numbers. A number n is a square number if and only if one can arrange the n points in a square. For example, the square numbers 4, 9 and 16 are represented in squares in Fig. 1.2.

The square of an odd number is odd whereas the square of an even number is even. This is clear since an even number is of the form $n = 2k$ for some k, and so $n^2 = 4k^2$ which is even. Similarly, an odd number is of the form $n = 2k + 1$ and so $n^2 = 4k^2 + 4k + 1$ which is odd.

A rectangular number n may be represented by a vertical and horizontal rectangle of n points. For example, the number 6 may be represented by a rectangle with length 3 and breadth 2, or a rectangle with length 2 and breadth 3. Similarly, the number 12 can be represented by a 4×3 or a 3×4 rectangle (Fig. 1.3).

A triangular number n may be represented by an equilateral triangle of n points. It is the sum of k natural numbers from 1 to k (Fig. 1.4).

That is, $n = 1 + 2 + \ldots + k$.

FIGURE 1.2
Square Numbers.

FIGURE 1.3
Rectangular Numbers.

FIGURE 1.4
Triangular Numbers.

Parity of Integers

The parity of an integer refers to whether the integer is odd or even. An integer n is odd if there is a remainder of 1 when it is divided by 2 (i.e., it is of the form $n = 2k + 1$). Otherwise, the number is even and of the form $n = 2k$.

The sum of two numbers is even if both are even or both are odd. The product of two numbers is even if at least one of the numbers is even. These properties are expressed as:

PROPERTIES OF PARITY

even ± even = even
even ± odd = odd
odd ± odd = even
even × even = even
even × odd = even
odd × odd = odd

Let a and b be integers with $a \neq 0$ then a is said to be a divisor of b (denoted by $a \mid b$) if there exists an integer k such that $b = ka$.

A divisor of n is called a *trivial divisor* if it is either 1 or n itself; otherwise, it is called a *non-trivial divisor*. A *proper divisor* of n is a divisor of n other than n itself.

Properties of Divisors

i. $a \mid b$ and $a \mid c$ then $a \mid b + c$
ii. $a \mid b$ then $a \mid bc$
iii. $a \mid b$ and $b \mid c$ then $a \mid c$

Proof (i)

Suppose $a \mid b$ and $a \mid c$ then $b = k_1a$ and $c = k_2a$.

Then $b + c = k_1a + k_2a = (k_1+k_2)a$ and so $a \mid b + c$.

Proof (iii)

Suppose $a \mid b$ and $b \mid c$ then $b = k_1a$ and $c = k_2b$.

Then $c = k_2b = (k_2k_1)\, a$ and thus $a \mid c$.

A *prime number* is a natural number (>1) whose only divisors are trivial. There are an infinite number of prime numbers.

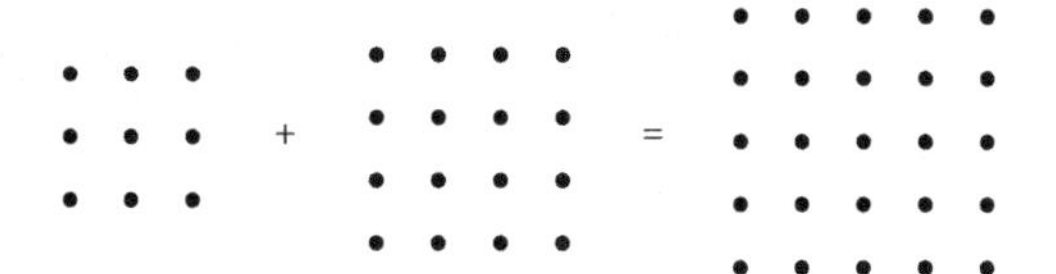

FIGURE 1.5
Pythagorean Triples.

The *fundamental theorem of arithmetic* states that every integer number can be factored as the product of prime numbers.

Pythagorean triples (Fig. 1.5) are combinations of three whole numbers that satisfy Pythagoras's equation $x^2 + y^2 = z^2$. There are an infinite number of such triples, and 3, 4 and 5 are an example since $3^2 + 4^2 = 5^2$.

Theorem 1.1: (Division Algorithm)

For any integer a and any positive integer b there exists unique integers q and r such that:

$$a = bq + r \quad 0 \leq r < b$$

Proof

The first part of the proof is to show the existence of integers q and r such that the equality holds, and the second part of the proof is to prove uniqueness of q and r. The details are in [ORg:20].

Theorem 1.2: (Irrationality of Square Root of 2)

The square root of 2 is an irrational number (i.e., it cannot be expressed as the quotient of two integer numbers).

Proof

The Pythagoreans[3] discovered this result and it led to a crisis in their community as number was considered to be the essence of everything in their world. The proof is indirect: i.e., the opposite of the desired result is assumed to be correct (i.e., that the square root of 2 is rational), and it is showed that this assumption leads to a contradiction. Therefore, the assumption that the square root of 2 is rational must be incorrect and so the result is proved.

Suppose $\sqrt{2}$ is rational then it can be put in the form $^p/_q$ where p and q are integers and $q \neq 0$. Therefore, we can choose p and q to be co-prime (i.e., without any common factors) and so

$$
\begin{aligned}
&(p/q)^2 = 2 \\
&\Rightarrow p^2/q^2 = 2 \\
&\Rightarrow p^2 = 2q^2 \\
&\Rightarrow 2 \mid p^2 \\
&\Rightarrow 2 \mid p \\
&\Rightarrow p = 2k \\
&\Rightarrow p^2 = 4k^2 \\
&\Rightarrow 4k^2 = 2q^2 \\
&\Rightarrow 2k^2 = q^2 \\
&\Rightarrow 2 \mid q^2 \\
&\Rightarrow 2 \mid q
\end{aligned}
$$

This is a contradiction as we have chosen p and q to be co-prime, and our assumption that there is a rational number that is the square root of 2 results in a contradiction. Therefore, this assumption must be false and we conclude that there is no rational number whose square is 2.

1.2 Fractions and Decimals

A simple fraction is of the form ${}^{a}/_{b}$ where a and b are integers, with the number a above the bar termed the *numerator* and the number b below the bar termed the *denominator*. Each fraction may be converted to a decimal representation by dividing the numerator by the denominator, and the decimal representation may be to an agreed number of decimal places (as the decimal representation may not terminate). For example, Fig. 1.6 describes how to convert the fraction ${}^{3}/_{8}$ to its decimal form 0.375 by long division.

The reverse operation of converting a decimal number to a fraction is straightforward, and involves determining the number of decimal places (n) of the number, and multiplying the number by the fraction $10^n/10^n$. The resulting fraction is then simplified.

For example, the conversion of the decimal number 0.25 to a fraction involves noting that we have two decimal places and so we multiply the decimal number 0.25 by $10^2/10^2$ (i.e., 100/100). This results in the fraction 25/100 which is then simplified to ${}^{1}/_{4}$.

The addition of two fractions with the same denominator is straightforward as all that is involved is adding the numerators of both fractions together and simplifying. For example, ${}^{1}/_{12} + {}^{5}/_{12} = {}^{(1+5)}/_{12} = {}^{6}/_{12} = {}^{1}/_{2}$.

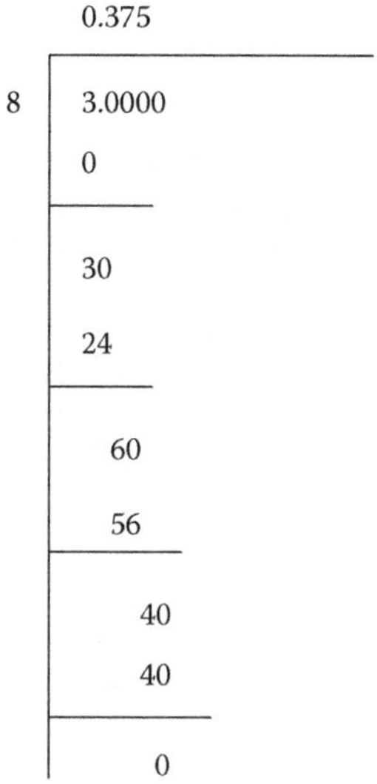

```
     0.375
   ________
8 | 3.0000
  | 0
  |____
  | 30
  | 24
  |____
  |  60
  |  56
  |_____
  |   40
  |   40
  |______
  |    0
```

FIGURE 1.6
Decimal Representation of Fraction.

The addition of fractions with different denominators is more difficult. One way to do this is to multiply the denominators of both fractions together to form a common denominator and then simplify. That is,

$$\frac{a}{m} + \frac{b}{n} = \frac{na + mb}{mn}$$

For example, $^1/_2 + ^1/_3 = {}^{(3.1\,+\,2.1)}/_{3.2} = {}^{(3\,+\,2)}/_6 = {}^5/_6$

However, the usual approach when adding two fractions is to determine the least common multiple of both denominators, and then to convert each fraction into the equivalent fraction with the common LCM denominator, and then to add both numerators together and simplify. For example, consider

$$\frac{3}{4} + \frac{5}{6} =$$

First, the LCM of 4 and 6 is determined (section 1.4) and the LCM (4, 6) is the smallest multiple of both 4 and 6 and this is 12. We then convert both fractions into the equivalent fractions under the common LCM (that is, we multiply the first fraction $^3/_4$ by $^3/_3$ and the second fraction $^5/_6$ by $^2/_2$ and this yields:

$$\frac{3}{4} + \frac{5}{6} = \frac{3.3}{12} + \frac{5.2}{12} = \frac{9 + 10}{12} = \frac{19}{12}$$

The multiplication of two numbers involves multiplying the numerators together and the denominators together and then simplifying:

$$\frac{a}{m} \times \frac{b}{n} = \frac{ab}{mn}$$

The division of one fraction by another involves inverting the divisor and multiplying and simplifying. That is,

$$\frac{a}{m} \div \frac{b}{n} = \frac{a}{m} \times \frac{n}{b} = \frac{an}{mb}$$

1.3 Prime Number Theory

A positive integer $n > 1$ is called prime if its only divisors are n and 1. A number that is not a prime is called composite.

Properties of Prime Numbers

i. There are an infinite number of primes.
ii. There is a prime number p between n and $n! + 1$ such that $n < p \leq n! + 1$.
iii. If n is composite then n has a prime divisor p such that $p \leq \sqrt{n}$.
iv. There are arbitrarily large gaps in the series of primes (given any $k > 0$ there exist k consecutive composite integers).

Proof (i)

Suppose there are a finite number of primes and they are listed as $p_1, p_2, p_3, \ldots, p_k$. Then consider the number N obtained by multiplying all known primes and adding 1. That is,

$$N = p_1 p_2 p_3 \cdots p_k + 1.$$

Clearly, N is not divisible by any of $p_1, p_2, p_3, \ldots, p_k$ since they all leave a remainder of 1. Therefore, N is either a new prime or divisible by a prime q (that is not in the list of $p_1, p_2, p_3, \ldots, p_k$).

This is a contradiction since this was the list of all the prime numbers, and so the assumption that there are a finite number of primes is false, and we deduce that there are an infinite number of primes. For a proof of the other properties, see [ORg:20].

Algorithm for Determining Primes

The *Sieve of Eratosthenes algorithm* was developed by the Hellenistic mathematician, Eratosthenes, and is used for determining the prime numbers up to a given number n.

The algorithm involves first listing all of the numbers from 2 to n. The first step is to remove all multiples of 2 up to n; the second step is to remove all multiples of 3 up to n; and so on.

The kth step involves removing multiples of the kth prime p_k up to n and the steps in the algorithm continue while $p \leq \sqrt{n}$. The numbers remaining in the list are the prime numbers from 2 to n. That is, the steps are:

1. List the integers from 2 to n.
2. For each prime p_k up to $\sqrt{n}$ remove all multiples of p_k.
3. The numbers remaining are the prime numbers between 2 and n.

The prime numbers between 1 and 50 are:

2, 3, 5, 7, 11, 13, 17, 19, 23, 29, 31, 37, 41, 43 and 47

Theorem 1.3: (Fundamental Theorem of Arithmetic)

Every natural number $n > 1$ may be written uniquely as the product of primes:

$$n = p_1^{\alpha_1} p_2^{\alpha_2} p_3^{\alpha_3} \ldots p_k^{\alpha_k}$$

Proof

There are two parts to the proof. The first part shows that there is a factorization and the second part shows that the factorization is unique. The details are in [ORg:20].

1.3.1 Greatest Common Divisors (GCD)

Let a and b be integers (not both 0) then the *greatest common divisor* d of a and b is a divisor of a and b (i.e., $d \mid a$ and $d \mid b$), and it is the largest such divisor (i.e., if $k \mid a$ and $k \mid b$ then $k \mid d$). It is denoted by gcd (a, b).

PROPERTIES OF GREATEST COMMON DIVISOR

i. Let a and b be integers (not both 0) then there exist integers x and y such that:

$$d = \gcd(a, b) = ax + by$$

ii. Let a and b be integers (not both 0) then the set $S = \{ax + by$ where $x, y \in \mathbb{Z}\}$ is the set of all multiples of $d = \gcd(a, b)$.

Proof (i)

Consider the set of all linear combinations of a and b, i.e., the set $\{ka + nb: k, n \in \mathbb{Z}\}$. We choose x and y such that $m = ax + by$ is the smallest positive integer in the set, and we show that m is the greatest common divisor [ORg:20].

Proof (ii)

This follows since $d \mid a$ and $d \mid b \Rightarrow d \mid ax + by$ for all integers x and y and so every element in the set $S = \{ax + by \text{ where } x, y \in \mathbb{Z}\}$ is a multiple of d.

Relatively Prime

Two integers a and b are relatively prime if gcd $(a, b) = 1$.

Properties

If p is a prime and $p \mid ab$ then $p \mid a$ or $p \mid b$.

Proof

Suppose $p \nmid a$ then from the results on the greatest common divisor we have gcd $(a, p) = 1$. That is,

$$\begin{aligned} & ra + sp = 1 \\ \Rightarrow\ & rab + spb = b \\ \Rightarrow\ & p \mid b (since\ \ p \mid rab \text{ and } p \mid spb \text{ and } so\ \ p \mid rab + spb) \end{aligned}$$

1.3.2 Euclid's Algorithm for GCD

Euclid's[4] algorithm is one of the oldest known algorithms, and it provides a procedure for finding the greatest common divisor of two numbers. It is described in Book VII of *Euclid's Elements* [Hea:56].

Lemma

Let a, b, q and r be integers with $b > 0$ and $0 \le r < b$ such that $a = bq + r$. Then gcd (a, b) = gcd (b, r).

Proof

Let K = gcd (a, b) and let L = gcd (b, r) and so we need to show that $K = L$. Suppose m is a divisor of a and b then as $a = bq + r$ we have m is a divisor of r and so any common divisor of a and b is a divisor of r.

Similarly, any common divisor n of b and r is a divisor of a, and so the set of common divisors of a and b is equal to the set of common divisors of b

and r. Therefore, the greatest common divisor of a and b is equal to the greatest common divisor of b and r.

Theorem 1.4: (Euclid's Algorithm)

Euclid's algorithm for finding the greatest common divisor of two positive integers a and b involves applying the division algorithm repeatedly as follows:

EUCLID'S ALGORITHM

$$\begin{array}{ll}
a = bq_0 + r_1 & 0 < r_1 < b \\
b = r_1q_1 + r_2 & 0 < r_2 < r_1 \\
r_1 = r_2q_2 + r_3 & 0 < r_3 < r_2 \\
\ldots\ldots\ldots\ldots & \\
\ldots\ldots\ldots\ldots & \\
r_{n-2} = r_{n-1}q_{n-1} + r_n & 0 < r_n < r_{n-1} \\
r_{n-1} = r_nq_n &
\end{array}$$

Then r_n is the greatest common divisor of a and b, i.e., gcd $(a,b) = r_n$.

Lemma

Let n be a positive integer greater than 1 then the positive divisors of n are precisely those integers of the form:

$$d = p_1^{\beta_1}p_2^{\beta_2}p_3^{\beta_3}\ldots p_k^{\beta_k} \quad \text{(where } 0 \le \beta_i \le \alpha_i\text{)}$$

where the unique factorization of n is given by

$$n = p_1^{\alpha_1}p_2^{\alpha_2}p_3^{\alpha_3}\ldots p_k^{\alpha_k}$$

Proof

Suppose d is a divisor of n then $n = dq$. By the unique factorization theorem (Theorem 1.3), the prime factorization of n is unique, and so the prime numbers in the factorization of d must appear in the prime factors p_1, p_2, p_3, ..., p_k of n.

Clearly, the power β_i of p_i must be less than or equal to α_i, i.e., $\beta_i \le \alpha_i$. Conversely, whenever $\beta_i \le \alpha_i$ then clearly d divides n.

1.3.3 Least Common Multiple (LCM)

LEAST COMMON DIVISOR

If m is a multiple of a and m is a multiple of b then it is said to be a *common multiple* of a and b. The least common multiple is the smallest of the common multiples of a and b and it is denoted by LCM (a, b).

Properties of LCM

If x is a common multiple of a and b then $m \mid x$. That is, every common multiple of a and b is a multiple of the least common multiple m.

Proof

We assume that both a and b are non-zero as otherwise the result is trivial (since all common multiples are 0). Clearly, by the division algorithm, we have:

$$x = mq + r \text{ where } 0 \leq r < m$$

Since x is a common multiple of a and b we have $a \mid x$ and $b \mid x$ and also that $a \mid m$ and $b \mid m$. Therefore, $a \mid r$ and $b \mid r$ and so r is a common multiple of a and b and since m is the least common multiple we have r is 0. Therefore, x is a multiple of the least common multiple m as required.

Example (LCM)

The LCM is calculated by determining the prime factors of each number, and then multiplying each factor by the greatest number of times that it occurs. The procedure may be seen more clearly with the calculation of the LCM of 8 and 12, as $8 = 2^3$ and $12 = 2^2.3$. Therefore, the LCM $(8,12) = 2^3.3 = 24$, since the greatest number of times that the factor 2 occurs is 3 and the greatest number of times that the factor 3 occurs is once.

1.3.4 Distribution of Primes

There are an infinite number of primes, but most integer numbers are composite. How many primes are there less than a certain number? The number of primes less than or equal to x is given by the prime distribution function (denoted by $\pi(x)$):

$$\pi(x) = \sum_{p \le x} 1 \quad \text{(where } p \text{ is prime)}$$

It satisfies the following properties:

PROPERTIES OF PRIME DISTRIBUTION FUNCTION

i. $\lim_{x \to \infty} \frac{\pi(x)}{x} = 0$

ii. $\lim_{x \to \infty} \pi(x) = \infty$

The first property expresses the fact that most integer numbers are composite, and the second property expresses the fact that there are an infinite number of prime numbers.

There is an approximation of the prime distribution function in terms of the logarithmic function ($^x/_{\ln x}$) as follows:

PRIME NUMBER THEOREM

$$\lim_{x \to \infty} \frac{\pi(x)}{x/\ln x} = 1$$

The approximation $x/\ln x$ to $\pi(x)$ gives an easy way to determine the approximate value of $\pi(x)$ for a given value of x. This result is known as the *Prime Number Theorem*, and it was originally conjectured by Gauss.

1.4 Ratios and Proportions

Ratios and proportions are used to solve business problems such as computing inflation, currency exchange and taxation.

A *ratio* is a comparison of the relative values of numbers or quantities where the quantities are expressed in the same units. Business information is often based on a comparison of related quantities stated in the form of a ratio, and a ratio is usually written in the form of *num* 1 to *num* 2 or *num*1:*num*2 (e.g., 3 to 4 or 3:4).

The numbers appearing in a ratio are called the terms of the ratio, and the ratio is generally reduced to the lowest terms, e.g., the term 80:20 would generally be reduced to the ratio 4:1 with the common factor of 20 used to reduce the terms. If the terms contain decimals then the terms are each multiplied by the same number to eliminate the decimals and the term is then simplified.

One application of ratios is to allocate a quantity into parts by a given ratio (i.e., allocating a portion of a whole into parts).

Example

Consider a company that makes a profit of €180,000 which is to be divided between its three partners A, B and C in the ratio 3:4:2 (where the ratio expresses the degree of ownership of the company among the partners). How much does each partner receive?

Solution

The total number of parts is 3 + 4 + 2 = 9. That is, for every nine parts A receives 3, B receives 4 and C receives 2. That is, A receives $^3/_9 = {}^1/_3$ of the profits, B receives $^4/_9$ of the profits and C receives $^2/_9$ of the profits. That is,

$$\text{A receives } 1/3 \times €180{,}000 = €60{,}000$$
$$\text{B receives } 4/9 \times €180{,}000 = €80{,}000$$
$$\text{C receives } 2/9 \times €180,000 = €40{,}000$$

A proportion is two ratios that are equal or equivalent (i.e., they have the same value and the same units). For example, the ratio 3:4 is the same as the ratio 6:8 and so they are the same proportion.

Often, an unknown term arises in a proportion and in such a case the proportions form a linear equation in one variable.

Example

Solve the proportion $2:5 = 8:x$

Solution

$$\frac{2}{5} = \frac{8}{x}$$

Cross-multiplying we get

$$2 \times x = 8 \times 5$$
$$\Rightarrow 2x = 40$$
$$\Rightarrow x = 20$$

Example

A car travels 384 km on 32 L of petrol. How far does it travel on 25 L of petrol?

Solution

We let x represent the unknown distance that is travelled on 25 L of petrol. We then write the two ratios as a proportion, and solve the simple equation to find the unknown value x.

$$\frac{384}{32} = \frac{x}{25}$$

Cross-multiplying we get:

$$25 \times 384 = x \times 32$$
$$\Rightarrow 9600 = 32x$$
$$\Rightarrow x = 300 \text{ km (for } 25\,\text{L of petrol)}$$

1.5 Percentages

Percent means "per hundred" and the symbol % indicates parts per hundred (i.e., a percentage is a fraction where the denominator is 100, which provides an easy way to compare two quantities). A percentage may also be represented as a decimal or as a fraction, and Table 1.1 shows the representation of 25% as a percentage, decimal and fraction.

Percentages are converted to decimals by moving the decimal point two places to the left (e.g., 25% = 0.25). Conversely, the conversion of a decimal to a percentage involves moving the decimal point two places to the right and adding the percentage symbol.

A percentage is converted to a fraction by dividing it by 100 and then simplifying (e.g., 25% = $^{25}/_{100} = {}^{1}/_{4}$). Similarly, a fraction can be converted to a decimal by dividing the numerator by the denominator, and a decimal

TABLE 1.1

Percentage, Decimal and Fraction

Percentage	Decimal	Fraction
25%	0.25	$^{25}/_{100} = {}^{1}/_{4}$

may be converted to a percentage by moving the decimal point two places to the right and adding the percent symbol.

The value of the percentage of a number is calculated by multiplying the rate by the number to yield the new value. For example, 80% of 50 is given by $0.8 \times 50 = 40$. That is, the value of the new number is given by

$$\text{New number} = \text{rate} \times \text{original number}$$

The rate (or percentage) that a new number is with respect to the original number is given by

$$\text{Rate} = \frac{\text{New Number}}{\text{Original Number}} \times 100$$

For example, to determine what percentage of €120 that €15 is, we apply the formula to get

$$\text{Rate} = \frac{15}{120} \times 100 = 12.5\%$$

Suppose that 30% of the original number is 15 and we wish to find the original number. Then we let x represent the original number and we form a simple equation with one unknown:

$$\begin{aligned} &0.3x = 15 \\ &\Rightarrow x = 15/0.3 \\ &\Rightarrow x = 50 \end{aligned}$$

In general, when we are given the rate and the new number we may determine the original number from the following formula:

$$\text{Original Number} = \frac{\text{New Number}}{\text{Rate}}$$

Example

Barbara is doing renovations on her apartment. She has budgeted 25% of the renovation costs for new furniture. If the cost of the new furniture is €2200 determine the total cost of the renovations to her apartment.

Solution

We let x represent the unknown cost of renovation and we form the simple equation with one unknown:

$$
\begin{aligned}
0.25x &= 2200 \\
\Rightarrow x &= €2200/0.25 \\
\Rightarrow x &= €8800
\end{aligned}
$$

Example

(i) What number is 25% greater than 40? (ii) What number is 20% less than 40?

Solution

We let x represent the new number.

For the first case, $x = 40 + 0.25(40) = 40 + 10 = 50$.

For the second case, $x = 40 - 0.2(40) = 40 - 8 = 32$

Often, we will wish to determine the rate of increase or decrease as in the following example.

Example

(i) Determine the percentage that 280 is greater than 200. (ii) Determine the percentage that 170 is less than 200?

Solution

i. The amount of change is $280 - 200 = 80$. The rate of change is therefore ${}^{80}/_{200} * 100 = 40\%$ (a 40% increase).
ii. The amount of change is $170 - 200 = -30$. The rate of change is therefore ${}^{-30}/_{200} * 100 = -15\%$ (a 15% decrease).

Example

Lilly increased her loan payments by 40% and now pays €700 back on her loan. What was her original payment?

Solution

Let the original amount be x and we form a simple equation of one unknown.

$$
\begin{aligned}
x + 0.4x &= 700 \\
1.4x &= 700 \\
x &= 700/1.4 \\
x &= 500
\end{aligned}
$$

For more detailed information on basic arithmetic, see [Bir:05].

1.6 Review Questions

1. Convert the fraction $^7/_8$ to a decimal.
2. Compute $^7/_8 + {}^5/_{12}$
3. Compute $^7/_8 * {}^5/_{12}$
4. Find the prime factorization of 18 and 24.
5. Find the least common multiple of 18 and 24.
6. Find the greatest common divisor of 18 and 24.
7. A company makes a profit of €120,000 which is to be divided between its three partners A, B and C in the ratio 2:7:6. Find the amount that each partner gets.
8. Solve the proportion $2:7 = 4:x$.
9. What number is 15% greater than 140?

1.7 Summary

Arithmetic is the branch of mathematics that is concerned with the study of numbers and their properties. It includes the study of the integer numbers, and operations on them, such as addition, subtraction, multiplication and division.

The natural numbers consist of the numbers $\{1, 2, 3, \ldots\}$. The integer numbers are a superset of the set of natural numbers, and the rational numbers are a superset of the set of integer numbers, which consist of all numbers of the form $\{^p/_q$ where p and q are integers and $q \neq 0\}$. A simple fraction is of the form $^a/_b$ where a and b are integers and $b \neq 0$, with the number a above the bar termed the numerator and the number b below the bar termed the denominator.

A positive integer $n > 1$ is called prime if its only divisors are n and 1, and a number that is not a prime is called composite. Prime numbers are the key building blocks in number theory, and the fundamental theorem of arithmetic states that every number may be written uniquely as the product of factors of prime numbers.

Euclid's algorithm is used for finding the greatest common divisor of two positive integers a and b. The least common multiple of two numbers is the smallest number that can be divided by both numbers.

Ratios and proportions are used to solve business problems where a ratio is a comparison of numbers where the quantities are expressed in the same units. The numbers appearing in a ratio are called the terms of the ratio, and the ratios are generally reduced to the lowest terms. One application of ratios is to allocate a quantity into parts by a given ratio (i.e., allocating a portion of a whole into parts).

Percent means "per hundred" and the symbol % indicates parts per hundred (i.e., a percentage is a fraction where the denominator is 100 which provides an easy way to compare two quantities).

Notes

1 Goldbach was an 18th-century German mathematician and Goldbach's Conjecture has been verified to be true for all integers $n < 12 * 10^{17}$.
2 Pierre de Fermat was a 17th French civil servant and amateur mathematician. He occasionally wrote to contemporary mathematicians announcing his latest theorem without providing the accompanying proof and inviting them to find the proof. The fact that he never revealed his proofs caused a lot of frustration among his contemporaries, and in his announcement of his famous last theorem, he stated that he had a wonderful proof that was too large to include in the margin. He corresponded with Pascal and they did some early work on the mathematical rules of games of chance and early probability theory.
3 Pythagoras of Samos (a Greek island in the Aegean sea) was an influential ancient mathematician and philosopher of the 6th century B.C. He gained his mathematical knowledge from his travels throughout the ancient world (especially in Egypt and Babylon). He became convinced that everything is number and he and his followers discovered the relationship between mathematics and the physical world as well as relationships between numbers and music. The Pythagorean brotherhood became a secret society, and focused on the study of mathematics. Pythagoras is mainly remembered today for the famous theorem named after him, which states that for a right-angled triangle the square of the hypotenuse is equal to the sum of the square of the other two sides. The Pythagoreans discovered the irrationality of the square root of 2 and as this result conflicted in a fundamental way with their philosophy that number is everything, and they suppressed the truth of this mathematical result.
4 Euclid was a 3rd-century B.C. Hellenistic mathematician and is considered the father of geometry.

2

Algebra

Algebra is the branch of mathematics that uses letters in the place of numbers, where the letters stand for variables or constants that are used in mathematical expressions. It is the study of such mathematical symbols and the rules for manipulating them, and it is a powerful tool for problem solving in science, engineering and business.

The origins of algebra are in work done by Islamic mathematicians during the Golden age of Islamic civilization, and the word "algebra" comes from the Arabic *"al-jabr,"* which appears as part of the title of a book by the Islamic mathematician, Al Khwarizmi, in the 9th century A.D. The third-century A.D. Hellenistic mathematician, Diophantus, also did early work on algebra.

Algebra covers many areas such as elementary algebra, linear algebra and abstract algebra. Elementary algebra includes the study of symbols and rules for manipulating them to form valid mathematical expressions, simultaneous equations, quadratic equations, polynomials, indices and logarithms. Linear algebra is concerned with the solution of a set of linear equations, and the study of matrices and vectors. Abstract algebra is concerned with the study of abstract algebraic structures such as groups, rings, fields and so on.

We show how to solve simple equations by bringing the unknown variable to one side of the equation and the values to the other side. We show how simultaneous equations may be solved by the method of substitution, the method of elimination and graphical techniques. We show how to solve quadratic equations by factorization, completing the square, the quadratic formula and graphical techniques. We show how simultaneous equations and quadratic equations may be used to solve practical problems.

We present the laws of indices and show the relationship between indices and logarithms. We discuss the exponential function e^x and the natural logarithm $\log_e x$ or $\ln x$.

2.1 Simplification of Algebraic Expressions

An algebraic expression is a combination of letters and symbols connected through various operations such as +, −, /, ×, (, and). Arithmetic expressions are formed from terms, and *like terms* (i.e., terms with the same

DOI: 10.1201/9781003308140-2

variables and exponents) may be added or subtracted. There are two terms in the algebraic expression below with the terms separated by a "–."

$$\underbrace{5x(2x^2 + y)}_{term\ 1} - \underbrace{4x(x + 2y - 1)}_{term\ 2}$$

The terms may include powers of some number (e.g., x^3 represents x raised to the power of 3, and 5^4 represents 5 raised to the power of 4).

In algebra, the like terms may be added or subtracted by adding or subtracting the numerical coefficients of the like terms. For example,

$$4x - 2x = 2x$$
$$5x^2 - 2x^2 = 3x^2$$
$$5x - 2y + 3x - 2y = 8x - 4y$$
$$4x^2 - 2y^3 - 3x^2 + 3y^3 = x^2 + y^3$$
$$-(4x - 3y) = -4x + 3y$$
$$5(3x) = (5 \times 3)x = 15x$$
$$5x.2x = 10x^2$$

Algebraic expressions may be simplified

$$\begin{aligned}(\mathrm{a}x + \mathrm{b}y)(\mathrm{a}x - \mathrm{b}y) &= \mathrm{aa}x^2 + (-\mathrm{ab} + \mathrm{ba})xy - \mathrm{bb}y^2 \\ &= \mathrm{a}^2x^2 - \mathrm{b}^2y^2\end{aligned}$$

Therefore,

$$(2x - 3y)(2x + 3y) = 2^2x^2 - 3^2y^2 = 4x^2 - 9y^2$$
$$(\mathrm{a} + \mathrm{b})(\mathrm{a} - \mathrm{b}) = \mathrm{a}^2 - \mathrm{b}^2$$

Let $P(x) = (\mathrm{a}x + \mathrm{b})$ and $Q(x) = (\mathrm{c}x + \mathrm{d})$. Then $P(x)Q(x) =$

$$(\mathrm{a}x+\mathrm{b})(\mathrm{c}x+\mathrm{d}) = (\mathrm{a}x+\mathrm{b})(\mathrm{c}x+\mathrm{d})$$

That is,

$$\begin{aligned}(\mathrm{a}x + \mathrm{b})(\mathrm{c}x + \mathrm{d}) &= \mathrm{a}x(\mathrm{c}x + \mathrm{d}) + \mathrm{b}(\mathrm{c}x + \mathrm{d}) \\ &= \mathrm{a}x\mathrm{c}x + \mathrm{a}x\mathrm{d} + \mathrm{b}\mathrm{c}x + \mathrm{bd} \\ &= \mathrm{ac}x^2 + (\mathrm{ad} + \mathrm{bc})x + \mathrm{bd}\end{aligned}$$

A polynomial $P(x)$ of degree n is defined as $P(x) = a_nx^n + a_{n-1}x^{n-1} + a_{n-2}x^{n-2} + \ldots + a_1x + a_0$. A polynomial may be multiplied by another, and when we multiply two polynomials $P(x)$ of degree n and $Q(x)$ of degree m together, the resulting polynomials is of degree $n + m$.

Example

Multiply $(2a + 3b)$ by $(a + b)$

Solution

This is given by $2a^2 + 2ab + 3ab + 3b^2 = 2a^2 + 5ab + 3b^2$.

2.2 Simple and Simultaneous Equations

A *simple equation* is an equation with one unknown, and the unknown may be on both the left-hand side and right-hand side of the equation. The method of solving such equations is to bring the unknowns to one side of the equation, and the values to the other side.

Simultaneous equations are equations with two (or more) unknowns. There are a number of methods to finding a solution to two simultaneous equations such as elimination, substitution and graphical techniques. The solution of n linear equations with n unknowns may be done using Gaussian elimination or matrix theory.

Example (Simple Equation)

Solve the simple equation $4 - 3x = 2x - 11$

Solution (Simple Equation)

$$\begin{aligned} 4 - 3x &= 2x - 11 \\ 4 - (-11) &= 2x - (3x) \\ 4 + 11 &= 2x + 3x \\ 15 &= 5x \\ 3 &= x \end{aligned}$$

Example (Simple Equation)

Solve the simple equation

$$\frac{2y}{5} + \frac{3}{4} + 5 = \frac{1}{20} - \frac{3y}{2}$$

Solution (Simple Equation)

The LCM of 4, 5 and 20 is 20. We multiply both sides by 20 to get:

$$20 * \frac{2y}{5} + 20 * \frac{3}{4} + 20 * 5 = 20 * \frac{1}{20} - 20 * \frac{3y}{2}$$
$$8y + 15 + 100 = 1 - 30y$$
$$38y = -114$$
$$y = -3$$

Simple equations may be used to solve practical problems where there is one unknown value to be determined. The information in the problem is converted into a simple equation with one unknown, and the equation is then solved.

Example (Practical Problem – Simple Equations)

The distance (in metres) travelled in time t seconds is given by the formula $s = ut + ½ at^2$, where u is the initial velocity in m/s and a is the acceleration in m/s^2. Find the acceleration of the body if it travels 168 m in 6 s, with an initial velocity of 10 m/s.

Solution

Using the formula $s = ut + {}^1/_2\, at^2$ we get:

$$168 = 10 * 6 + 1/2a * 6^2$$
$$168 = 60 + 18a$$
$$108 = 18a$$
$$a = 6m/s^2$$

2.2.1 Simultaneous Equations

Simultaneous equations are equations with two (or more) unknowns. There are several methods available to find a solution to the simultaneous equations including the method of substitution, the method of elimination and graphical techniques. We start with the substitution method where we express one of the unknowns in terms of the other. The method of substitution involves expressing x in terms of y and substituting it in the other equation (or vice versa expressing y in terms of x and substituting it in the other equation).

Example (Simultaneous Equation – Substitution Method)

Solve the following simultaneous equations by the method of substitution.

$$x + 2y = -1$$
$$4x - 3y = 18$$

Solution (Simultaneous Equation – Substitution Method)

For this example, we use the first equation to *express* x in terms of y.

$$x + 2y = -1$$
$$x = -1 - 2y$$

We then substitute for x, i.e., instead of writing x we write $(-1 - 2y)$ for x in the second equation, and we get a simple equation with just the unknown value y.

$$4(-1 - 2y) - 3y = 18$$
$$\Rightarrow\ -4 - 8y - 3y = 18$$
$$\Rightarrow\ -11y = 18 + 4$$
$$\Rightarrow\ -11y = 22$$
$$\Rightarrow y = -2$$

We then obtain the value of x from the substitution:

$$x = -1 - 2y$$
$$\Rightarrow x = -1 - 2(-2)$$
$$\Rightarrow x = -1 + 4$$
$$\Rightarrow x = 3$$

We can then verify that our solution is correct by checking our answer for both equations.

$$3 + 2(-2) = -1 \quad \surd$$
$$4(3) - 3(-2) = 18 \quad \surd$$

The approach of the method of elimination is to manipulate both equations so that we may eliminate either x or y, and so reduce the equations to a simple equation of one unknown value of x or y. This is best seen with an example.

Example (Simultaneous Equation – Method of Elimination)

Solve the following simultaneous equations by the method of elimination.

$$3x + 4y = 5$$
$$2x - 5y = -12$$

Solution (Simultaneous Equation – Method of Elimination)

We will use the method of elimination in this example to eliminate x, and so we multiply equation (2.1) by 2 and equation (2.2) by –3, and this yields two equations which have equal but opposite coefficients of x.

$$\begin{aligned} 6x + 8y &= 10 \\ -6x + 15y &= 36 \\ \hline 0x + 23y &= 46 \\ y &= 2 \end{aligned}$$

We then add both equations together and conclude that $y = 2$. We then determine the value of x by replacing y with 2 in the first equation.

$$\begin{aligned} 3x + 4(2) &= 5 \\ 3x + 8 &= 5 \\ 3x &= 5 - 8 \\ 3x &= -3 \\ x &= -1 \end{aligned}$$

We can then verify that our solution is correct as before by checking our answer for both equations.

Each simultaneous equation represents a straight line, and so the solution to the two simultaneous equations satisfies both equations and so is on both lines, i.e., the solution is the point of intersection of both lines (if there is such a point). Therefore, the solution involves drawing each line and finding the point of intersection of both lines (Fig. 2.1).

Example (Simultaneous Equation – Graphical Techniques)

Find the solution to the following simultaneous equations using graphical techniques:

$$\begin{aligned} x + 2y &= -1 \\ 4x - 3y &= 18 \end{aligned}$$

Solution (Simultaneous Equation – Graphical Techniques)

First, we find two points on line 1, e.g., (0, –0.5) and (–1, 0) are on line 1, since when $x = 0$ we have $2y = -1$ and so $y = -0.5$. Similarly, when $y = 0$ we have $x = -1$. Next, we find two points on line 2, e.g., when x is 0 y is –6 and when y is 0 we have $x = 4.5$ and so the points (0, –6) and (4.5, 0) are on line 2.

We then draw the X-axis and the Y-axis, draw the scales on the axes, label the axes, plot the points and draw both lines. Finally, we find the point of

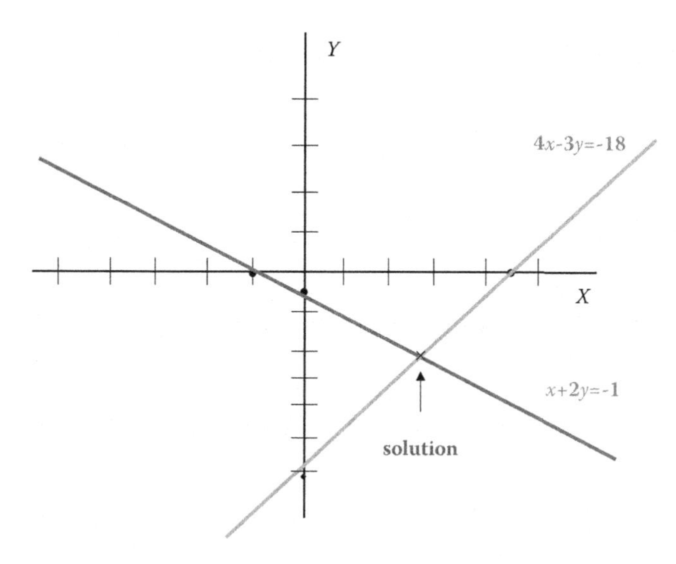

FIGURE 2.1
Graphical Solution to Simultaneous Equations.

intersection of both lines (if there is such a point), and this is our solution to the simultaneous equations.

The graph shows that the two lines intersect, and so we need to determine the point of intersection, and this involves determining the x and y co-ordinate of the solution which is given by $x = 3$ and $y = -2$. The solution using graphical techniques requires care (as inaccuracies may be introduced from poor drawing) and graph paper is required for accuracy.

The solution to practical problems often involves solving two simultaneous equations.

Example (Simultaneous Equation – Problem Solving)

Three new cars and four new vans supplied to a dealer together cost £97,000, and five new cars and two new vans of the same models cost £103,100. Find the cost of a car and a van.

Solution (Simultaneous Equation – Problem Solving)

We let C represent the cost of a car and V represent the cost of a van. We convert the information provided into two equations with two unknowns and then solve for V and C.

$$3C + 4V = 97000 \tag{2.1}$$

$$5C + 2V = 103100 \tag{2.2}$$

We multiply equation (2.2) by −2 and add to equation (2.1) to eliminate V

$$\begin{aligned} 3C + 4V &= 97000 \\ -10C - 4V &= -206200 \\ \hline -7C &= -109200 \\ C &= £15600 \end{aligned}$$

We then calculate the cost of a van by substituting the value of C in equation (2.2) to reduce it to an equation of one unknown.

$$\begin{aligned} 5C + 2V &= 103100 \\ 78000 + 2V &= 103100 \\ 2V &= 25100 \\ V &= £12550 \end{aligned}$$

Therefore, the cost of a car is £15,600 and the cost of a van is £12,550.

2.3 Quadratic Equations

A quadratic equation is an equation of the form $ax^2 + bx + c = 0$, and solving the quadratic equation is concerned with finding the unknown value x (roots of the quadratic equation). There may be no solution, one solution (a double root) or two solutions. There are several techniques for solving quadratic equations such as factorization, completing the square, the quadratic formula and graphical techniques.

Example (Quadratic Equations – Factorization)

Solve the quadratic equation $3x^2 - 11x - 4 = 0$ by factorization.

Solution (Quadratic Equations – Factorization)

The approach taken is to find the factors of the quadratic equation. Sometimes this is easy, but often other techniques will need to be employed. For the above quadratic equation, we note immediately that its factors are $(3x + 1)(x - 4)$ since

$$\begin{aligned} &(3x + 1)(x - 4) \\ &= 3x^2 - 12x + x - 4 \\ &= 3x^2 - 11x - 4 \end{aligned}$$

Next, we note the property that if the product of two numbers A and B is 0 then either A is 0 or B is 0. Another words, $AB = 0 \Rightarrow A = 0$ or $B = 0$. We conclude from this property that:

$$
\begin{aligned}
&3x^2 - 11x - 4 = 0 \\
&\Rightarrow (3x + 1)(x - 4) = 0 \\
&\Rightarrow (3x + 1) = 0 \text{ or } (x - 4) = 0 \\
&\Rightarrow 3x = -1 \text{ or } x = 4 \\
&\Rightarrow x = -0.33 \text{ or } x = 4
\end{aligned}
$$

Therefore, the solution (or roots) of the quadratic equation $3x^2 - 11x - 4 = 0$ are $x = -0.33$ or $x = 4$.

Example (Quadratic Equations – Completing the Square)

Solve the quadratic equation $2x^2 + 5x - 3 = 0$ by completing the square.

Solution (Quadratic Equations – Completing the Square)

First, we convert the quadratic equation to an equivalent quadratic with a unary coefficient of x^2. This involves division by 2. Next, we examine the coefficient of x (in this case $^5/_2$) and we add the square of half the coefficient of x to both sides. This allows us to complete the square, and we then take the square root of both sides. Finally, we solve for x.

$$
\begin{aligned}
&2x^2 + 5x - 3 = 0 \\
&\Rightarrow x^2 + 5/2x - 3/2 = 0 \\
&\Rightarrow x^2 + 5/2x = 3/2 \\
&\Rightarrow x^2 + 5/2x + (5/4)^2 = 3/2 + (5/4)^2 \\
&\Rightarrow (x + 5/4)^2 = 3/2 + (25/16) \\
&\Rightarrow (x + 5/4)^2 = 24/16 + (25/16) \\
&\Rightarrow (x + 5/4)^2 = 49/16 \\
&\Rightarrow (x + 5/4) = \pm 7/4 \\
&\Rightarrow x = -5/4 \pm 7/4 \\
&\Rightarrow x = -5/4 - 7/4 \text{ or } x = -5/4 + 7/4 \\
&\Rightarrow x = -12/4 \text{ or } x = 2/4 \\
&\Rightarrow x = -3 \text{ or } x = 0.5
\end{aligned}
$$

Example 1 (Quadratic Equations – Quadratic Formula)

Establish the quadratic formula for solving quadratic equations.

Solution (Quadratic Equations – Quadratic Formula)

We complete the square and the result will follow.

$$ax^2 + bx + c = 0$$
$$\Rightarrow x^2 + b/ax + c/a = 0$$
$$\Rightarrow x^2 + b/ax = -c/a$$
$$\Rightarrow x^2 + b/ax + (b/2a)^2 = -c/a + (b/2a)^2$$
$$\Rightarrow (x + b/2a)^2 = -c/a + (b/2a)^2$$
$$\Rightarrow (x + b/2a)^2 = \frac{-4ac}{4a^2} + \frac{b^2}{4a^2}$$
$$\Rightarrow (x + b/2a)^2 = \frac{b^2 - 4ac}{4a^2}$$
$$\Rightarrow (x + b/2a) = \pm\frac{\sqrt{b^2 - 4ac}}{2a}$$
$$\Rightarrow x = \frac{-b \pm \sqrt{b^2 - 4ac}}{2a}$$

Example 2 (Quadratic Equations – Quadratic Formula)

Solve the quadratic equation $2x^2 + 5x - 3 = 0$ using the quadratic formula.

Solution (Quadratic Equations – Quadratic Formula)

For this example $a = 2$; $b = 5$; and $c = -3$, and we put these values into the quadratic formula.

$$x = \frac{-5 \pm \sqrt{5^2 - 4.2(-3)}}{2.2} = \frac{-5 \pm \sqrt{25 + 24}}{4}$$
$$x = \frac{-5 \pm \sqrt{49}}{4} = \frac{-5 \pm 7}{4}$$
$$x = 0.5 \text{ or } x = -3.$$

Example (Quadratic Equations – Graphical Techniques)

Solve the quadratic equation $2x^2 - x - 6 = 0$ using graphical techniques given that the roots of the quadratic equation lie between $x = -3$ and $x = 3$.

Solution (Quadratic Equations – Graphical Techniques)

The approach is first to create a table of values for the curve $y = 2x^2 - x - 6$ (Table 2.1), and to draw the X-axis and Y-axis and scales, and then to plot the points from the table of values, and to join the points together to form the curve (Fig. 2.2).

TABLE 2.1

Table of Values for Quadratic Equation

x	−3	−2	−1	0	1	2	3
$y = 2x^2 - x - 6$	15	4	−3	−6	−5	0	9

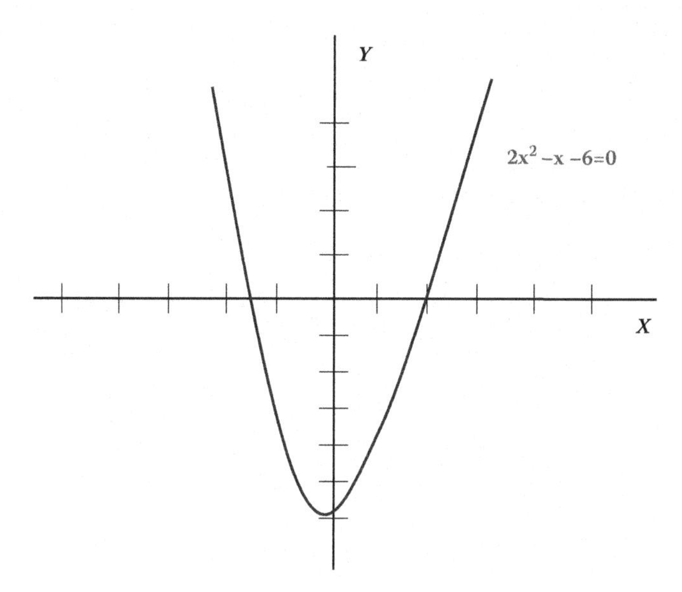

FIGURE 2.2
Graphical Solution to Quadratic Equation.

The graphical solution to the quadratic equation is then given by the points where the curve intersects the X-axis (i.e., $y = 0$ on the X-axis). There may be no solution (i.e., the curve does not intersect the X-axis), one solution (a double root) or two solutions.

The graph for the curve $y = 2x^2 - x - 6$ is given in Table 2.1, and so the points where the curve intersects the X-axis are determined. We note from the graph that the curve intersects the X-axis at two distinct points, and we see from the graph that the roots of the quadratic equation are given by $x = -1.5$ and $x = 2$.

The solution to quadratic equations using graphical techniques requires care in the plotting the points (as in determining the solution to simultaneous equations using graphical techniques), and graph paper is required for accuracy.

Quadratic equations often arise in solving practical problems as the following example shows.

Example (Quadratic Equations – Practical Problem)

A shed is 7.0-m long and 5.0-m wide. A concrete path of constant width x is laid all the way around the shed, and the area of the path is 30 m^2. Calculate its width x to the nearest centimetre (and use the quadratic formula).

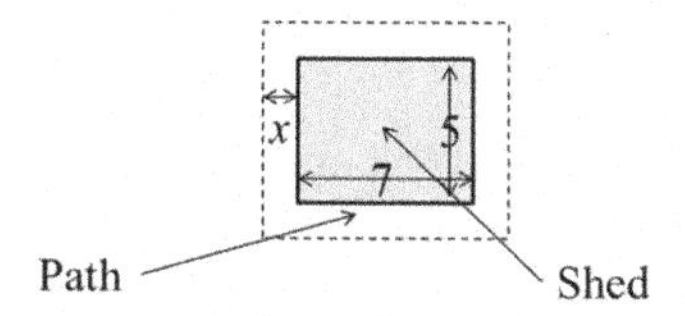

Solution (Quadratic Equations – Practical Problem)

Let x be the width of the path. Then the area of the path is the difference in area between the area of the large rectangle and the shed. We let A_S denote the area of the shed and let A_{S+P} denote the area of the large rectangle (i.e., the area of the shed + the area of the path)

$$
\begin{aligned}
A_S &= 7 * 5 = 35 \\
A_{S+P} &= (7 + 2x)(5 + 2x) \\
&= 35 + 14x + 10x + 4x^2 \\
&= 35 + 24x + 4x^2
\end{aligned}
$$

$$
\begin{aligned}
A_P &= 30 \\
&= A_{S+P} - A_S \\
&\Rightarrow 35 + 24x + 4x^2 - 35 = 30 \\
&\Rightarrow -30 + 24x + 4x^2 = 0 \\
&\Rightarrow 4x^2 + 24x - 30 = 0 \\
&\Rightarrow x = 0.91 \text{ m (from the quadratic formula)}
\end{aligned}
$$

2.4 Indices and Logarithms

The product $a.a.a.a\ldots a$ (n times) is denoted by a^n, and the number n is the index of a. The following are properties of indices.

$$
\begin{aligned}
a^o &= 1 \\
a^{m+n} &= a^m.\ a^n \\
a^{mn} &= (a^m)^n \\
a^{-n} &= \frac{1}{a^n} \\
a^{\frac{1}{n}} &= \sqrt[n]{a}
\end{aligned}
$$

Logarithms are closely related to indices, and if the number b can be written in the form $b = a^x$, then we say that log to the base a of b is x, i.e., $\log_a b = x \Leftrightarrow a^x = b$. Clearly, $\log_{10} 100 = 2$ since $10^2 = 100$. The following are the properties of logarithms:

$$
\begin{aligned}
\log_a AB &= \log_a A + \log_a B \\
log_a A^n &= n\ \log_a A \\
\log \tfrac{A}{B} &= \log A - \log B
\end{aligned}
$$

We will prove the first property of logarithms. Suppose $\log_a A = x$ and $\log_a B = y$. Then, $A = a^x$ and $B = a^y$ and so $AB = a^x a^y = a^{x+y}$ and so $\log_a AB = x + y = \log_a A + \log_a B$.

Example

Solve $\log_2{}^1/_{64}$ without a calculator.

Solution

$$\begin{aligned}
&\log_2 1/64 = x \\
&\Rightarrow 2^x = 1/64 \\
&\Rightarrow 2^x = 1/4 \times 4 \times 4 \\
&\Rightarrow 2^x = 1/2 \times 2 \times 2 \times 2 \times 2 \times 2 \\
&\Rightarrow 2^x = 1/2^6 \\
&\Rightarrow 2^x = 2^{-6} \\
&\Rightarrow x = -6
\end{aligned}$$

Example

Write $\log\left(\frac{16 \times \sqrt[3]{5}}{81}\right)$ in terms of log 2, log 3 and log 5 to any base.

Solution

$$\begin{aligned}
&\log\left(\frac{16 \times \sqrt[3]{5}}{81}\right) \\
&= \log\ 16 + 1/3 \log\ 5 - \log\ 81 \\
&= \log\ 2^4 + 1/3 \log\ 5 - \log\ 3^4 \\
&= 4\ \log\ 2 + 1/3 \log\ 5 - 4 \log\ 3
\end{aligned}$$

The law of logarithms may be used to solve certain indicial equations and we illustrate this with two examples.

Example (Indicial Equations)

Solve the equation $\log(x - 4) + \log(x + 2) = 2 \log (x - 2)$

Solution

$$\begin{aligned}
&\log(x - 4) + \log(x + 2) = 2 \log(x - 2) \\
&\Rightarrow \log(x - 4)(x + 2) = \log (x - 2)^2 \\
&\Rightarrow \log(x^2 - 2x - 8) = \log(x^2 - 4x + 4) \\
&\Rightarrow x^2 - 2x - 8 = x^2 - 4x + 4 \\
&\Rightarrow -2x - 8 = -4x + 4 \\
&\Rightarrow 2x = 12 \\
&\Rightarrow x = 6
\end{aligned}$$

Example (Indicial Equations)

Solve the equation $2^x = 3$, correct to four significant places.

Solution

$$\begin{aligned}
&2^x = 3 \\
&\Rightarrow \log_{10}2^x = \log_{10}3 \\
&\Rightarrow x\log_{10}2 = \log_{10}3 \\
&\Rightarrow x = \frac{\log_{10}3}{\log_{10}2} \\
&= \frac{0.4771}{0.3010} \\
&\Rightarrow x = 1.585
\end{aligned}$$

2.5 Exponentials and Natural Logarithms

The number e is a mathematical constant that occurs frequently in mathematics and its value is approximately equal to 2.7183 (it is an irrational number). The exponential function e^x (where e is the base and x is the exponent) is widely used in mathematics and science, and especially in problems involving growth and decay. The exponential function has the property that it is the unique function that is equal to its own derivative (i.e., $d/dx\ e^x = e^x$). The number e is the base of the natural logarithm, and e is sometimes called Euler's number or Euler's constant.

The value of the exponential function may be determined from its power series, and the value of $e^{0.1}$ may be determined by substituting 0.1 for x in the power series. However, it is more common to determine its value by using a scientific calculator which contains the e^x function.

$$e^x = 1 + x + \frac{x^2}{2!} + \ldots + \frac{x^n}{n!} + \ldots$$

Logarithms to the base e are termed natural logarithms (or Naperian logs). The natural logarithm of x is written as $\log_e x$ and it is more commonly written as $\ln x$.

Example (Natural Logs)

Solve the equation $7 = 4e^{-3x}$ to find x, correct to four decimal places.

Solution (Natural Logs)

$$
\begin{aligned}
&7 = 4e^{-3x} \\
&\Rightarrow 7/4 = e^{-3x} \\
&\Rightarrow \ln(1.75) = \ln(e^{-3x}) = -3x \\
&0.55966 = -3x \\
&x = -0.1865
\end{aligned}
$$

Example (Practical Problem)

The length of a bar, l, at a temperature θ is given by $l = l_0 e^{\alpha\theta}$, where l_0 and α are constants. Evaluate l, correct to four significant figures, when $l_0 = 2.587$, $\theta = 321.7$ and $\alpha = 1.771 \times 10^{-4}$.

Solution (Practical Problem)

$$
\begin{aligned}
l &= l_0 e^{\alpha\theta}, \\
&= 2.587 e^{1.771\times 10-4 \,*\, 321.7} \\
&= 2.587 * 0.56973 \\
&= 1.4739
\end{aligned}
$$

2.6 Review Questions

1. Solve the simple equation: $4(3x + 1) = 7(x + 4) - 2(x + 5)$
2. Solve the following simultaneous equations by

$$
\begin{aligned}
x + 2y &= -1 \\
4x - 3y &= 18
\end{aligned}
$$

   a. Graphical techniques
   b. Method of substitution
   c. Method of elimination
3. Solve the quadratic equation $3x^2 + 5x - 2 = 0$ given that the solution is between $x = -3$ and $x = 3$ by:
   a. Graphical techniques
   b. Factorization
   c. Quadratic formula

4. Solve the following indicial equation using logarithms:

$$2^{x-1} = 3^{2x-1}$$

5. Solve $\log_3{}^1/_{81}$ without a calculator.
6. Solve the equation $\log(x - 3) + \log(x + 3) = 2 \log (x - 1)$
7. Solve the equation $4 = 15e^{-2x}$ to find x, correct to four decimal places.

2.7 Summary

This chapter provided a brief introduction to algebra, which is the branch of mathematics that studies mathematical symbols and the rules for manipulating them. Algebra is a powerful tool for problem solving in science, engineering and business.

Elementary algebra includes the study of simple equations, where a simple equation is an equation with one unknown; simultaneous equations involve two or more equations with two or more unknowns, and may be solved by the method of substitution, the method of elimination or graphical techniques; quadratic equations are of the form $ax^2 + bx + c = 0$, and they may be solved by factorization, completing the square, the quadratic formula or graphical techniques.

Algebra involves the study of indices and logarithms, and we discussed the various rules of indices and logarithms. We discussed the exponential constant e, and the natural logarithm $\ln x$. The number e occurs frequently in mathematics.

3

Sets, Relations and Functions

This chapter introduces fundamental building blocks in mathematics such as sets, relations and functions. Sets are collections of well-defined objects; relations indicate relationships between members of two sets A and B; and functions are a special type of relation where there is exactly (or at most)[1] one relationship for each element $a \in A$ with an element in B.

A set is a collection of well-defined objects that contains no duplicates. The term "well defined" means that for a given value it is possible to determine whether or not it is a member of the set. There are many examples of sets such as the set of natural numbers $\mathbb{N}$; the set of integer numbers $\mathbb{Z}$; and the set of rational numbers $\mathbb{Q}$. The natural number $\mathbb{N}$ is an infinite set consisting of the numbers $\{1, 2, \ldots\}$. Venn diagrams may be used to represent sets pictorially.

A binary relation $R(A, B)$ where A and B are sets is a subset of the Cartesian product $(A \times B)$ of A and B. The domain of the relation is A and the co-domain of the relation is B. The notation aRb signifies that there is a relation between a and b and that $(a, b) \in R$. An n-ary relation $R\ (A_1, A_2, \ldots, A_n)$ is a subset of $(A_1 \times A_2 \times \ldots \times A_n)$. However, an n-ary relation may also be regarded as a binary relation $R(A, B)$ with $A = A_1 \times A_2 \times \ldots \times A_{n-1}$ and $B = A_n$.

Functions may be total or partial. A total function $f{:}A \to B$ is a special relation such that for each element $a \in A$ there is exactly one element $b \in B$. This is written as $f(a) = b$. A partial function differs from a total function in that the function may be undefined for one or more values of A. The domain of a partial function (denoted by **dom** f) is the set of values in A for which the partial function is defined. The domain of the function is A if f is a total function. The co-domain of the function is B.

3.1 Set Theory

A set is a fundamental building block in mathematics, and it is defined as a collection of well-defined objects. The elements in a set are of the same kind, and they are distinct with no repetition of the same element in the set.[2] Most sets encountered in the computer field are finite, as computers can only deal with finite entities. Venn diagrams[3] are often employed to give a pictorial representation of a set, and to illustrate various set operations such as set union, intersection and set difference.

DOI: 10.1201/9781003308140-3

There are many well-known examples of sets including the set of natural numbers denoted by $\mathbb{N}$; the set of integers denoted by $\mathbb{Z}$; the set of rational numbers denoted by $\mathbb{Q}$; the set of real numbers denoted by $\mathbb{R}$; and the set of complex numbers denoted by $\mathbb{C}$.

Example 3.1

The following are examples of sets:

- The books on the shelves in a library
- The books that are currently overdue from the library
- The customers of a bank
- The bank accounts in a bank
- The set of natural numbers $\mathbb{N} = \{1, 2, 3, \ldots\}$
- The set of prime numbers = $\{2, 3, 5, 7, 11, 13, 17, \ldots\}$
- The integer numbers $\mathbb{Z} = \{\ldots, -3, -2, -1, 0, 1, 2, 3, \ldots\}$
- The non-negative integers $\mathbb{Z}^+ = \{0, 1, 2, 3, \ldots\}$
- The rational numbers are the set of quotients of integers

$$\mathbb{Q} = \{p/q\colon p,\ \ q \in \mathbb{Z} \text{ and } q \neq 0\}$$

A finite set may be defined by listing all its elements. For example, the set $A = \{2, 4, 6, 8, 10\}$ is the set of all even natural numbers less than or equal to 10. The order in which the elements are listed is not relevant, i.e., the set $\{2, 4, 6, 8, 10\}$ is the same as the set $\{8, 4, 2, 10, 6\}$.

Sets may be defined by using a predicate to constrain set membership. For example, the set $S = \{n\colon \mathbb{N}\colon n \leq 10 \wedge n \bmod 2 = 0\}$ also represents the set $\{2, 4, 6, 8, 10\}$. That is, the use of a predicate allows a new set to be created from an existing set by using the predicate to restrict membership of the set. The set of even natural numbers may be defined by a predicate over the set of natural numbers that restricts membership to the even numbers. It is defined by:

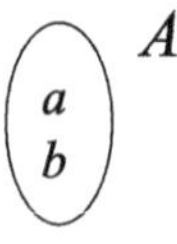

$$\text{Evens} = \{x \mid x \in \mathbb{N} \wedge even(x)\}.$$

In this example, $even(x)$ is a predicate that is true if x is even and false otherwise. In general, $A = \{x \in E \mid P(x)\}$ denotes a set A formed from a set E

using the predicate P to restrict membership of A to those elements of E for which the predicate is true.

The elements of a finite set S are denoted by $\{x_1, x_2, \ldots, x_n\}$. The expression $x \in S$ denotes that the element x is a member of the set S, whereas the expression $x \notin S$ indicates that x is not a member of the set S.

A set S is a subset of a set T (denoted $S \subseteq T$) if whenever $s \in S$ then $s \in T$, and in this case the set T is said to be a superset of S (denoted $T \supseteq S$). Two sets S and T are said to be equal if they contain identical elements, i.e., $S = T$ if and only if $S \subseteq T$ and $T \subseteq S$. A set S is a proper subset of a set T (denoted $S \subset T$) if $S \subseteq T$ and $S \neq T$. That is, every element of S is an element of T and there is at least one element in T that is not an element of S. In this case, T is a proper superset of S (denoted $T \supset S$).

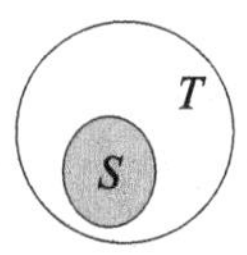

The empty set (denoted by $\varnothing$ or $\{\}$) represents the set that has no elements. Clearly, $\varnothing$ is a subset of every set. The singleton set containing just one element x is denoted by $\{x\}$, and clearly $x \in \{x\}$ and $x \neq \{x\}$. Clearly, $y \in \{x\}$ if and only if $x = y$.

Example 3.2

i. $\{1,2\} \subseteq \{1,2,3\}$
ii. $\varnothing \subset \mathbb{N} \subset \mathbb{Z} \subset \mathbb{Q} \subset \mathbb{R} \subset \mathbb{C}$

The cardinality (or size) of a finite set S defines the number of elements present in the set. It is denoted by $|S|$. The cardinality of an infinite[4] set S is written as $|S| = \infty$.

Example 3.3

i. Given $A = \{2, 4, 5, 8, 10\}$ then $|A| = 5$.
ii. Given $A = \{x \in \mathbb{Z}: x^2 = 9\}$ then $|A| = 2$
iii. Given $A = \{x \in \mathbb{Z}: x^2 = -9\}$ then $|A| = 0$.

3.1.1 Set Theoretical Operations

Several set theoretical operations are considered in this section. These include the Cartesian product operation; the power set of a set; the set union operation; the set intersection operation; the set difference operation; and the symmetric difference operation.

Cartesian Product

The Cartesian product allows a new set to be created from existing sets. The Cartesian[5] product of two sets S and T (denoted $S \times T$) is the set of ordered pairs $\{(s, t) \mid s \in S, t \in T\}$. Clearly, $S \times T \neq T \times S$ and so the Cartesian product of two sets is not commutative. Two ordered pairs (s_1, t_1) and (s_2, t_2) are considered equal if and only if $s_1 = s_2$ and $t_1 = t_2$.

The Cartesian product may be extended to that of n sets $S_1, S_2, \ldots, S_n$. The Cartesian product $S_1 \times S_2 \times \ldots \times S_n$ is the set of ordered n-tuples $\{(s_1, s_2, \ldots, s_n) \mid s_1 \in S_1, s_2 \in S_2, \ldots, s_n \in S_n\}$. Two ordered n-tuples $(s_1, s_2, \ldots, s_n)$ and $(s_1', s_2', \ldots, s_n')$ are considered equal if and only if $s_1 = s_1', s_2, = s_2', \ldots, s_n = s_n'$.

The Cartesian product may also be applied to a single set S to create ordered n-tuples of S, i.e., $S^n = S \times S \times \ldots \times S$ (n times).

Power Set

The power set of a set A (denoted $\mathbb{P}A$) denotes the set of subsets of A. For example, the power set of the set $A = \{1, 2, 3\}$ has eight elements and is given by:

$$\mathbb{P}A = \{\emptyset, \ \{1\}, \ \{2\}, \ \{3\}, \ \{1, 2\}, \ \{1, 3\}, \ \{2, 3\}, \ \{1, 2, 3\}\}.$$

There are $2^3 = 8$ elements in the power set of $A = \{1, 2, 3\}$ where the cardinality of A is 3. In general, there are $2^{|A|}$ elements in the power set of A.

Theorem 3.1: (Cardinality of Power Set of A)

There are $2^{|A|}$ elements in the power set of A.

Proof

Let $|A| = n$ then the cardinality of the subsets of A are subsets of size 0, 1, …, n. There are $\binom{n}{k}$ subsets of A of size k[6]. Therefore, the total number of subsets of A is the total number of subsets of size 0, 1, 2, … up to n. That is,

$$|\mathbb{P}A| = \sum_{k=0}^{n} \binom{n}{k}$$

The Binomial Theorem states that

$$(1 + x)^n = \sum_{k=0}^{n} \binom{n}{k} x^k$$

Therefore, putting $x = 1$ we get that

$$2^n = (1 + 1)^n = \sum_{k=0}^{n} \binom{n}{k} 1^k = |\mathbb{P}A|$$

Union and Intersection Operations

The union of two sets A and B is denoted by $A \cup B$. It results in a set that contains all of the members of A and of B and is defined by:

$$A \cup B = \{r \mid r \in A \ \text{ or } \ r \in B\}.$$

For example, suppose $A = \{1, 2, 3\}$ and $B = \{2, 3, 4\}$ then $A \cup B = \{1, 2, 3, 4\}$. Set union is a commutative operation, i.e., $A \cup B = B \cup A$. Venn diagrams are used to illustrate these operations pictorially.

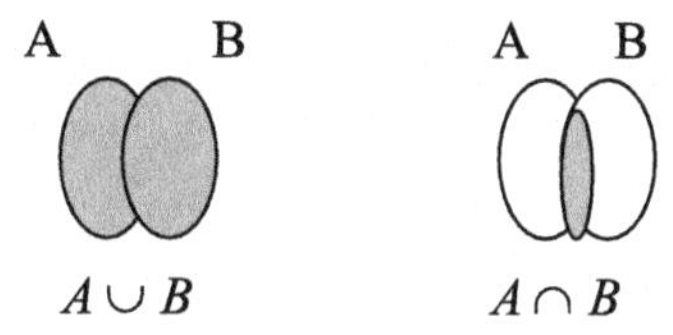

The intersection of two sets A and B is denoted by $A \cap B$. It results in a set containing the elements that A and B have in common and is defined by:

$$A \cap B = \{r \mid r \in A \ \text{ and } \ r \in B\}.$$

Suppose $A = \{1, 2, 3\}$ and $B = \{2, 3, 4\}$ then $A \cap B = \{2, 3\}$. Set intersection is a commutative operation, i.e., $A \cap B = B \cap A$.

Union and intersection may be extended to more generalized union and intersection operations. For example:

$\cup_{i=1}^{n} A_i$ denotes the union of n sets.

$\cap_{i=1}^{n} A_i$ denotes the intersection of n sets

Set Difference Operations

The set difference operation $A \setminus B$ yields the elements in A that are not in B. It is defined by

$$A \setminus B = \{a \mid a \in A \ \text{ and } \ a \notin B\}$$

For A and B defined as $A = \{1, 2\}$ and $B = \{2, 3\}$ we have $A \backslash B = \{1\}$ and $B \backslash A = \{3\}$. Clearly, set difference is not commutative, i.e., $A \backslash B \neq B \backslash A$. Clearly, $A \backslash A = \varnothing$ and $A \backslash \varnothing = A$.

The symmetric difference of two sets A and B is denoted by $A \Delta B$ and is given by:

$$A \ \Delta \ B = A \backslash B \cup B \backslash A$$

The symmetric difference operation is commutative, i.e., $A \Delta B = B \Delta A$. Venn diagrams are used to illustrate these operations pictorially.

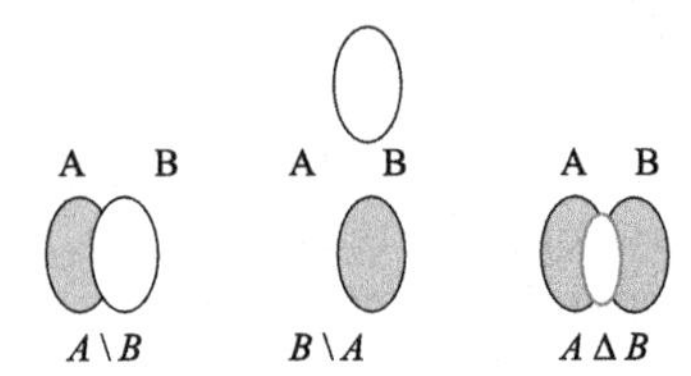

The complement of a set A (with respect to the universal set U) is the elements in the universal set that are not in A. It is denoted by A^c (or A') and is defined as:

$$A^c = \{u \mid u \in U \text{ and } u \notin A\} = U \backslash A$$

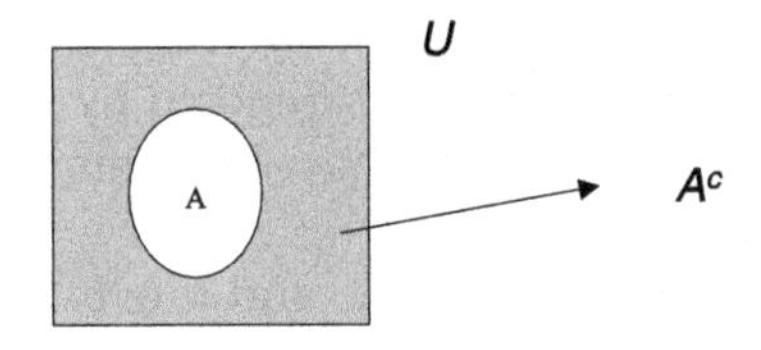

The shaded area illustrates the complement of the set A.

3.1.2 Properties of Set Theoretical Operations

The set union and set intersection properties are commutative and associative. Their properties are listed in Table 3.1.

These properties may be seen to be true with Venn diagrams or with a mathematical proof. We give a proof of the distributive property.

Proof of Properties (Distributive Property)

To show $A \cap (B \cup C) = (A \cap B) \cup (A \cap C)$

Suppose $x \in A \cap (B \cup C)$ then

$$x \in A \wedge x \in (B \cup C)$$

$$\Rightarrow x \in A \wedge (x \in B \vee x \in C)$$

TABLE 3.1

Properties of Set Operations

Property	Description
Commutative	Union and intersection operations are commutative, i.e., $S \cup T = T \cup S$ $S \cap T = T \cap S$
Associative	Union and intersection operations are associative, i.e., $R \cup (S \cup T) = (R \cup S) \cup T$ $R \cap (S \cap T) = (R \cap S) \cap T$
Identity	The identity under set union is the empty set ∅, and the identity under intersection is the universal set U. $S \cup \varnothing = \varnothing \cup S = S$ $S \cap U = U \cap S = S$
Distributive	The union operator distributes over the intersection operator and vice versa. $R \cap (S \cup T) = (R \cap S) \cup (R \cap T)$ $R \cup (S \cap T) = (R \cup S) \cap (R \cup T)$.
DeMorgan's[7] Law	The complement of $S \cup T$ is given by $(S \cup T)^c = S^c \cap T^c$ The complement of $S \cap T$ is given by $(S \cap T)^c = S^c \cup T^c$

$$\Rightarrow (x \in A \wedge x \in B) \vee (x \in A \wedge x \in C)$$
$$\Rightarrow x \in (A \cap B) \vee x \in (A \cap C)$$
$$\Rightarrow x \in (A \cap B) \cup (A \cap C)$$

Therefore, $A \cap (B \cup C) \subseteq (A \cap B) \cup (A \cap C)$
Similarly, $(A \cap B) \cup (A \cap C) \subseteq A \cap (B \cup C)$
Therefore, $A \cap (B \cup C) = (A \cap B) \cup (A \cap C)$

3.1.3 Russell's Paradox

Bertrand Russell (Fig. 3.1) was a famous British logician, mathematician and philosopher. He was the co-author with Alfred Whitehead of *Principia Mathematica*, which aimed to derive all the truths of mathematics from logic. Russell's Paradox was discovered by Bertrand Russell in 1901, and showed that the system of logicism being proposed by Frege contained a contradiction.

Question (Posed by Russell to Frege)

Is the set of all sets that do not contain themselves as members a set?

Russell's Paradox

Let $A = \{S$ a set and $S \notin S\}$. Is $A \in A$? Then $A \in A \Rightarrow A \notin A$ and vice versa. Therefore, a contradiction arises in either case and there is no such set A.

FIGURE 3.1
Bertrand Russell.

Two ways of avoiding the paradox were developed in 1908, and these were Russell's theory of types and Zermelo set theory [ORg:20].

3.2 Relations

A binary relation $R(A, B)$ where A and B are sets is a subset of $A \times B$, i.e., $R \subseteq A \times B$. The domain of the relation is A and the co-domain of the relation is B. The notation aRb signifies that $(a, b) \in R$.

A binary relation $R(A, A)$ is a relation between A and A (or a relation on A). This type of relation may always be composed with itself, and its inverse is also a binary relation on A. The identity relation on A is defined by $a\ i_A a$ for all $a \in A$.

Example 3.4

There are many examples of relations:

i. The relation on a set of students in a class where $(a, b) \in R$ if the height of a is greater than the height of b.
ii. The relation between A and B where $A = \{0, 1, 2\}$ and $B = \{3, 4, 5\}$ with R given by:

$$R = \{(0, 3),\ \ (0, 4),\ \ (1, 4)\}$$

iii. The relation less than (<) between and $\mathbb{R}$ and $\mathbb{R}$ is given by:

$$\{(x, y) \in \mathbb{R}^2 : x < y\}$$

iv. A bank may represent the relationship between the set of accounts and the set of customers by a relation. The implementation of a bank account may be a positive integer with at most eight decimal digits.

The relationship between accounts and customers may be done with a relation $R \subseteq A \times B$, with the set A chosen to be the set of natural numbers, and the set B chosen to be the set of all human beings alive or dead. The set

$$A \text{ could also be chosen to be } A = \{n \in \mathbb{N}: n < 10^8\}$$

A relation $R(A, B)$ may be represented pictorially. This is referred to as the graph of the relation, and it is illustrated in the diagram below. An arrow from x to y is drawn if (x, y) is in the relation. Thus for the height relation R given by $\{(a, p), (a, r), (b, q)\}$ an arrow is drawn from a to p, from a to r and from b to q to indicate that (a, p), (a, r) and (b, q) are in the relation R.

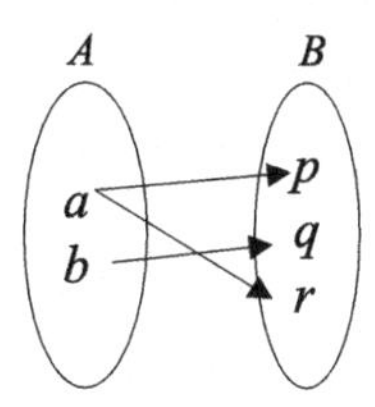

The pictorial representation of the relation makes it easy to see that the height of a is greater than the height of p and r; and that the height of b is greater than the height of q.

An n-ary relation $R(A_1, A_2, \ldots, A_n)$ is a subset of $(A_1 \times A_2 \times \ldots \times A_n)$. However, an n-ary relation may also be regarded as a binary relation $R(A, B)$ with $A = A_1 \times A_2 \times \ldots \times A_{n-1}$ and $B = A_n$.

3.2.1 Reflexive, Symmetric and Transitive Relations

A binary relation on A may have additional properties such as being reflexive, symmetric or transitive.

PROPERTIES OF RELATIONS

A relation on a set A is *reflexive* if $(a, a) \in R$ for all $a \in A$.

A relation R is *symmetric* if whenever $(a, b) \in R$ then $(b, a) \in R$.

A relation is *transitive* if whenever $(a, b) \in R$ and $(b, c) \in R$ then $(a, c) \in R$.

A relation that is reflexive, symmetric and transitive is termed an *equivalence relation*.

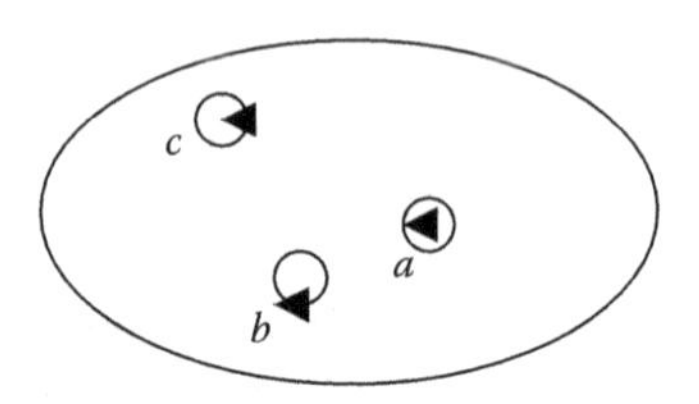

FIGURE 3.2
Reflexive Relation.

Example 3.5 (Reflexive Relation)

A relation is reflexive if each element possesses an edge looping around on itself. The relation in Fig. 3.2 is reflexive.

Example 3.6 (Symmetric Relation)

The graph of a symmetric relation will show for every arrow from a to b an opposite arrow from b to a. The relation in Fig. 3.3 is symmetric, i.e., whenever $(a, b) \in R$ then $(b, a) \in R$.

Example 3.7 (Transitive Relation)

The graph of a transitive relation will show that whenever there is an arrow from a to b and an arrow from b to c that there is an arrow from a to c. The relation in Fig. 3.4 is transitive, i.e., whenever $(a, b) \in R$ and $(b, c) \in R$ then $(a, c) \in R$.

Example 3.8 (Equivalence Relation)

The relation on the set of integers $\mathbb{Z}$ defined by $(a, b) \in R$ if $a - b = 2k$ for some $k \in \mathbb{Z}$ is an equivalence relation, and it partitions the set of integers into two equivalence classes, i.e., the even and odd integers.

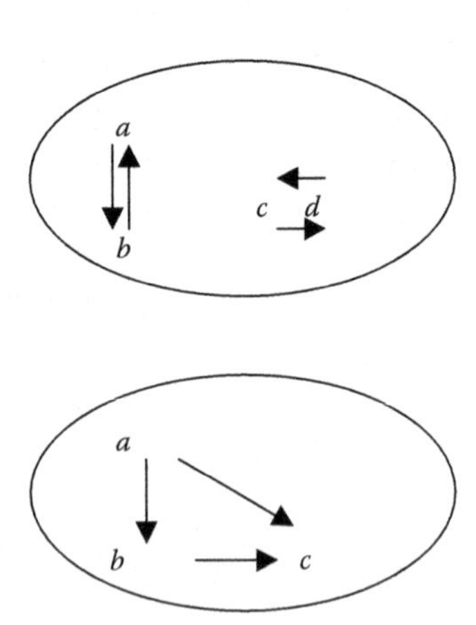

FIGURE 3.3
Symmetric Relation.

FIGURE 3.4
Transitive Relation.

Domain and Range of Relation

The domain of a relation R (A, B) is given by $\{a \in A \mid \exists\, b \in B$ and $(a, b) \in R\}$. It is denoted by **dom** R. The domain of the relation $R = \{(a, p), (a, r), (b, q)\}$ is $\{a, b\}$.

The range of a relation R (A, B) is given by $\{b \in B \mid \exists\, a \in A$ and $(a, b) \in \mathrm{R}\}$. It is denoted by **rng** R. The range of the relation $R = \{(a, p), (a, r), (b, q)\}$ is $\{p, q, r\}$.

Inverse of a Relation

Suppose $R \subseteq A \times B$ is a relation between A and B then the inverse relation $R^{-1} \subseteq B \times A$ is defined as the relation between B and A and is given by:

$$b \; R^{-1} \, a \text{ if and only if } a \; R \; b$$

That is,

$$R^{-1} = \{(b, a) \in B \times A\colon (a, b) \in R\}$$

Example 3.9

Let R be the relation between $\mathbb{Z}$ and $\mathbb{Z}^+$ defined by mRn if and only if $m^2 = n$. Then $R = \{(m,n) \in \mathbb{Z} \times \mathbb{Z}^+\colon m^2 = n\}$ and $R^{-1} = \{(n, m) \in \mathbb{Z}^+ \times \mathbb{Z}\colon m^2 = n\}$.

For example, $-3\ R\ 9$, $-4\ R\ 16$, $0\ R\ 0$, $16\ R^{-1}\ {-4}$, $9\ R^{-1}\ {-3}$, etc.

Partitions and Equivalence Relations

An equivalence relation on A leads to a partition of A, and vice versa for every partition of A there is a corresponding equivalence relation.

Let A be a finite set and let $A_1, A_2, \ldots, A_n$ be subsets of A such $A_i \neq \varnothing$ for all i, $A_i \cap A_j = \varnothing$ if $i \neq j$ and $A = \cup_i^n A_i = A_1 \cup A_2 \cup \ldots \cup A_n$.

The sets A_i partition the set A, and these sets are called the classes of the partition (Fig. 3.5).

Theorem 3.2: (Equivalence Relation and Partitions)

An equivalence relation on A gives rise to a partition of A where the equivalence classes are given by Class(a) $= \{x \mid x \in A$ and $(a, x) \in R\}$. Similarly, a partition gives rise to an equivalence relation R, where $(a, b) \in R$ if and only if a and b are in the same partition.

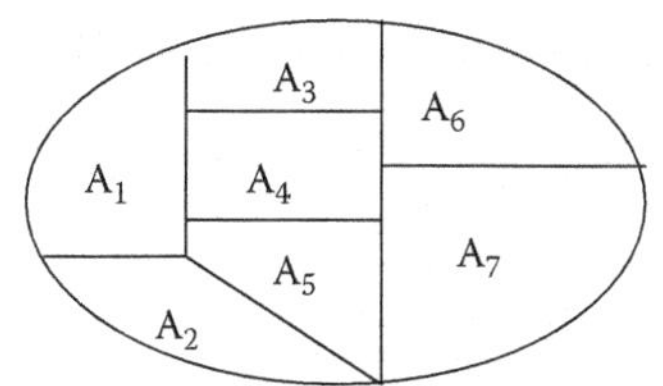

FIGURE 3.5
Partitions of A.

Proof

The proof involves showing that the equivalence classes are disjoint and that a partition yields an equivalence relation [ORg:20].

3.2.2 Composition of Relations

The composition of two relations $R_1(A, B)$ and $R_2(B, C)$ is given by $R_2 \text{ o } R_1$ where $(a, c) \in R_2 \text{ o } R_1$ if and only there exists $b \in B$ such that $(a, b) \in R_1$ and $(b, c) \in R_2$. The composition of relations is associative, i.e.,

$$(R_3 \text{ o } R_2) \text{ o } R_1 = R_3 \text{ o } (R_2 \text{ o } R_1)$$

Example 3.10 (Composition of Relations)

Consider a library that maintains two files. The first file maintains the serial number s of each book as well as the details of the author a of the book. This may be represented by the relation $R_1 = sR_1a$. The second file maintains the library card number c of its borrowers and the serial number s of any books that they have borrowed. This may be represented by the relation $R_2 = c\ R_2s$.

The library wishes to issue a reminder to its borrowers of the authors of all books currently on loan to them. This may be determined by the composition of $R_1 \text{ o } R_2$, i.e., $c\ R_1 \text{ o } R_2\ a$ if there is book with serial number s such that $c\ R_2s$ and $s\ R_1a$.

Example 3.11 (Composition of Relations)

Consider sets $A = \{a, b, c\}$, $B = \{d, e, f\}$, $C = \{g, h, i\}$ and relations $R(A, B) = \{(a, d), (a, f), (b, d), (c, e)\}$ and $S(B, C) = \{(d, h), (d, i), (e, g), (e, h)\}$. Then we graph these relations and show how to determine the composition pictorially.

$S \text{ o } R$ is determined by choosing $x \in A$ and $y \in C$ and checking if there is a route from x to y in the graph. If so, we join x to y in $S \text{ o } R$. For example, if we consider a and h we see that there is a path from a to d and from d to h and therefore (a, h) is in the composition of S and R (Fig. 3.6).

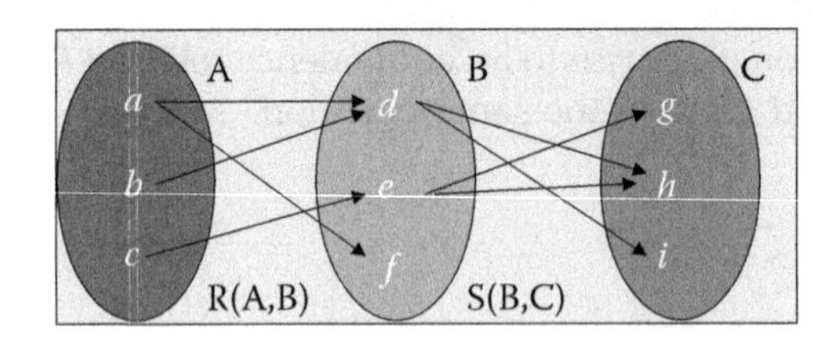

FIGURE 3.6
Composition of Relations $S \text{ o } R$.

The union of two relations $R_1(A, B)$ and $R_2(A, B)$ is meaningful (as these are both subsets of $A \times B$). The union $R_1 \cup R_2$ is defined as $(a, b) \in R_1 \cup R_2$ if and only if $(a, b) \in R_1$ or $(a, b) \in R_2$.

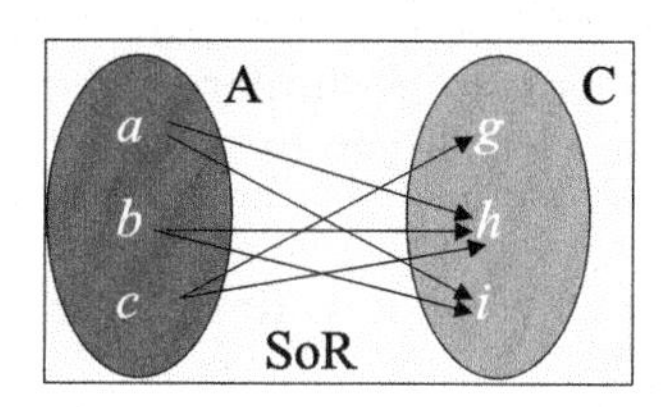

Similarly, the intersection of R_1 and R_2 ($R_1 \cap R_2$) is meaningful and is defined as $(a, b) \in R_1 \cap R_2$ if and only if $(a, b) \in R_1$ and $(a, b) \in R_2$. The relation R_1 is a subset of R_2 ($R_1 \subseteq R_2$) if whenever $(a, b) \in R_1$ then $(a, b) \in R_2$.

The inverse of the relation R was discussed earlier and is given by the relation R^{-1} where $R^{-1} = \{(b, a) \mid (a, b) \in R\}$.

The composition of R and R^{-1} yields: $R^{-1} \circ R = \{(a, a) \mid a \in \text{dom } R\} = i_A$ and $R \circ R^{-1} = \{(b, b) \mid b \in \textbf{dom } R^{-1}\} = i_B$.

3.2.3 Binary Relations

A binary relation R on A is a relation between A and A, and a binary relation can always be composed with itself. Its inverse is a binary relation on the same set. The following are all relations on A:

$$R^2 = R \circ R$$
$$R^3 = (R \circ R) \circ R$$
$$R^0 = i_A \text{ (identity relation)}$$
$$R^{-2} = R^{-1} \circ R^{-1}$$

Example 3.12

Let R be the binary relation on the set of all people P such that $(a, b) \in R$ if a is a parent of b. Then, the relation R^n is interpreted as:

R is the parent relationship.
R^2 is the grandparent relationship.
R^3 is the great grandparent relationship.
R^{-1} is the child relationship.
R^{-2} is the grandchild relationship.
R^{-3} is the great grandchild relationship.

This can be generalized to a relation R^n on A where $R^n = R \circ R \circ \ldots \circ R$ (n-times). The transitive closure of the relation R on A is given by:

$$R^* = \cup_{i=0}^{\infty} R^i = R^0 \cup R^1 \cup R^2 \cup \ldots R^n \cup ..$$

where R^0 is the reflexive relation containing only each element in the domain of R, i.e., $R^0 = i_A = \{(a, a) \mid a \in \mathbf{dom}\ R\}$.

The positive transitive closure is similar to the transitive closure except that it does not contain R^0. It is given by:

$$R^+ = \cup_{i=1}^{\infty} R^i = R^1 \cup R^2 \cup \ldots \cup R^n \cup ..$$

$a\ R^+\ b$ if and only if $a\ R^n\ b$ for some $n > 0$, i.e., there exists $c_1, c_2, \ldots, c_n \in A$ such that

$$aRc_1, \quad c_1Rc_2, \ \ldots, \ c_nRb$$

3.2.4 Applications of Relations to Databases

A Relational Database Management System (RDBMS) is a system that manages data using the relational model, where a relation is defined as a set of tuples that is usually represented by a table. A table is data organized in rows and columns, and the data in each column are of the same data type. Constraints may be employed to provide restrictions on the kinds of data that may be stored in the relations, and these Boolean expressions are a way of implementing business rules in the database.

Relations have one or more keys associated with them, and the *key uniquely identifies the row of the table.* An index is a way of providing fast access to the data in a relational database, as it allows the tuple in a relation to be looked up directly (using the index) rather than checking all tuples in the relation.

The concept of a relational database was first described in a paper "A Relational Model of Data for Large Shared Data Banks" by Codd [Cod:70]. A relational database is a database that conforms to the relational model, and it may be defined as a set of relations (or tables).

Codd (Fig. 3.7) developed the *relational database model* in the late 1960s, and today, this is the standard way that information is organized and retrieved from computers. Relational databases are at the heart of systems from hospitals' patient records to airline flight and schedule information.

An n-ary relation R $(A_1, A_2, \ldots, A_n)$ is a subset of the Cartesian product of the n sets, i.e., a subset of $(A_1 \times A_2 \times \ldots \times A_n)$, and the data in the relational model are defined as a set of n-tuples, and are represented by a table which is a visual representation with the data organized in rows and columns. There is more detailed information on the relational model and databases in [ORg:20].

FIGURE 3.7
Edgar Codd.

3.3 Functions

A function $f: A \rightarrow B$ is a special relation such that for each element $a \in A$ there is exactly (or at most)[8] one element $b \in B$. This is written as $f(a) = b$.

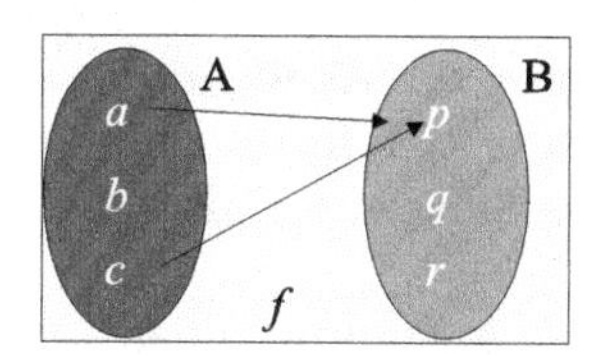

A function is a relation but not every relation is a function. For example, the relation in Fig. 3.8 is not a function since there are two arrows from the element $a \in A$.

The domain of the function (denoted by **dom** f) is the set of values in A for which the function is defined. The domain of the function is A if f is a total function. The co-domain of the function is B. The range of the function (denoted **rng** f) is a subset of the co-domain and consists of:

$$\textbf{rng}\ f = \{r \mid r \in B \text{ such that } f(a) = r \text{ for some } a \in A\}.$$

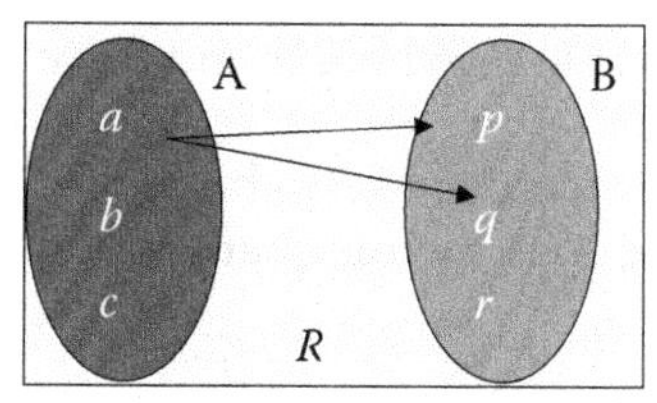

FIGURE 3.8
A Relation That Is Not a Function.

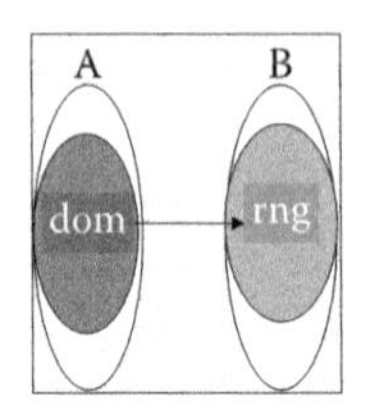

FIGURE 3.9
Domain and Range of a Partial Function.

Functions may be partial or total. A *partial function* (or partial mapping) may be undefined for some values of A, and partial functions arise regularly in the computing field (Fig. 3.9). *Total functions* are defined for every value in A and most functions encountered in mathematics are total.

Example 3.13 (Functions)

Functions are an essential part of mathematics and computer science, and there are many well-known functions such as the trigonometric functions $\sin x$, $\cos x$ and $\tan x$; the logarithmic function $\ln(x)$; the exponential functions e^x; and polynomial functions.

i. Consider the partial function $f: \mathbb{R} \to \mathbb{R}$ $f(x) = {}^1/_x$ (where $x \neq 0$). Then, this partial function is defined everywhere except for $x = 0$.
ii. Consider the function $f: \mathbb{R} \to \mathbb{R}$ where
$$f(x) = x^2$$
Then, this function is defined for all $x \in \mathbb{R}$.

Partial functions often arise in computing as a program may be undefined or fail to terminate for several values of its arguments (e.g., infinite loops). Care is required to ensure that the partial function is defined for the argument to which it is to be applied.

Consider a program P that has one natural number as its input and which fails to terminate for some input values. It prints a single real result and halts if it terminates. Then P can be regarded as a partial mapping from $\mathbb{N}$ to $\mathbb{R}$.

$$P: \mathbb{N} \to \mathbb{R}$$

Example 3.14

How many total functions $f: A \to B$ are there from A to B (where A and B are finite sets)?

Each element of A maps to any element of B, i.e., there are $|B|$ choices for each element $a \in A$. Since there are $|A|$ elements in A the number of total functions is given by:

$$|B|\ |B| \ldots |B| \quad (|A| \text{ times})$$
$$= |B|^{|A|} \quad \text{total functions between } A \text{ and } B.$$

Example 3.15

How many partial functions $f: A \to B$ are there from A to B (where A and B are finite sets)?

Each element of A may map to any element of B or to no element of B (as it may be undefined for that element of A). In other words, there are $|B| + 1$ choices for each element of A. As there are $|A|$ elements in A, the number of distinct partial functions between A and B is given by:

$$(|B| + 1)\ (|B| + 1) \ldots (|B| + 1) \quad (|A| \text{ times})$$
$$= (|B| + 1)^{|A|}$$

Two partial functions f and g are equal if:

1. $\text{dom } f = \text{dom } g$
2. $f(a) = g(a)$ for all $a \in \text{dom } f$.

A function f is less defined than a function g ($f \subseteq g$) if the domain of f is a subset of the domain of g, and the functions agree for every value on the domain of f.

1. $\text{dom } f \subseteq \text{dom } g$
2. $f(a) = g(a)$ for all $a \in \text{dom } f$.

The composition of functions is similar to the composition of relations. Suppose $f: A \to B$ and $g: B \to C$ then $g \circ f: A \to C$ is a function, and it is written as $g \circ f(x)$ or $g(f(x))$ for $x \in A$.

The composition of functions is not commutative and this can be seen by an example. Consider the function $f: \mathbb{R} \to \mathbb{R}$ such that $f(x)= x^2$ and the function $g: \mathbb{R} \to \mathbb{R}$ such that $g(x) = x + 2$. Then

$$g \circ f(x) = g(x^2) = x^2 + 2.$$
$$f \circ g(x) = f(x + 2) = (x + 2)^2 = x^2 + 4x + 4.$$

Clearly, $g \circ f(x) \neq f \circ g(x)$ and so the composition of functions is not commutative. The composition of functions is associative, as the composition of relations is associative and every function is a relation. For $f: A \to B$, $g: B \to C$, and $h: C \to D$ we have:

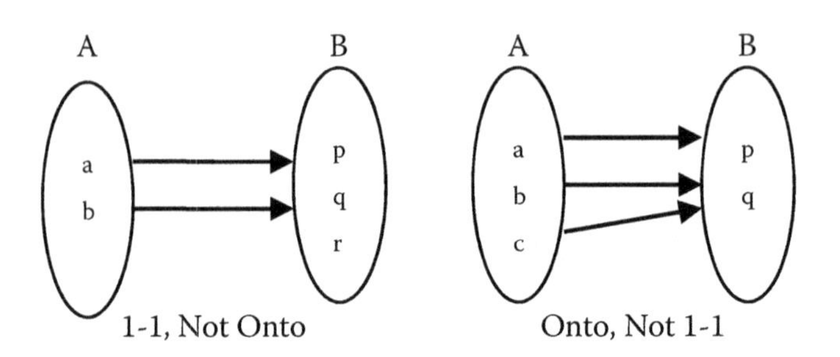

FIGURE 3.10
Injective and Surjective Functions.

$$h \ o(g \ o \ f) = (h \ o \ g)o \ f$$

A function f: $A \rightarrow B$ is *injective* (*one to one*) if

$$f(a_1) = f(a_2) \Rightarrow a_1 = a_2.$$

For example, consider the function f: $\mathbb{R} \rightarrow \mathbb{R}$ with $f(x) = x^2$. Then $f(3) = f(-3) = 9$ and so this function is not one to one.

A function f: $A \rightarrow B$ is *surjective* (*onto*) if given any $b \in B$ there exists an $a \in A$ such that $f(a) = b$. Consider the function f: $\mathbb{R} \rightarrow \mathbb{R}$ with $f(x) = x + 1$. Clearly, given any $r \in \mathbb{R}$ then $f(r - 1) = r$ and so f is onto (Fig. 3.10).

A function is *bijective* if it is one to one and onto (Fig. 3.11). That is, there is a one to one correspondence between the elements in A and B, and for each $b \in B$ there is a unique $a \in A$ such that $f(a) = b$.

The inverse of a relation was discussed earlier and the relational inverse of a function f: $A \rightarrow B$ clearly exists. The relational inverse of the function may or may not be a function.

However, if the relational inverse is a function it is denoted by f^{-1}: $B \rightarrow A$. A total function has an inverse if and only if it is bijective whereas a partial function has an inverse if and only if it is injective.

The identity function 1_A: $A \rightarrow A$ is a function such that $1_A(a) = a$ for all $a \in A$. Clearly, when the inverse of the function exists then we have that $f^{-1} o f = 1_A$ and $f^{-} o f^{-1} = 1_B$.

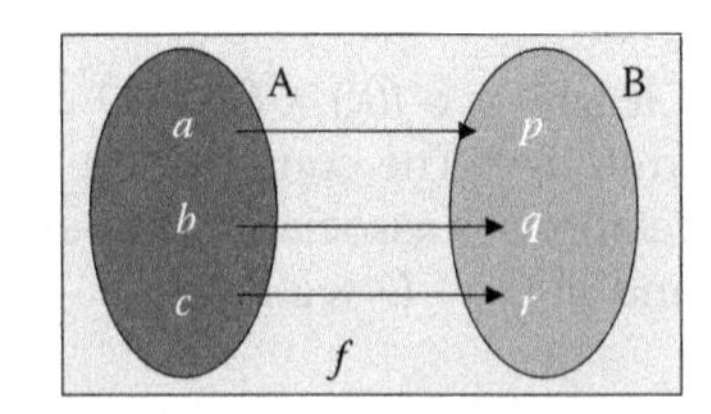

FIGURE 3.11
Bijective Function (One to One and Onto).

Theorem 3.3: (Inverse of Function)

A total function has an inverse if and only if it is bijective.

Proof

Suppose f: $A \rightarrow B$ has an inverse f^{-1}. Then we show that f is bijective.

We first show that f is one to one.

Suppose $f(x_1) = f(x_2)$ then

$$\begin{aligned} f^{-1}(f(x_1)) &= f^{-1}(f(x_2)) \\ \Rightarrow f^{-1} \circ f(x_1) &= f^{-1} \circ f(x_2) \\ \Rightarrow 1_A(x_1) &= 1_A(x_2) \\ \Rightarrow x_1 &= x_2 \end{aligned}$$

Next, we first show that f is onto. Let

$b \in B$ and let $a = f^{-1}(b)$ then

$$f(a) = f(f^{-1}(b)) = 1_B(b) = b \text{ and so } f \text{ is surjective}$$

The second part of the proof is concerned with showing that if f: $A \rightarrow B$ is bijective then it has an inverse f^{-1}. Clearly, since f is bijective we have that for each $a \in A$ there exists a unique $b \in B$ such that $f(a) = b$.

Define g: $B \rightarrow A$ by letting $g(b)$ be the unique a in A such that $f(a) = b$. Then we have that:

$$g \circ f(a) = g(b) = a \text{ and } f \circ g(b) = f(a) = b.$$

Therefore, g is the inverse of f.

3.4 Review Questions

1. What is a set? A relation? A function?
2. Explain the difference between a partial and a total function.
3. Explain the difference between a relation and a function.
4. Determine $A \times B$ where $A = \{a, b, c, d\}$ and $B = \{1, 2, 3\}$.
5. Determine the symmetric difference $A \Delta B$ where $A = \{a, b, c, d\}$ and $B = \{c, d, e\}$.

6. What is the graph of the relation $\leq$ on the set $A = \{2, 3, 4\}$?
7. What is the composition of S and R (i.e., $S \circ R$), where R is a relation between A and B, and S is a relation between B and C. The sets A, B, C are defined as $A = \{a, b, c, d\}$, $B = \{e, f, g\}$, $C = \{h, i, j, k\}$ and $R = \{(a, e), (b, e), (b, g), (c, e), (d, f)\}$ with $S = \{(e, h), (e, k), (f, j), (f, k), (g, h)\}$.
8. What is the domain and range of the relation R where $R = \{(a, p), (a, r), (b, q)\}$.
9. Deternine the inverse relation R^{-1} where $R = \{(a, 2), (a, 5), (b, 3), (b, 4), (c, 1)\}$.
10. Determine the inverse of the function $f: \mathbb{R} \times \mathbb{R} \rightarrow \mathbb{R}$ defined by

$$f(x) = \frac{x-2}{x-3} (x \neq 3) \quad \text{and } f(3) = 1$$

11. Give examples of injective, surjective and bijective functions.
12. Let $n \geq 2$ be a fixed integer. Consider the relation $\equiv$ defined by $\{(p, q): p, q \in \mathbb{Z}, n \mid (q - p)\}$
    a. Show $\equiv$ is an equivalence relation.
    b. What are the equivalence classes of this relation?

3.5 Summary

This chapter provided an introduction to set theory, relations and functions. Sets are collections of well-defined objects; a relation between A and B indicates relationships between members of the sets A and B; and functions are a special type of relation where there is at most one relationship for each element $a \in A$ with an element in B.

A set is a collection of well-defined objects that contains no duplicates. The Cartesian product of two sets allows a new set to be created from existing sets. The Cartesian product of two sets S and T (denoted $S \times T$) is the set of ordered pairs $\{(s,t) \mid s \in S, t \in T\}$.

A binary relation $R(A, B)$ is a subset of the Cartesian product $(A \times B)$ of A and B where A and B are sets. The domain of the relation is A and the codomain of the relation is B. The notation aRb signifies that there is a relation between a and b and that $(a, b) \in R$. An n-ary relation $R(A_1, A_2, \ldots A,_n)$ is a subset of $(A_1 \times A_2 \times \ldots \times A_n)$.

A total function $f: A \rightarrow B$ is a special relation such that for each element $a \in A$ there is exactly one element $b \in$ B. This is written as $f(a) = b$. A function is a relation but not every relation is a function.

The domain of the function (denoted by **dom** f) is the set of values in A for which the function is defined. The domain of the function is A provided that f is a total function. The co-domain of the function is B.

Notes

1. We distinguish between total and partial functions. A total function $f: A \rightarrow B$ is defined for every element in A whereas a partial function may be undefined for one or more values in A.
2. There are mathematical objects known as *multi-sets* or *bags* that allow duplication of elements. For example, a bag of marbles may contain three green marbles, two blue marbles and one red marble.
3. The British logician, John Venn, invented the Venn diagram. It provides a visual representation of a set and the various set theoretical operations. Their use is limited to the representation of two or three sets as they become cumbersome with a larger number of sets.
4. The natural numbers, integers and rational numbers are countable sets (i.e., they may be put into a one-to-one correspondence with the natural numbers), whereas the real and complex numbers are uncountable sets.
5. Cartesian product is named after René Descartes who was a famous 17th French mathematician and philosopher. He invented the Cartesian coordinates system that links geometry and algebra, and allows geometric shapes to be defined by algebraic equations.
6. Permutations and combinations are discussed in Chapter 4.
7. De Morgan's law is named after Augustus De Morgan, a 19th-century English mathematician who was a contemporary of George Boole.
8. We distinguish between total and partial functions. A total function is defined for all elements in the domain, whereas a partial function may be undefined for one or more elements in the domain.

4

Sequences, Series and Permutations and Combinations

The goal of this chapter is to provide an introduction to sequences and series, including arithmetic and geometric sequences and arithmetic and geometric series. Sequences and series are ubiquitous in mathematics and they may be finite or infinite. The odd numbers 1, 3, 5, 7, 9, … and the even numbers 2, 4, 6, 8, 10, … are examples of sequences. Sequences arise naturally in business mathematics as may be seen in the example of £1000 is invested at an interest rate of 10% compounded annually, and where the sequence lists the value of the investment each year:

£1000, £1100, £1210, £1331, £1464.10, ….

A finite series is the sum of the terms of a finite sequence, for example, 1 + 3 + 5 + 7 + 9 is the sum of the finite sequence 1, 3, 5, 7 and 9. An infinite series is the sum of the terms of an infinite sequence, and in some cases (e.g., in the case of certain geometric series) the sum approaches a limit called the limiting sum of the series. Otherwise, the series is divergent.

We derive formulae for the sum of an *arithmetic series* and *geometric series*, and we discuss the convergence of a geometric series when $|r| < 1$, and the limit of its sum as n gets larger and larger.

We consider the *counting principle* where one operation has m possible outcomes and a second operation has n possible outcomes. We determine that the total number of outcomes after performing the first operation followed by the second operation is $m \times n$.

A *permutation* is an arrangement of a given number of objects, by taking some or all of them at a time. The order of the arrangement is important, as the arrangement "abc" is different from "cba." A *combination* is a selection of a number of objects in any order, where the order of the selection is unimportant. That is, the selection "abc" is the same as the selection "cba." We determine the number of permutations (${}^nP_r$) of r objects from n objects, and the number of combinations (${}^nC_r$) of r objects from n objects.

DOI: 10.1201/9781003308140-4

4.1 Sequences and Series

A sequence $a_1, a_2, \ldots, a_n$ is any succession of terms (usually numbers), and the nth term of the sequence is written as a_n. The *Fibonacci sequence* is a famous sequence discovered in the 12th century and which has several applications in science and mathematics. Each term in the Fibonacci sequence (apart from the first two terms) is obtained from the sum of the previous two terms in the sequence.

$$1,\ 1,\ 2,\ 3,\ 5,\ 8,\ 13,\ 21,\ 34\ldots.$$

A sequence may be finite (with a finite number of terms) or infinite. The Fibonacci sequence is infinite whereas the sequence 2, 4, 6, 8, 10 is finite. We distinguish between convergent and divergent sequences, where a *convergent* sequence approaches a certain limiting value, as n gets larger and larger. Technically, we say that $\lim_{n\to\infty} a_n$ exists (i.e., the limit of a_n exists). Otherwise, the sequence is said to be *divergent*.

Often, there is a mathematical expression for the nth term in a sequence (e.g., for the sequence of even integers 2, 4, 6, 8, … the general expression for a_n is given by $a_n = 2n$). Clearly, the sequence of the even integers is divergent, as it does not approach a particular value, as n gets larger and larger. Consider the following sequence:

$$1,\ -1,\ 1,\ -1,\ 1,\ -1$$

Then, this sequence is divergent since it does not approach a particular value, as n gets larger and larger, since it continues to alternate between 1 and −1. The formula for the nth term in the sequence may be given by

$$(-1)^{n+1}$$

The sequence 1, ½, ⅓, ¼, …, $^1/_n$ … is convergent and it converges to 0. The nth term a_n in the sequence is given by $^1/_n$, and as n gets larger and larger a_n gets closer and closer to 0.

A series is the sum of the terms in a sequence, and the sum of the first n terms of the sequence $a_1, a_2, \ldots, a_n$ is given by $a_1 + a_2 + \ldots + a_n$ which is denoted by

$$\sum_{k=1}^{n} a_k$$

A series is convergent if its sum approaches a certain limiting value S as n gets larger and larger, and this is written formally as

$$\lim_{n \to \infty} \sum_{k=1}^{n} a_k = S$$

Otherwise, the series is said to be divergent.

4.2 Arithmetic and Geometric Sequences

An *arithmetic sequence* is a sequence of numbers such that the difference between consecutive terms is a constant. Consider the sequence 1, 4, 7, 10, … where each term is obtained from the previous term by adding the constant value 3. This is an example of an arithmetic sequence, and there is a difference of 3 between any term and the previous one. The general form of a term in this sequence is $a_n = 3n - 2$.

The general form of an *arithmetic sequence* is given by

$$a, \quad a + d, \quad a + 2d, \quad a + 3d, \ \ldots\ldots\ldots \ a + (n - 1)d, \ \ldots\ldots$$

The value a is the initial term in the sequence, and the value d is the constant difference between a term and its successor. If the common difference d is positive then the terms in the sequence are increasing and grow towards positive infinity. If the common difference d is negative then the terms in the sequence are decreasing and grow towards negative infinity.

The nth term of an arithmetic sequence is given by $a + (n-1)d$. For the sequence, 1, 4, 7, …, we have $a = 1$ and $d = 3$, and the sequence is not convergent. In fact, all arithmetic sequences (apart from the constant sequence $a, a, \ldots, a$ which converges to a) are divergent.

A *geometric sequence* is a sequence of numbers where each term after the first is determined by multiplying the previous term by a common ratio (i.e., a fixed non-zero number) r. Consider the sequence 1, 3, 9, 27, 81, … where each term (apart from the first) is achieved from the previous term by multiplying it by the constant value 3. This is an example of a geometric sequence, and the general form of a geometric sequence is given by

$$a, \quad ar, \quad ar^2, \quad ar^3, \ \ldots, \ ar^{n-1}$$

The first term of the geometric sequence is a and r is the common ratio, and the nth term of a geometric sequence is ar^{n-1}. Each term is obtained from the previous one by multiplying it by the common ratio r. For the sequence 1, 3, 9, 27, the value of a is 1 and the value of r is 3.

A geometric sequence is convergent if $-1 < r < 1$, and in this case there is exponential decay of the sequence towards 0, and the sequence converges to 0. A geometric sequence is also convergent if $r = 1$, as for this case it is simply the constant sequence $a, a, a, \ldots$, which converges to a. For the case when $r = -1$ we have the alternating sequence $a, -a, a, -a, \ldots$, which is divergent.

For the case where $r > 1$, there will be exponential growth towards positive or negative infinity (depending on the sign of the initial term), and the sequence is divergent. For where $r < -1$ there will be exponential growth towards positive or negative infinity due to the alternating sign, and the sequence is divergent.

4.3 Arithmetic and Geometric Series

An arithmetic series is the sum of the terms in an arithmetic sequence, and a geometric sequence is the sum of the terms in a geometric sequence. It is possible to derive a simple formula for the sum of the first n terms in arithmetic and geometric series.

Formula for Arithmetic Series

We write the series in two ways: first, the normal left to right addition of the terms in the series, and then we write the addition in reverse, and then we add both series together.

$$Sn = a + (a + d) + (a + 2d) + (a + 3d) + \ldots\ldots + (a + (n - 1)d)$$

$$Sn = a + (n - 1)d + a + (n - 2)d + \ldots\ldots + \quad + (a + d) + a$$

$$2Sn = [2a + (n - 1)d] + [2a + (n - 1)d] + \ldots. + [2a + (n - 1)d] \quad (n \textit{ times})$$

$$2Sn = n \times [2a + (n - 1)d]$$

Therefore, we conclude that

$$S_n = \frac{n}{2}[2a + (n - 1)d]$$

Example (Arithmetic Series)

Find the sum of the first n terms in the following arithmetic series 1, 3, 5, 7, 9.

Solution

Clearly, $a = 1$ and $d = 2$. Therefore, applying the formula we get

$$S_n = \frac{n}{2}[2.1 + (n - 1)2] = \frac{2n^2}{2} = n^2$$

Formula for Geometric Series

We develop a formula for the sum of a geometric series by noting that for a geometric series we have

$$\begin{aligned} S_n &= a + ar + ar^2 + ar^3 + \ldots\ldots..+ar^{n-1} \\ \Rightarrow rS_n &= \quad ar + ar^2 + ar^3 + \ldots\ldots..+ar^{n-1} + ar^n \\ \hline \Rightarrow rS_n - S_n &= ar^n - a \\ &= a(r^n - 1) \\ \Rightarrow (r-1)S_n &= a(r^n - 1) \end{aligned}$$

Therefore, we conclude that (where $r \neq 1$) that

$$S_n = \frac{a(r^n - 1)}{r - 1} = \frac{a(1 - r^n)}{1 - r}$$

The case of when $r = 1$ corresponds to the arithmetic series $a + a + \ldots + a$, and the sum of this series is simply na, and the sum of the infinite series is divergent. The sum of the first n terms of the geometric series when $r = -1$ is either a or 0 (depending on whether n is odd or even as it is an alternating sequence). The sum of the infinite geometric series is therefore divergent when $r = -1$. The geometric series converges when $|r| < 1$ as $r^n \to 0$ as $n \to \infty$, and so:

$$S_n \to \frac{a}{1 - r} \quad \text{as } n \to \infty$$

Example (Geometric Series)

Find the sum of the first n terms in the following geometric series $1, {}^1/_2, {}^1/_4, {}^1/_8, \ldots$ What is the sum of the series?

Solution

Clearly, $a = 1$ and $r = {}^1/_2$. Therefore, applying the formula we get

$$S_n = \frac{1(1 - 1/2^n)}{1 - 1/2} = \frac{(1 - 1/2^n)}{1 - 1/2} = 2(1 - 1/2^n)$$

The sum of the infinite series is the limit of the sum of the first n terms as n approaches infinity. This is given by

$$\lim_{n \to \infty} S_n = \lim_{n \to \infty} 2(1 - 1/2^n) = 2$$

4.4 Permutations and Combinations

A permutation is an arrangement of a given number of objects, by taking some or all of them at a time. A combination is a selection of a number of objects where the order of the selection is unimportant. The number of permutations and combinations are defined in terms of the factorial function $n!$, where $n! = n(n-1) \dots 3.2.1$.

The difference between permutations and combinations may be seen in a simple example. Suppose we wish to select a pair of objects from four objects A, B, C and D. Then the number of permutations (selections where the order is relevant) is 12 and is given by

AB AC AD BC BD CD BA CA DA CB DB DC

Each of the 12 selections is a permutation and they are a permutation of four objects taken two at a time. In general, if there are n objects available from which to select and k objects are selected then the number of permutations is given by ${}^nP_k$.

The number of combinations (selections where the order of the selection is irrelevant) is six and is given by

AB AC AD BC BD CD

The combination AB is the same as BA (i.e., the order is irrelevant) and each combination is a selection from four objects taken two at a time. In general, if there are n objects available from which to select and k objects are selected then the number of combinations is given by ${}^nC_k$.

PRINCIPLES OF COUNTING

a. Suppose one operation has m possible outcomes and a second operation has n possible outcomes, then the total number of possible outcomes when performing the first operation **followed by** the second operation is $m \times n$. (***Product Rule***).

b. Suppose one operation has m possible outcomes and a second operation has n possible outcomes then the possible outcomes of the first operation or the second operation is given by $m + n$. (*Sum Rule*)

Example (Counting Principle)

Suppose a dice is thrown and a coin is then tossed. How many different outcomes are there and what are they?

Solution

There are six possible outcomes from a throw of the dice: 1, 2, 3, 4, 5 or 6, and there are two possible outcomes from the toss of a coin: H or T. Therefore, the total number of outcomes is determined from the product rule as $6 \times 2 = 12$. The outcomes are given by

(1, H), (2, H), (3, H), (4, H), (5, H), (6, H), (1, T), (2, T), (3, T), (4, T), (5, T), (6, T)

Example (Counting Principle (b))

Suppose a dice is thrown and if the number is even a coin is tossed and if it is odd then there is a second throw of the dice. How many different outcomes are there?

Solution

There are two experiments involved with the first experiment involving an even number and a toss of a coin. There are three possible outcomes that result in an even number and two outcomes from the toss of a coin. Therefore, there are $3 \times 2 = 6$ outcomes from the first experiment.

The second experiment involves an odd number from the throw of a dice and the further throw of the dice. There are three possible outcomes that result in an odd number and six outcomes from the throw of a dice. Therefore, there are $3 \times 6 = 18$ outcomes from the second experiment.

Finally, there are six outcomes from the first experiment and 18 outcomes from the second experiment, and so from the sum rule there are a total of $6 + 18 = 24$ outcomes.

Pigeonhole Principle

The pigeonhole principle states that if n items are placed into m containers (with $n > m$) then at least one container must contain more than one item (Fig. 4.1).

FIGURE 4.1
Illustration of Pigeonhole Principle.

Examples (Pigeonhole Principle)

a. Suppose there is a group of 367 people then there must be at least two people with the same birthday.
   This is clear since there are 365 days in a year (with 366 days in a leap year), and so there are at most 366 possible distinct birthdays in a year. The group size is 367 people, and so there must be at least two people with the same birthday.
b. Suppose that a class of 102 students is assessed in an examination (the outcome from the exam is a mark between 0 and 100). Then, there are at least two students who receive the same mark.
   This is clear as there are 101 possible outcomes from the test (as the mark that a student may achieve is between 0 and 100), and as there are 102 students in the class and 101 possible outcomes from the test, then there must be at least two students who receive the same mark.

4.4.1 Permutations

A permutation is an arrangement of a number of objects (taken k at a time) in a definite order.

Consider the three letters A, B and C. If these letters are written in a row then there are six possible arrangements:

ABC or ACB or BAC or BCA or CAB or CBA

There is a choice of three letters for the first place, then there is a choice of two letters for the second place, and there is only one choice for the third place. Therefore, there are $3 \times 2 \times 1 = 6$ possible arrangements.

In general, the total number of different arrangements (permutations) of n objects (taken n at a time) is given by $n! = n(n-1)(n-2) \ldots 3.2.1$.

Consider the four letters A, B, C and D. How many arrangements are there of the four letters (taking two letters at a time with no repetition of the letters)?

There are four choices for the first letter and three choices for the second letter, and so there are 12 possible arrangements. These are given by

AB AC AD BA BC BD CA CB CD DA DB DC

The total number of arrangements of n different objects taking r at a time ($r \leq n$) is given by

$$^{n}P_{r} = n(n-1)(n-2) \ldots (n-r+1).$$

This may also be written as

$$^{n}P_{r} = \frac{n!}{(n-r)}$$

Example (Permutations)

Suppose A, B, C, D, E and F are six students. How many ways can they be seated in a row if:

i. There is no restriction on the seating.
ii. A and B must sit next to one another.
iii. A and B must not sit next to one another.

Solution

For unrestricted seating the number of arrangements is given by 6.5.4.3.2.1 = 6! = 720.

For the case where A and B must be seated next to one another, then consider A and B as one person, and then the five people may be arranged in 5! = 120 ways. There are 2! = 2 ways in which AB may be arranged, and so there are 2! × 5! = 240 arrangements.

AB	C	D	E	F

For the case where A and B must not be seated next to one another, then this is given by the difference between the total number of arrangements and the number of arrangements with A and B together, i.e., 720 – 240 = 480.

4.4.2 Combinations

A combination is a selection of a number of objects in any order, as the order of the selection is unimportant. That is, both AB and BA represent the same selection since the order is irrelevant.

The total number of permutations or arrangements of n different objects taking r at a time is given by ^nP_r, and we can determine the total number of combinations ^nC_r from this.

The number of combinations is the number of ways that r objects can be selected from n different objects (where the order of the selection is irrelevant), and as each selection may be permuted $r!$ times to form the permutations, the total number of permutations is $r!$ × total number of combinations. In another words:

$$^nP_r = r! \times {}^nC_r,$$

We may also write this as

$$nC_r = {}^nP_r/r!$$

$$\binom{n}{r} = \frac{n!}{r!\,(n-r)!} = \frac{n(n-1)\ldots(n-r+1)}{r!}$$

It is clear from the definition that

$$\binom{n}{r} = \binom{n}{n-r}$$

Example 1 (Combinations)

How many ways are there to choose a team of 11 players from a panel of 15 players?

Solution

Clearly, the number of ways is given by $\binom{15}{11} = \binom{15}{4}$
That is, 15.14.13.12/4.3.2.1 = 1365.

Example 2 (Combinations)

How many ways can a committee of four people be chosen from a panel of 10 people where

i. There is no restriction on membership of the panel.
ii. A certain person must be a member.
iii. A certain person must not be a member.

Solution

For (i) with no restrictions on membership the number of selections of a committee of four people from a panel of 10 people is given by

$$\binom{10}{4} = 210$$

For (ii) where one person must be a member of the committee then this involves choosing three people from a panel of nine people and is given by

$$\binom{9}{3} = 84$$

For (iii) where one person must not be a member of the committee then this involves choosing four people from a panel of nine people, and is given by

$$\binom{9}{4} = 126$$

4.5 Review Questions

1. Determine the formula for the general term and the sum of the following arithmetic sequence

$$1, \ 4, \ 7, \ 10, \ldots.$$

2. Write down the formula for the nth term in the following sequence:

$$1/4, 1/12, 1/36, 1/108, ...$$

3. Find the sum of the following geometric sequence:

$$1/3, 1/6, 1/12, 1/24, ...$$

4. How many different five-digit numbers can be formed from the digits 1, 2, 3, 4, 5 where
   i. No restrictions on digits and repetitions allowed.
   ii. The number is odd and no repetitions are allowed.
   iii. The number is even and repetitions are allowed.
5. i. How many ways can a group of five people be selected from nine people?
   ii. How many ways can a group be selected if two particular people are always included?
   iii. How many ways can a group be selected if two particular people are always excluded?

4.6 Summary

This chapter provided a brief introduction to sequences and series, including arithmetic and geometric sequences and arithmetic and geometric series. We derived formulae for the sum of an arithmetic series and geometric series, and we discussed the convergence of a geometric series when $|r| < 1$.

We considered counting principles including the product and sum rules. The product rule is concerned with where one operation has m possible outcomes and a second operation has n possible outcomes then the total number of possible outcomes when performing the first operation followed by the second operation is $m \times n$.

We discussed the pigeonhole principle, which states that if n items are placed into m containers (with $n > m$) then at least one container must contain more than one item. We discussed permutations and combinations where permutations are an arrangement of a given number of objects, by taking some or all of them at a time. A combination is a selection of a number of objects in any order, and the order of the selection is unimportant.

5

Simple Interest and Applications

Interest is the additional payment that a borrower makes to the lender and is separate to the repayment of the principal. We distinguish between simple and compound interest, with simple interest is calculated on the principal only, whereas compound interest is calculated on both the principal and the accumulated interest from previous compounding periods. That is, *simple interest* is always calculated on the original principal, whereas for *compound interest*, the interest is added to the principal sum, so that interest is also earned on the accumulated interest for the next compounding period. In other words, compound interest includes interest on the interest earned in previous compounding periods whereas simple interest does not.

A customer pays interest to borrow from the bank and so the amount that is repaid by the customer is always greater than the amount borrowed (unless there is a period of negative interest rates in the economy). The rate of interest is the interest amount paid/received over a period divided by the principal amount borrowed or lent (expressed as a percentage).

The interest paid to a saver depends on the type of bank account held with most accounts paying compound rather than simple interest. In practice, interest is most generally calculated on a daily, monthly or yearly basis.

We discuss future and present values where the *future value* is the amount that the principal will grow to at a given rate of interest over a period of time, whereas the *present value* of an amount to be received at a given date in the future is the principal that will grow to that amount at a given rate of interest over that period of time.

We discuss *promissory notes* that are a written promise by one party to pay a certain sum of money (with or without interest) on a particular date to another party. A promissory note may be interest bearing (where the rate of interest is stated in the note) or non-interest-bearing (where there is no rate of interest specified) and these are termed treasury bills.

5.1 Simple Interest

Savers receive interest for placing deposits at the bank for a period of time, whereas lenders pay interest on their loans to the bank. Simple interest is

DOI: 10.1201/9781003308140-5

generally paid on *term deposits* (these are usually short-term fixed-term deposits for 3, 6 or 12 months) or short-term investments or loans. The interest earned on a savings account depends on the principal amount placed on deposit at the bank, the period of time that it will remain on deposit, and the specified rate of interest for the period.

For example, if €1000 is placed on deposit at a bank with an interest rate of 10% per annum for 2 years, then it will earn a total of €200 in simple interest for the period. The interest amount is calculated by

$$\frac{1000 * 10 * 2}{100} = \text{Euro } 200$$

SIMPLE INTEREST FORMULA

The general formula for calculating the amount of simple interest A due for placing principal P on deposit at a rate r of interest (where r is expressed as a percentage) for a period of time T (in years) is given by

$$I = \frac{P \times r \times t}{100}$$

If the rate of interest r is expressed as a decimal then the formula for the interest earned is simply

$$I = P \times r \times t$$

It is essential in using the interest rate formula that the units for time and rate of interest are the same.

Example (Simple Interest)

Calculate the simple interest payable for the following short-term investments:

1. £5000 placed on deposit for 6 months ($^1/_2$ year) at an interest rate of 4%.
2. £3000 placed on deposit for 1 month ($^1/_{12}$ year) at an interest rate of 5%.
3. £10,000 placed on deposit for 1 day ($^1/_{365}$ year) at an interest rate of 7%.

Solution (Simple Interest)

1. $A = 5000 * 0.04 * 0.5 =$ £100
2. $A = 3000 * 0.05 * 0.08333 =$ £12.50
3. $A = 10000 * 0.07 * 0.00274 =$ £1.92

We may derive various formulae from the simple interest formula $A = P \times r \times T$.

$$P = \frac{I}{rt} \quad r = \frac{I}{Pt} \quad t = \frac{I}{Pr}$$

Example (Finding the Principal, Rate or Time)

Find the value of the principal or rate or time in the following:

1. What principal will earn interest of €24.00 at 4.00% in 8 months?
2. Find the annual rate of interest for a principal of €800 to earn €50 in interest in 9 months.
3. Determine the number of months required for a principal of €2000 to earn €22 in interest at a rate of 5%.

Solution

We use the formulae derived from the simple interest formula to determine these.

1. $P = \frac{I}{rt} = \frac{24}{0.04 \times 0.6666} = €\,900$
2. $r = \frac{I}{Pt} = \frac{50}{800 \times 0.75} = 0.0833 = 8.33\%$
3. $t = \frac{I}{Pr} = \frac{22}{2000 \times 0.05} = 0.22 \text{ years} = 2.64 \text{ months}$

5.2 Computing Future and Present Values

The future value is what the principal will amount to in the future at a given rate of interest, whereas the present value of an amount to be received in the future is the principal that would grow to that amount at a given rate of interest.

5.2.1 Computing Future Value

A fixed-term account is an account that is opened for a fixed period of time (typically 3, 6 or 12 months). The interest rate is fixed during the term and thus the interest due at the maturity date is known in advance. That is, the customer knows what the *future value* (FV) of the investment is, and knows what is due on the maturity date of the account (this is termed the *maturity value*).

On the *maturity date*, both the interest due and the principal are paid to the customer, and the account is generally closed on this date. In some cases, the customer may agree to rollover the principal, or the principal and interest for a further fixed period of time, but there is no obligation on the customer to do so. The account is said to mature on the maturity date, and the maturity value (*MV*) or future value (*FV*) is given by the sum of the principal and interest.

COMPUTING FUTURE VALUE – (USING SIMPLE INTEREST)

$$MV = FV = P + I$$

Further, since $I = prt$ we can write this as $MV = P + prt$ Or

$$FV = MV = P(1 + rt)$$

Example (Computing Maturity Value)

Jo invests €10,000 in a short-term investment for 3 months at an interest rate of 9%. What is the maturity value of her investment?

Solution (Computing Maturity Value)

$$\begin{aligned} MV &= P(1 + rt) \\ &= 10000(1 + 0.09 * 0.25) \\ &= €10{,}225 \end{aligned}$$

5.2.2 Computing Present Values

The present value of an amount to be received at a given date in the future is the principal that will grow to that amount at a given rate of interest over that period of time. We computed the maturity value of a given principal at a given rate of interest r over a period of time t as:

COMPUTING PRESENT VALUE – (USING SIMPLE INTEREST)

$$MV = P(1 + rt)$$

Therefore, the present value ($PV = P$) of an amount V to be received t years in the future at an interest rate r (simple interest) is given by

$$PV = P = \frac{V}{(1 + rt)}$$

Example (Computing Present Value)

Compute the present value of an investment 8 months prior to the maturity date, where the investment earns interest of 7% per annum and has a maturity value of €920.

Solution (Computing Present Value)

$$V = 920, \quad r = 0.07, \quad t = 8/12 = 0.66$$

$$PV = P = \frac{V}{(1 + rt)} = \frac{920}{(1 + 0.07 * 0.66)} = €879.37$$

Example (Computing Present and Future Values)

Compute the present and future values of the following:

1. The present value of £3000 due in 3 months where the interest rate is 12%.
2. The future value of £3000 due in 6 months where the interest rate is 12%.

Solution (Computing Present and Future Values)

1. The present value of £3000 due in 3 months at an interest rate of 12% is given by

$$PV = P = \frac{V}{(1 + rt)} = \frac{3000}{(1 + 0.12 * 0.25)} = €2912.62$$

2. The future value of £3000 in 6 months where the rate is 12% is

$$\begin{aligned} FV &= P(1 + rt) \\ &= 3000(1 + 0.12 * 0.5) \\ &= 3000(1.06) \\ &= £3180 \end{aligned}$$

5.2.3 Equivalent Values and Payment

We showed in the previous example that the present value of £3000 to be received in 3 months at an interest rate of 12% is £2912.62, and that the future value of £3000 in 6 months time at an interest rate of 12% is £3180.

That is, at an interest rate of 12%, we have that £3000 to be received in 3 months is equivalent to receiving £2912.62 now, and that £3000 received now is equivalent to £3180 to be received in 6 months.

That is, when computing equivalent values we first determine the relevant date, and depending on whether the payment date is before or after this date we apply the present value or future value formula, and the calculations are as follows:

Payment Date	Formula
Late	Use future value formula $FV = P(1 + rt)$
Early	Use present value formula $PV = P = \frac{V}{(1+rt)}$

Example (Equivalent Values)

Compute the payment due for the following:

1. A payment of \$2000 is due in 1 year. It has been decided to repay early and payment is to be made today. What is the equivalent payment that should be made given that the interest rate is 10%?
2. A payment of \$2000 is due today. It has been agreed that payment will instead be made 6 months later. What is the equivalent payment that will be made at that time given that the interest rate is 10%?

Solution (Equivalent Values)

1. The original payment date is 12 months from today and is now being made 12 months earlier than original date. Therefore, we compute the present value of \$2000 for 12 months at an interest rate of 10%.

$$PV = P = \frac{V}{(1 + rt)} = \frac{2000}{(1 + 0.1 * 1)} = €1818.18$$

2. The original payment date is today but has been changed to 6 months later, and so we compute the future value of \$2000 for 6 months at an interest rate of 10%.

$$FV = P(1 + rt) = 2000(1 + 0.1 * 0.5) = 2000(1.05) = \$2100$$

Example (Equivalent Values)

A payment of €5000 that is due today is to be replaced by two equal payments (we call the unknown payments value x) due in 4 and 8 months, where the interest rate is 10%. Find the value of the replacement payments.

Solution (Equivalent Values)

The sum of the present value of the two (equal but unknown) payments is €5000.

The present value of x (received in 4 months) is

$$PV = P = \frac{V}{(1+rt)} = \frac{x}{(1 + 0.1 * 0.33)} = \frac{x}{1.033} = 0.9678\,x$$

The present value of x (received in 8 months) is

$$PV = P = \frac{V}{(1+rt)} = \frac{x}{(1 + 0.1 * 0.66)} = \frac{x}{1.066} = 0.9375\,x$$

Therefore,

$$\begin{aligned} 0.9678x + 0.9375x &= €5000 \\ \Rightarrow 1.9053x &= €5000 \\ \Rightarrow x &= €2624.26 \end{aligned}$$

5.3 Promissory Notes

A *promissory note* is a written promise by one party to pay a certain sum of money (with or without interest) on a particular date to another party. A promissory note may be interest bearing (where the rate of interest is stated in the note), or non-interest-bearing (where there is no rate of interest specified) and the latter are termed treasury bills.

The maturity value of an interest-bearing promissory note MV where P is the face value of the note, r is the interest rate and t is the interest rate period from the date of issue to the date of maturity is given by

$$MV = P + I = P(1 + rt)$$

Example (Maturity Value – Interest-Bearing Promissory Note)

Calculate the maturity value of a 3-month €10,000 promissory note with interest rate of 6% dated July 1, 2019.

Solution (Maturity Value – Interest-Bearing Promissory Note)

The date of issue is July 1, 2019 and as the term is 3 months the maturity date is September 30, 2019. The number of elapsed days is 92 days (or $^{92}/_{365}$ = 0.252 years). The maturity value is

$$MV = P(1 + rt) = 10000(1 + 0.06 * 0.252) = 10000(1.01512) = €10,151.20$$

The present value of a promissory note is its value at any time prior to the maturity date of the note. It is calculated by determining the present date and the maturity date, the maturity value MV, the prevailing interest rate r and the duration t between today's date and the maturity date. It is given by the present value formula:

$$PV = \frac{MV}{(1 + rt)}$$

Example (Present Value – Interest-Bearing Promissory Note)

A 90-day promissory note with an interest rate of 10% and face value of €5000 is issued on July 15, 2019.

1. Calculate the maturity value of the note.
2. If the time value of money is 7% find the present value of the note on September 1, 2019.
3. Calculate what the present value of the note on September 1, 2019 would be if the time value of money is 10%.
4. Explain the difference between the value in (3) and (4).

Solution (Present Value – Interest-Bearing Promissory Note)

The date of issue is July 15, 2019 and as the term of the note is 90 days the maturity date is October 14, 2019.

1. The maturity value of the note is

$$\begin{aligned} MV &= P(1 + rt) \\ &= 5000(1 + 0.1 * 90/365) \\ &= 5000(1 + 0.1 * 0.2466) \\ &= 5000(1.02466) \\ &= €5123.30 \end{aligned}$$

2. The maturity date of the note is October 14, 2019 and so the number of days from 1 September to 14 October is 43 days ($^{43}/_{365} = 0.1178$).

$$PV = \frac{MV}{(1 + rt)} = \frac{5123.30}{(1 + 0.07 * 0.1178)} = \frac{5123.30}{1.00825} = € \, 5081.39$$

3. The present value when the time value of money is 10% is given by

$$PV = \frac{MV}{(1 + rt)} = \frac{5123.30}{(1 + 0.1 * 0.1178)} = \frac{5123.30}{1.01178} = € \, 5063.65$$

4. An interest-bearing promissory note becomes more valuable when the interest rate of the note exceeds the rate at which money is worth. It becomes less valuable where the interest rate of the note is less than the rate at which money is worth.

5.4 Treasury Bills

A *treasury bill* (T-bill) is a promissory note that does not carry an interest rate as such, and instead the bill is issued at a discount to the investor, and the face value of the bill is paid to the investor on the maturity date. The term of a treasury bill is generally 91,182 or 364 days (Fig. 5.1).

Treasury bills are short-term debt that are issued by a government to potential investors at an auction, and the investors are in effect lending money to the government, and so treasury bills are backed by the state. The investors are mainly banks and financial institutions, and they purchase a treasury bill at a discount to its face value. For example, a 182-day treasury bill with a face value of $10,000 may be sold at the auction at a discount of 5% from the face value (i.e., for $9500), and as the government pays the investor $10,000 at the maturity date, and the investor makes a profit of $500 (which is considered the interest on the bill).

The government pays the face value of the bill on the maturity date, and the discounted price is a discount on the face value of the bill. In the example where the investor pays $9,500 for a $10,000 182-day T-bill the discount is 5% on the face value of the bill, and the investor earns $500 in interest at an interest rate of $^{500}/_{9500} * 100 = 5.62\%$ for the term of the treasury bill (this is the interest rate applied for the term and the annual rate of interest is 11.27%, i.e., $^{5.62}/_{182} * 365$).

There are often secondary markets where the investor may buy or sell treasury bills prior to their maturity date, and the value of the bill may

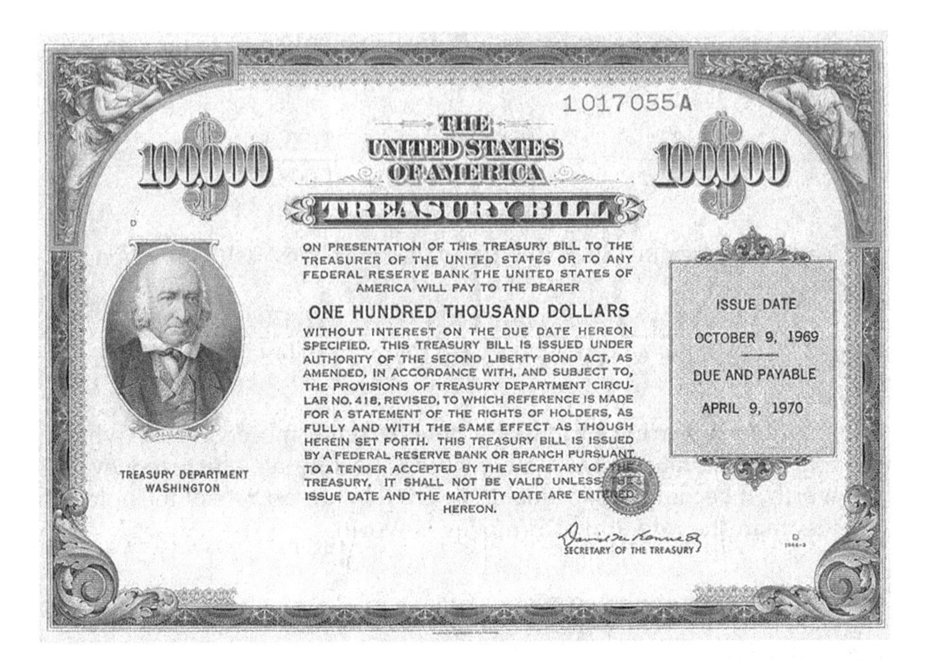

FIGURE 5.1
US Treasury Bill.

increase/decrease depending on factors such as government fiscal policies, the rate of inflation and interest rates (the rate that money is worth). The present value of the treasury bill is determined from the maturity value, the number of days to the maturity date and the rate that money is worth (interest rate):

$$PV = \frac{MV}{(1 + rt)}$$

Example (Treasury Bill)

An investor bought a 182-day treasury bill with a face value of $10,000 to yield an annual return of 4%. What was the purchase price of the bill?

Solution (Treasury Bill)

The purchase price is the present value of $10,000 where the rate is 4% and the time t is $^{182}/_{365} = 0.4986$.

$$PV = \frac{MV}{(1 + rt)} = \frac{10000}{(1 + 0.04 * 0.4986)} = \frac{10000}{1.0199} = €9804.46$$

Example (Treasury Bill)

An investor purchased a 364-day treasury bill 315 days before the maturity date to yield an annual rate of 5%, where the T-bill has a face value of €20,000. He sells the bill 90 days later to yield 6%.

1. How much did the investor pay for the treasury bills?
2. How much did the investor sell the treasury bills for?
3. What rate of return did the investor obtain?

Solution (Treasury Bill)

1. The purchase price of the treasury bill is the present value of €20,000 where the rate r is 5% and the time t is $^{315}/_{365} = 0.863$.

$$PV = \frac{MV}{(1 + rt)} = \frac{20000}{(1 + 0.05 * 0.863)} = \frac{20000}{1.04315} = \text{€ } 19{,}172.27$$

2. The bill is sold 90 days later (i.e., 315 – 90 = 225 days before the maturity date) and so $t = {}^{225}/_{365} = 0.6164$ and the yield is 6%

$$PV = \frac{MV}{(1 + rt)} = \frac{20000}{(1 + 0.06 * 0.6164)} = \frac{20000}{1.0369} = \text{€}19{,}286.66$$

3. The investor bought the T-bill for €19,172.27 and sold it 90 days later for €19,286.66 and so the gain is €114.39.
   We recall the formula $I = P \times r \times t$ where $I = 114.39$, $P = \text{€}19{,}172.27$, $t = {}^{90}/_{365} = 0.2466$ and so we calculate r as

$$r = \frac{I}{Pt} = \frac{114.39}{19{,}172.27 * 0.2466} = 0.0242 = 2.42\%$$

For more information on banking and financial services, see [RoH:12].

5.5 Review Questions

1. What is simple interest?
2. Calculate the simple interest payable on an investment of £12,000 placed on deposit for 9 months at an interest rate of 8%.

3. What principal will earn interest of €45.00 at an interest rate of 6.00% in 6 months?
4. What is the future value of the investment of €5000 invested for 8 months at an interest rate of 9%?
5. Explain the terms present value and future value.
6. What is the present value of £5000 due in 4 months where the interest rate is 12%?
7. What is the future value of £5000 in 6 months at an interest rate of 12%?
8. What is meant by equivalent value?
9. A payment of $10000 is due in 1 year and it has been decided to repay early with payment today. What is the equivalent payment that should be made given that the interest rate is 6%?
10. Explain the difference between a promissory note and a treasury bill.
11. Calculate the maturity value of a 90-day promissory note with an interest rate of 6% and face value of €6000 is issued on March 15, 2019.
12. An investor bought a 91-day treasury bill with a face value of $5000 to yield an annual return of 5%. What was the purchase price of the bill?

5.6 Summary

This chapter discussed simple interest, which is calculated on the principle only, whereas compound interest includes interest that is paid on the interest earned in previous compounding periods. The rate of simple interest is determined from the amount of interest paid/received over a period divided by the principal amount borrowed or lent (expressed as a percentage).

The interest paid to a saver depends on the type of bank account held with most accounts paying compound rather than simple interest. In practice, interest is most commonly calculated on a daily, monthly or yearly basis.

We discussed the future value that a principal will grow to at a given rate of interest over a period of time, and the present value of an amount to be received in the future is the principal that will grow to that amount at a given rate of interest over that period.

We discussed promissory notes that are a written promise by one party to pay a certain sum of money (with or without interest) on a particular date to another party. A promissory note may be interest bearing or non-interest-bearing.

6

Compound Interest and Applications

Simple interest is calculated on the principal only, whereas compound interest is calculated on both the principal and the accumulated interest of previous periods. That is, *simple interest* is always calculated on the original principal and is used in the calculation of interest on short-term deposits, loans and investments. However, for *compound interest*, the interest is added to the principal sum so that interest is also earned on both principal and accumulated interest in the next compounding period.

In other words, compound interest always includes interest on the accumulated interest earned in previous periods (with the exception of the first interest period) whereas simple interest does not. Compound interest is generally used for long-term investments and loans, and its calculation is more complicated than that of simple interest.

The time value of money is the concept that the earlier that cash is received the greater its value to the recipient. Similarly, the later that a cash payment is made, the lower its value to the payee and the lower its cost to the payer. This is due to the fact that money received now can be invested and earns interest, and also that the purchasing power of money received later is impacted by the inflation during the period. The future value of a principal P invested for n years at a compound rate r of interest per compounding period is given by $A = P(1 + r)^n$.

The present value of a given amount A that will be received n years in the future at an interest rate r for each compounding period is the principal that will grow to that amount and is given by $P = A(1 + r)^{-n}$.

A long-term promissory note has a term greater than 1 year and may be bought or sold prior to its maturity date. The calculation of its value is similar to that used in the calculation of the value of short-term promissory notes (except that the compounding formula is employed).

6.1 Compound Interest

The calculation of compound interest is more complicated as may be seen from the following example:

DOI: 10.1201/9781003308140-6

Example (Compound Interest)

Calculate the interest earned and what the new principal will be on €1000, which is placed on deposit at a bank, with an interest rate of 10% per annum (compound) for 3 years.

Solution

At the end of year 1, €100 of interest is earned, and this is added to the existing principal making the new principal (at the start of year 2) €1000 + €100 = €1100. At the end of year 2, €110 is earned in interest, and this is added to the principal making the new principal (at the start of year 3) €1100 + €110 = €1210. Finally, at the end of year 3, a further €121 is earned in interest, and so the new principal is €1331 and the total interest earned for the 3 years is the sum of the interest earned for each year (i.e., €331). This may be seen in Table 6.1.

The new principal each year is given by a geometric sequence (recall a geometric sequence is a sequence in the form $a, ar, ar^2, \ldots, ar^n$). For this example, we have $a = 1000$, and as the interest rate is 10% = $^1/_{10}$ = 0.1 we have $r = (1 + 0.1)$, and so the sequence is:

$$1000,\ 1000(1.1),\ 1000(1.1)^2,\ 1000(1.1)^3,\ \ldots\ldots$$

COMPOUND INTEREST FORMULA

If a principal amount P is invested for n years at a rate r of interest (r is expressed as a decimal) then it will amount to:

$$A = FV = P(1 + r)^n$$

For our example above, $A = 1000$, $n = 3$ and $r = 0.1$. Therefore,

$$\begin{aligned} A &= 1000(1.1)^3 \\ &= €1331 \text{ (as before)} \end{aligned}$$

TABLE 6.1

Calculation of Compound Interest

Year	Principal	Interest Earned	New Principal
1	€1000	€100	€1100
2	€1100	€110	€1210
3	€1210	€121	€1331

A principal amount P invested for n years at a rate r of simple interest (r is expressed as a decimal) will amount to:

$$A = FV = P(1 + rt)$$

The principal €1000 invested for 3 years at a rate of interest of 10% (simple interest) will amount to:

$$A = 1000(1 + 0.1 * 3) = 1000(1.3) = €1300$$

There are variants of the compound interest formula to cover situations where there are m-compounding periods per year. For example, interest may be compounded annually, semi-annually (with two compounding periods per year), quarterly (with four compounding periods per year), monthly (with 12 compounding periods per year), or daily (with 365 compounding periods per year).

MULTIPLE COMPOUNDING PERIODS – INTEREST RATE

The periodic rate of interest (i) per compound period is given by the nominal annual rate of interest (r) divided by the number of compounding periods (m) per year:

$$i = \frac{\text{Nominal Rate}}{\text{\#compounding periods}} = \frac{r}{m}$$

For example, if the nominal annual rate is 10% and interest is compounded quarterly then the period rate of interest per quarter is $^{10}/_{4} = 2.5\%$. That is, compound interest of 2.5% is calculated at the end of each quarter and applied to the account.

The number of compounding periods for the total term of a loan or investment is given by the number of compounding periods per year (m) multiplied by the number of years of the investment or loan.

$$n = \#years \times m$$

Example (Compound Interest – Multiple Compounding Periods)

An investor places £10,000 on a term deposit that earns interest at 8% per annum compounded quarterly for 3 years and 9 months. At the end of the term, the interest rate changes to 6% compounded monthly and it is invested for a further term of 2 years and 3 months.

1. How many compounding periods are there for 3 years and 9 months?
2. What is the value of the investment at the end of 3 years and 9 months?
3. How many compounding periods are there for 2 years and 3 months?
4. What is the final value of the investment at the end of the 6 years?

Solution (Compound Interest – Multiple Compounding Periods)

1. The initial term is for 3 years and 9 months (i.e., 3.75 years), and so the total number of compounding periods is given by n = #years * m, where #years = 3.75 and $m = 4$. Therefore, $n = 3.75 * 4 = 15$.
2. The nominal rate of interest r is 8% = 0.08, and so the interest rate i per quarterly compounding period is 0.08/4 = 0.02.
Therefore at the end of the term, the principal amounts to

$$\begin{aligned} A &= FV_1 = P(1+i)^n \\ &= 10000(1+0.02)^{15} \\ &= 10000(1.02)^{15} \\ &= 10000(1.3458) \\ &= £13{,}458 \end{aligned}$$

3. The term is for 2 years and 3 months (i.e., 2.25 years), and so the total number of compounding periods is given by n = #years * m, where #years = 2.25 and $m = 12$. Therefore, $n = 2.25 * 12 = 27$.
4. The new nominal interest rate is 6% = 0.06 and so the interest rate i per compounding period is ${}^{0.06}/_{12} = 0.005$. Therefore, at the end of the term, the principal amounts to

$$\begin{aligned} A &= FV_2 = FV_1(1+i)^n \\ &= 13458(1+0.005)^{27} \\ &= 13458(1.005)^{27} \\ &= 13458 * 1.14415 \\ &= £15{,}398 \end{aligned}$$

6.2 Present Value

The time value of money is the concept that the earlier that cash is received the greater its value to the recipient. Similarly, the later that a cash payment is made, the lower its value to the payee and the lower its cost to the payer.

This is clear if we consider the example of a person who receives $1000 now and a person who receives $1000 in 5 years from now. The person who receives $1000 now is able to invest it and to receive compound interest on the principal, whereas the other person who receives $1000 in 5 years earns no interest during the period. Further, the inflation during the period means that the purchasing power of $1000 is less in 5 years' time than it is today.

We presented the general formula for what the future value of a principal P invested for n compounding periods at a compound rate r of interest per compounding period as

$$A = P(1 + r)^n$$

The present value of a given amount A that will be received in the future is the principal ($P = PV$) that will grow to that amount where there are n compounding periods and the rate of interest is r for each compounding period.

COMPOUND INTEREST FORMULA

The present value of an amount A received in n compounding periods at an interest rate r for the compounding period is given by

$$P = \frac{A}{(1 + r)^n}$$

We can also write the present value formula as $PV = P = A\ (1 + r)^{-n}$.

Example (Present Value)

Find the principal that will amount to $10,000 in 5 years at 8% per annum compounded quarterly.

Solution (Present Value)

The term is 5 years = 5 ∗ 4 = 20 compounding period. The nominal rate of interest is 8% = 0.08 and so the interest rate i per compounding period is $^{0.08}/_{4} = 0.02$. The present value is then given by

$$\begin{aligned} PV &= A(1 + i)^{-n} = FV(1 + i)^{-n} \\ &= 10000(1.02)^{-20} \\ &= \$6729.71 \end{aligned}$$

The difference between the known future value of \$10,000, and the computed present value (i.e., the principal of \$6729.71) is termed the compound discount and represents the compound interest that accumulates on the principal over the period of time. It is given by

$$\text{Compound Discount} = FV - PV$$

For this example the compound discount is \$10,000 – 6729.71 = \$3270.29.

Example (Present Value)

Elodie is planning to buy a home entertainment system for her apartment. She can pay £1500 now, or pay £250 now and £1600 in 2 years' time. Which option is better if the nominal rate of interest is 9% compounded monthly?

Solution (Present Value)

There are $2 * 12 = 24$ compounding periods and the interest rate i for the compounding period is ${}^{0.09}/_{12} = 0.0075$. The present value of £1600 in 2 years' time at an interest rate of 9% compounded monthly is

$$\begin{aligned} PV &= FV(1+i)^{-n} \\ &= 1600(1.0075)^{-24} \\ &= 1600/1.1964 \\ &= £1337.33 \end{aligned}$$

The total cost of the second option is £250 + 1337.33 = £1587.33

Therefore, Elodie should choose the first option since it is cheaper by £87.33 (i.e., £1587.33 – 1500).

6.3 Long-Term Promissory Notes

We discussed short-term promissory notes in Chapter 5, where a promissory note is a written promise by one party to pay a certain sum of money (with or without interest) to another party. A long-term promissory note has a term greater than 1 year and may be bought or sold before its maturity date. The principles involved in calculating the value of a promissory note are similar to those used in the calculation of the value of short-term promissory notes except that the compounding formula is employed (Fig. 6.1).

The maturity value of a promissory note is first determined (for a non-interest-bearing note, this is the same as the face value of the note). The value of a promissory note at a point in time is determined from the

FIGURE 6.1
A $1000 Promissory Note in 1840. Public Domain.

maturity value MV of the note, the current date and the maturity date of the note, the number of compound periods n to the maturity date of the note, and the discount rate r (or more commonly i) is determined from r by dividing by the number of compounding periods m. The value of the promissory note at a given point in time when there are n compounding periods remaining to maturity is given by the present value formula:

$$PV = MV(1 + i)^{-n}$$

Example (Non-Interest-Bearing Long-Term Promissory Note)

Find the value of a non-interest-bearing note of $5,000 that is discounted 2 years prior to maturity where the interest rate is 8% per annum compounded monthly.

Solution (Non-Interest-Bearing Long-Term Promissory Note)

First, the maturity value of the note is $5000 (as the note is non-interest-bearing with the maturity value equal to the face value). The period of time to maturity is 2 years, and as the interest is compounded monthly the number of compounding periods n is $2 * 12 = 24$. The interest rate per period $i = 0.08/12 = 0.00667$ and so the value of the promissory note is

$$\begin{aligned} PV &= MV(1 + i)^{-n} \\ &= 5000(1 + 0.00667)^{-24} \\ &= 5000(1.00667)^{-24} \\ &= 5000 * 0.8525 \\ &= \$4262.64 \end{aligned}$$

Next, we consider the calculation of the present value of an interest-bearing promissory note, and this involves calculating the maturity value of the note using its associated interest rate and then determining its present value at the discount date at the discount rate of interest.

Example (Interest-Bearing Long-Term Promissory Note)

A 3-year interest-bearing promissory note of $10,000 was issued on June 1, 2016 at an interest rate of 8% compounded quarterly. The note is discounted on June 1, 2018 at a discount rate of 6% compounded semi-annually. Find the discount value on this date.

Solution (Interest-Bearing Long-Term Promissory Note)

First, the maturity value of the note is determined and the period of time to maturity is 3 years and so the number of compounding periods is given by $n = 3 * 4 = 12$. The interest rate i per quarterly period is $8\%/4 = 2\% = 0.02$. Therefore, the maturity value of the promissory note is:

$$\begin{aligned} MV &= P(1+i)^n \\ &= 10000(1+0.02)^{12} \\ &= 10000(1.02)^{12} \\ &= 10000 * 1.2682 \\ &= \$12{,}682 \end{aligned}$$

Next, we determine the present value of $12,682 1-year prior to its maturity at a discount rate of 6% compounded semi-annually. The number of compounding periods is $n = 1 * 2 = 2$ and the interest rate i is $6\%/2 = 3\% = 0.03$. Therefore, the present value of the promissory note on the discount date is given by

$$\begin{aligned} PV &= MV(1+i)^{-n} \\ &= 12682(1+0.03)^{-2} \\ &= 12682(1.03)^{-2} \\ &= 12682 * 0.94259 \\ &= \$11{,}954 \end{aligned}$$

6.4 Equivalent Values

We discussed equivalent values and the time value of money in Chapter 5. When two sums of money are to be paid/received at different times they

FIGURE 6.2
Equivalent Weights.

are not directly comparable as such, and a point in time (the focal date) must be chosen to make the comparison (Fig. 6.2).

The choice of focal date determines whether the present value or future value formula will be used. That is, when computing equivalent values we first determine the focal date, and then depending on whether the payment date is before or after this reference date we apply the present value or future value formula.

If the due date of the payment is before the focal date, then we apply the future value *FV* formula:

$$FV = P(1 + i)^n$$

If the due date of the payment is after the focal date, then we apply the present value *PV* formula:

$$PV = FV(1 + i)^{-n}$$

Example (Equivalent Values)

A debt value of €1000 that was due 3 months ago, €2000 that is due today, and €1500 that is due in 18 months are to be combined into one payment due 6 months from today at 8% compounded monthly. Determine the amount of the single payment.

Solution (Equivalent Values)

The focal date is 6 months from today and so we need to determine the equivalent values E_1, E_2 and E_3 of the three payments on this date, and we then replace the three payments with one single payment $E = E_1 + E_2 + E_3$ that is payable 6 months from today.

The equivalent value E_1 of €1000 which was due 3 months ago in 6 months from today is determined from the future value formula where the number of interest periods $n = 6 + 3 = 9$. The interest rate per period is 8%/12 = 0.66% = 0.00667.

$$\begin{aligned} FV &= P(1+i)^n \\ &= 1000(1+0.00667)^9 \\ &= 1000(1.00667)^9 \\ &= 1000 * 1.0616 \\ &= €1061.60 \end{aligned}$$

The equivalent value E_2 of €2000 which is due today in 6 months is determined from the future value formula, where the number of interest periods $n = 6$. The interest rate per period is = 0.00667.

$$\begin{aligned} FV &= P(1+i)^n \\ &= 2000(1+0.00667)^6 \\ &= 2000(1.00667)^6 \\ &= 2000 * 1.0407 \\ &= €2081.40 \end{aligned}$$

The equivalent value E_3 of €1500 which is due in 18 months from today is determined from the present value formula, where the number of interest periods $n = 18 - 6 = 12$. The interest rate per period is = 0.00667.

$$\begin{aligned} PV &= FV(1+i)^{-n} \\ &= 1500(1+0.0067)^{-12} \\ &= 1500(1.0067)^{-12} \\ &= €1384.49 \end{aligned}$$

$$\begin{aligned} E &= E_1 + E_2 + E_3 \\ &= 1061.60 + 2081.40 + 1384.49 \\ &= €4527.49 \end{aligned}$$

Example (Equivalent Values – Replacement Payments)

Liz was due to make a payment of £2000 today. However, she has negotiated a deal to make two equal payments: the first payment is to be made 1 year from now and the second payment 2 years from now. Determine the amount of the equal payments where the interest rate is 9% compounded quarterly and the focal date is today.

Solution (Equivalent Values – Replacement Payments)

Let x be the value of the equal payments. The first payment is made in 1 year and so there are $n = 1 * 4 = 4$ compounding periods and the second payment is made in 2 years and so there are $n = 2 * 4 = 8$ compounding periods. The interest rate i is $9\%/4 = 2.25\% = 0.0225$.

The present value E_1 of a sum x received in 1 year is given by

$$\begin{aligned} PV &= FV(1+i)^{-n} \\ &= x(1+0.0225)^{-4} \\ &= x(1.0225)^{-4} \\ &= 0.9148x \end{aligned}$$

The present value E_2 of a sum x received in 2 years is given by

$$\begin{aligned} PV &= FV(1+i)^{-n} \\ &= x(1+0.0225)^{-8} \\ &= x(1.0225)^{-8} \\ &= 0.8369x \end{aligned}$$

The sum of the present value of E_1 and E_2 is £2000 and so we have

$$\begin{aligned} 0.9148x + 0.8369x &= 2000 \\ 1.7517x &= 2000 \\ x &= £1141.75 \end{aligned}$$

6.5 Analysing the Business Case for a Project

A project is generally carried out to achieve tangible benefits such as greater sales or cost savings, and many companies have a rigorous process to evaluate potential projects to ensure that only those projects that will deliver the greatest business return are carried out.

The business case describes the reason and justification for the project, and it needs to be aligned to the business strategy. It is based on the expected costs of the project, the associated risks and the expected business benefits and savings. A project should proceed only if it has a valid business case, and the project should be terminated should its business case cease to exist.

The business case describes the problem or opportunity, the options that are available to solve the problem or take advantage of the opportunity, and the preferred solution. The recommended solution is then subjected to an investment appraisal to ensure that resources are not wasted, and that there is an acceptable return on the investment.

It is therefore essential that the project makes business sense and provides a financial return to the organization. A project will generally result in extra income (or savings) being earned by the company in future years, but it requires the commitment of financial resources now for its implementation. There may be risks that could prevent the project from achieving its objectives, and these need to be identified and managed appropriately. The project should cover its costs in the sense that the present value of the expected future payments or savings (NPV) should exceed the financial cost of the project.

There should always be a business case for a project, and the business case should be monitored during project execution to ensure that it still remains valid, and should the business case cease to exist during project execution the project should immediately be terminated to avoid wasting resources as there is no longer any sound rational for the project.

The business case process describes the problem or business opportunity, the options that are available as a solution, the preferred solution, an estimate of the cost and timelines, the initial risks and an investment appraisal of the preferred solution (Table 6.2). It may include the following:

- Description of problem/business opportunity
- List of potential solutions
- Preferred solution
- Analysis (costs, timescales, risks and benefits)
- Investment analysis.

Example (Investment Analysis)

A business is evaluating whether to embark on a process improvement project that will deliver financial savings over 3 years. The costs and savings associated with the project are summarized in Table 6.3, and the business needs to determine whether it should authorize or reject the project based on its projected costs and savings. The discount rate used by the company is 8% (compounded annually). Should the project be authorized or rejected? Would the project cover its costs if the discount rate was adjusted to 10%?

TABLE 6.2

Terminology in Investment Analysis

Section	Description
Net benefit/net benefit after tax	Net benefit reflects the total benefits less the total cost in any 1 year. The Net Cash benefits reflect the net cash benefit after tax has been deducted.
Discount rate %	The discount rate reflects two measures namely risk and interest. It reflects the cost of borrowing money and the company's attitude to risk (e.g., the interest rate charged on a mortgage). Discounting is used to convert money expected in the future back to a single common period, the present (using the present value formula and is based on the fact that a sum of money received now is worth more than a sum of money received in the future).
Net present value (NPV) of net benefit	The net present value (NPV) of a project is the financial value of the project in today's money. It takes the time value of money into account using the discount rate. The sum of the discounted benefits is computed and the project costs deducted to give the extra profit that the company will make. A negative NPV means that the project does not cover its costs, and an NPV of zero means that the project breaks even.
Cost of capital	The cost of capital refers to the rate that investors expect (in the long term) to receive in return for investing in the company. It may be determined by calculating the cost of debt, which is the after-tax interest rate on loans and bonds. An alternative more complex approach involves estimating the cost of equity from analysing shareholder's expected return implicit in the price they have paid to buy or hold their shares.
Economic value added (EVA)	The EVA reflects the value of the project and is the net benefit after tax minus the total cost of capital. It determines if a business is earning more than its true cost of capital. A negative EVA means that the project doesn't generate any real profit whereas a positive EVA generates a profit. The higher the EVA, the more attractive the project will be.
Payback period (years)	The payback time is the period of time (expressed in years) that it takes to recover all project costs. This may be done on a discounted/non-discounted basis.
Internal rate of return (IRR)	The internal rate of return (IRR) is the discount rate (interest rate) at which the present value of the future cash flow of an investment equals the cost of the investment. Alternatively, it can be seen as the % to which the discount rate needs to rise for the project to only break even. The higher the IRR the more attractive the project.

Solution (Investment Analysis)

The total cost of the project is £74,000 and so we need to determine if the total savings cover the costs. That is, we determine the present value of savings E_1, E_2 and E_3 where $E = E_1 + E_2 + E_3$ and compare this to the costs to make the decision.

TABLE 6.3
Projected Costs and Savings

Cost Area	Amount	Savings	Amount
Consultancy	£15,000	Year 1	£35,000
Training	£12,000	Year 2	£30,000
Materials	£6,000	Year 3	£25,000
Effort	£37,000		
Expenses	£4,000		
Total costs	£74,000	Total savings	

$$E_1 = 35000(1.08)^{-1} = 35000 * 0.9259 = 32{,}407$$
$$E_2 = 30000(1.08)^{-2} = 30000 * 0.8573 = 25{,}720$$
$$E_3 = 25000(1.08)^{-3} = 25000 * 0.7938 = 19{,}845$$

$$E = E_1 + E_2 + E_3 = 32{,}407 + 25{,}720 + 19{,}845 = £77{,}972$$

The NPV is the total savings less the total costs = $S - V$ = £3,972 and so the project should be authorized.

For the second part, we proceed in a similar manner and calculate that

$$E_1 = 31{,}818$$
$$E_2 = 24{,}793$$
$$E_3 = 18{,}783$$

$$E = E_1 + E_2 + E_3 = £75{,}394$$

The total savings are still in excess of the projected costs (£1394), and so the project should proceed. For more information on banking and financial services, see [RoH:12].

6.6 Review Questions

1. An investor places £5000 on a term deposit that earns interest at 8% per annum compounded quarterly for 2 years and 3 months. Find the value of the investment on the maturity date.
2. Explain the difference between present and future value when interest is compounded.

3. Find the principal that will amount to \$12,000 in 3 years at 6% per annum compounded quarterly.
4. How many years will it take a principal of \$5000 to exceed \$10,000 at a constant annual growth rate of 6% compound interest?
5. What is the present value of \$5000 to be received in 5 years' time at a discount rate of 7% compounded annually?
6. What is a long-term promissory note?
7. Find the value of a non-interest-bearing note of \$10,000 that is discounted 3 years prior to maturity where the interest rate is 6% per annum compounded monthly.
8. Explain the concept of equivalent values and when to use the present value/future value in its calculation.
9. A debt value of €2000 due 6 months ago, €5000 due today and €3000 due in 18 months are to be combined into one payment due 3 months from today at a rate of 6% compounded monthly. Determine the amount of the single payment.

6.7 Summary

Simple interest is calculated on the principal only, whereas compound interest is calculated on both the principal and the accumulated interest of previous periods. Compound interest is generally used for long-term investments and loans, and its calculation is more complicated than that of simple interest.

The time value of money is the concept that the earlier that cash is received the greater its value to the recipient, and vice versa for late payments. The future value of a principal P invested for n compounding periods at a compound rate r of interest per compounding period is given by $A = P\,(1 + r)^n$. The present value of a given amount A that will be received n compounding periods in the future, at an interest rate r for each compounding period is the principal that will grow to that amount and is given by $P = A(1 + r)^{-n}$.

A long-term promissory note has a term greater than 1 year and may be bought or sold prior to its maturity date. The calculation of the value of a promissory note is similar to that used in the calculation of the value of short-term promissory notes.

7

Banking and Financial Services

Banks are at the heart of the financial system and they play a key role in facilitating the flow of money throughout the economy. Banks provide a service where those with excess funds (i.e., the savers) may lend to those who need funds (i.e., the borrowers). Banks pay interest to its savers[1] and earn interest from its borrowers, and the spread between the interest rate given to savers and the interest rate charged to borrowers provides banks with the revenue to provide service to their clients, and to earn a profit from the provision of the services.

Bank lending provides funding to consumers and business to enable them to make investments and purchases that they may otherwise be unable to make, and so the banking system is an integral part of the modern economy. In another words, banks through their provision of services such as savings and lending contribute to economic well-being and growth in the state, by ensuring that those who have a need for funding and who have appropriate financial means to repay the loan are able to access the required funding.

Money is a medium of exchange that is accepted as a means for paying for goods and services. Money acts as a *medium of exchange* (it can be exchanged for goods or services), a *unit of account* (it is measured in a specific currency which allows the price of things to be compared) and a *store of value* (it can be saved and used at a later time). Money includes cash, demand deposits, checking accounts and so on. There are various definitions of the money supply in the economy, including the more formal M_1, M_2 and M_3 measures, where $M_1 \leq M_2 \leq M_3$. M_1 consists of all the cash in circulation; M_2 consists of M_1 plus small deposits, short-term deposits and money market funds; M_3 consists of M_2 plus large deposits, long-term deposits and money market funds.

A bank is a financial intermediary between savers and borrowers and is responsible for the safeguarding, transfer, exchanging and lending of money. Banking today is a rapid moving, 24-hour, around the world activity, where banks compete with each other and with other financial institutions in the provision of services to meet the needs of their clients. There are several types of banks which are included in Table 7.1 (Fig. 7.1).

Most of a bank's income comes from the interest that consumers or business make in loan repayments, and the difference between what the bank pays in interest to its depositors and the interest that it receives from its borrowers is the net income or revenue of the bank. The profit (or loss) of the bank is given by the deduction of the bank's expenses from the revenue.

DOI: 10.1201/9781003308140-7

TABLE 7.1
Types of Banks

Type	Description
Central banks	Central banks manage the money supply in a specific country (or in several countries – e.g., the Eurozone). They are generally independent of the government and implement the appropriate monetary policy for the state, and supervise banks, set interest rates, control inflation and deflation, and control the flow of currency. The Federal Reserve in the United States, the ECB in Europe and the Bank of England are all examples of a central bank.
Retail banks	Retail banking is the provision of banking services to the general public rather than to large companies. These banks provide services such as savings accounts, current accounts, personal loans and mortgages, debit and credit cards and insurance.
Commercial banks	Commercial banks are focused on supporting the needs of business, and they provide services such as bank accounts and loans to large corporations and small businesses.
Investment banks	Investment banks manage the trading of stocks, bonds and other securities between companies and investors, and they also provide financial advice to individuals and corporations, as well as managing investment portfolios and participating in an advisory role in mergers and acquisitions.
Credit unions	Credit unions provide similar services to banks, except that their fees tend to be lower as they are not-for-profit institutions owned by their members.

FIGURE 7.1
US Federal Reserve.

Banks may have other sources of income such as maintenance fees for current accounts, ATM transaction fees, fees for online transactions, commission from foreign exchange and fees from investments.

The main assets of the bank are the loans that the bank has given to its clients, but the bank may also have property, investments and other assets. The main liabilities of the banks are its customer deposit accounts, and as the customer could make large withdrawals or close the account at any time the liabilities are more liquid than the assets (a liquid asset is an asset that can be readily exchanged for cash). The bank requires sufficient reserves so that it can meet any large or unexpected demand for withdrawals, and so it cannot lend out its reserves and these are maintained onsite or at the central bank. The reserves consist of the required reserves as set by the central bank (typically 10% of deposits), and the bank may then issue loans with the excess funds above the required level of reserves.

7.1 Central Bank

A central bank is a national institution that is independent of the government. It is responsible for monetary policy, the regulation of the banks and financial institutions in the state, keeping inflation under control and within appropriate limits, providing financial services to banks and the government and supporting economic development.

Central banks have several tools to influence monetary policy such as reserve requirements, setting interest rates and quantitative easing. The *reserve requirement* is the amount of money that the banks must have on hand at the end of each day, and this ratio is used by the central bank to control the amount that may be lent by the banks. The reserves are there to protect against potential loan losses[2] as well as unexpected demand for withdrawals. The effect of an increase in the central bank reserves ratio is that the banks need to keep more money on hand, and so they have less money to lend out to borrowers. That is, its effects are a contraction in economic activity, whereas a decrease in the reserve ratio is expansionary as banks are required to keep less money on hand and so they may lend out more to their clients leading to an expansion in economic activity.

Central banks set the interest rates that they charge banks, and this directly influences the rates that banks charge consumers for loans and mortgages. An increase in interest rates makes money more expensive, and leads to a contraction in economic activity, thereby keeping inflation under control (a contractionary monetary policy). Similarly, a decrease in interest rate makes money less expensive, and leads to an expansion in economic activity, thereby preventing a recession (an expansionary monetary policy). If interest rates are too high, it can lead to a recession and even deflation, and if interest rates are too low then it can overstimulate the economy and lead to inflation. Therefore, the central bank needs to monitor the key economic data very carefully to ensure it makes the right decisions on interest rate policy.

Quantitative easing is where the central bank purchases government securities or other securities from the market place, with the goal of increasing the money supply in the financial system and thereby stimulating the economy by encouraging lending and investment.

Banks play an essential role in the economy, and in view of the financial crises that arise from time to time (e.g., the 2008–2009 banking crisis) it is important to manage financial risks in the banking sector, and especially to manage the risks of a liquidity or insolvency crisis. The role of the central bank is to monitor the banks to ensure that these serious risks do not materialize.

There are several well-known central banks such as the Bank of England in the UK, the Federal Reserve in the United States, the European Central Bank and the Bank of Japan. Sweden created the first central bank, the Riksbank, in 1668 and the Bank of England was created in 1694.

7.2 Retail Banks

Retail banks provide various financial services to individual customers rather than to companies and corporations. Retail banking is the visible face of banking to the general public, with bank branches located in the urban areas throughout the country. Retail banks provide the following services:

- Current accounts
- Savings accounts
- Personal loans
- Mortgages
- Business loans
- Overdrafts
- ATM banking
- Debit cards
- Credit cards
- Standing orders and direct debits
- Internet and mobile banking
- Electronic funds transfer

A *current account* is a bank account that allows access to everyday financial services such as payment of salary, setting up direct debits for the payment of bills and setting up standing orders to make regular and equal payments of rent or mortgage. A current account generally provides a debit card and a cheque book and may have an agreed overdraft facility. An *overdraft* is an additional amount of funds that the bank allows the customer to use when there is no money left in their account, and the overdraft is essentially a short-term loan up to a defined limit. The bank charges interest for any funds used in the overdraft and there may also be fees to set up the overdraft. Some current accounts offer interest and the accounts may be accessed online via the Internet or phone. There are often transaction and maintenance fees for a current account.

A *savings account* is an interest-bearing account held at a bank or some other financial institution. It generally pays a modest rate of interest and there may be a limit to the number of withdrawals that may be made per year, but the funds are generally easy to access.

A *personal loan* allows a set amount of money to be borrowed from a bank (usually €2000–€20,000) for a fixed number of years (usually 1–5 years). Personal loans may be used for short-term or medium-term borrowing (e.g., to embark on a home improvement project or to buy a new car), and once the loan and its term have been approved the repayment amount per month is calculated. The borrower must make this monthly payment (via direct debit or standing order) to ensure that the loan is paid back on time. It may be possible to pay off the loan early or to extend the loan for a longer period of time, and this depends on the terms and conditions associated with the loan.

A *mortgage* is a long-term loan (usually 20–25 years) that is used to finance the purchase of a property, and the mortgage is a secured loan which uses the property as security on the amount borrowed. This reduces the risk for the lender, but it means that the borrower is exposed to the risk of losing the property should difficulties arise in meeting the monthly repayments. A mortgage may be a fixed rate or variable rate, with a fixed-rate mortgage providing certainty on the amount of the monthly repayment.

A *business loan* is a loan that is taken out specifically for business purposes, and it may be either secured or unsecured. The approval of a business loan may be dependent on an examination of the company's business plan and accounts. The company will need to provide collateral for secured loans, which could potentially be lost if the business has difficulty in making the monthly repayments.

An *automated teller machine* (ATM) is an electronic machine that allows customers to carry out withdrawals from the ATM and to perform basic banking without the assistance of a bank teller. *Credit* and *debit cards* look identical and are 16-digit card numbers with an expiry date and a private personal identification number (PIN). Credit cards allow the customer to borrow money from the card issuer up to a defined limit in order to purchase items or withdraw cash from an ATM machine. A debit card allows the customer to draw on funds that they already have with the bank, and so the money used to purchase items or withdraw cash from an ATM machine is deducted directly from the customer's bank account. This helps to prevent the user from building up debt, as they are spending what they already have in their bank account. Credit cards generally provide better consumer protection than debit cards.

A *standing order* is an instruction that the customer makes to their bank to make regular and equal payments (e.g., weekly, monthly or quarterly) to another party. A *direct debit* is where the customer authorizes another party to collect money from their bank account whenever a payment is due (e.g., a utility bill).

Internet and *mobile banking* allow consumers to access their accounts online and to carry out basic banking operations such as checking their account balance, paying bills and transferring funds to another party within or outside the country. *Electronic funds transfer* is concerned with the transfer of funds from one party to another (usually by Internet or mobile banking), where the name, bank, IBAN and SWIFT bank code of the recipient need to be known.

7.3 Financial Instruments

A *financial instrument* is a financial contract between two parties, and it may be created, modified, traded and settled. Financial instruments may be either cash instruments whose values are directly determined by markets, or derivative instruments that derive their value from one or more underlying entities such as an asset, index or interest rate. The financial instruments discussed in this section include the following:

- Debentures
- Bonds
- Bills
- Derivatives
- Foreign exchange

Securities include debentures, bonds, bills, options and shares that may be traded in financial markets. A *debenture* is a long-term (typically more than 10 years) debt instrument that is unsecured by collateral, and it relies on the reputation and creditworthiness of the issuer. Debentures are generally issued by large corporations or governments, and they pay periodic interest at a rate called the coupon rate. They may have an interest rate exposure in a rising interest rate environment and there is also the risk of default should the issuer go through financial difficulties. Convertible debentures may be converted to equity after a specified period, whereas non-convertible debentures may not be converted and have a defined maturity date when the investor is repaid.

A *bond* is a common debt instrument used by corporations and governments, and it serves as an IOU between the issuer and the investor (bondholder). It is a long-term debt (greater than 1 year) and the investor receives periodic interest payments during the bond's term. The bondholder is repaid in full on the maturity date, and the debt is secured in the sense that it is backed by the assets of the issuer. There are minimal risks with bonds issued by the government as they are guaranteed by the state, and so treasury bonds generally pay a lower rate of interest than corporate bonds.

A *treasury bill* (T-bill) is a short-term financial instrument issued by the government (central bank) or corporations to deal with short-term funding needs such as expanding business activities or managing short-term government expenditure. Government-issued treasury bills are a very safe investment as they are backed by the state, and the investor is essentially lending money to the government. Treasury bills are sold at a discount to the face (*par*) value of the bill, where the issuer promises to pay the investor the face value of the bill on the maturity date of the bill. For example, a treasury bill with a par value of €5000 may be sold for €4750, where the discount rate is 5% (i.e., the investor pays €4750 for the bill and receives €5000 on the maturity date), and so the discount to the face value is €250 and the discount rate is 5%. Investors may buy or sell bills on the secondary market prior to their maturity, and the purchasing price of the bill is influenced by factors such as the interest rate set by the central bank and the rate of inflation.

A *derivative* is a contract between two or more parties, and it is a secondary security whose value is based on (derived from) the primary security that it is linked to. The primary financial securities include bonds, currencies, commodities, interest rates, market indices and stocks. A derivative's value is based on the underlying primary security, but ownership of the derivative does not imply ownership of the primary asset. Common derivatives that are used include future contracts, forward contracts, options, swaps and warrants (similar to options). For example, a future contract is a derivative since its value is affected by the performance of the underlying asset from the issue date to the exercise date.

Foreign exchange is concerned with the purchase or sale of foreign currency, and it is a common operation that tourists perform when visiting other countries. Similarly, companies that are active in importing or exporting will need to purchase/sell currency as part of international trade. However, such transactions are subject to currency exchange risk, since the transaction is generally in the future and may be adversely affected by currency exchange rate movements during the period. A change in the exchange rate may result in increased costs in purchasing the currency, or receiving less when selling the currency, and so the management of exchange rate risk is essential to a business. There are several foreign exchange tools that may be employed to manage (*hedge*) FX risk, and these include FX instruments for purchasing/selling currency using spot, forward, swaps and options. Foreign exchange is discussed in more detail in section 7.6.

7.4 Basic Mathematics of Annuities

An *annuity* is a sequence of equal payments made over a period of time, and the payment is usually made at the end of the payment interval. For example, for a hire purchase contract the buyer makes an initial deposit for an item, and then pays an equal amount per month (the payment is generally at the end of the month) up to a fixed end date. Personal loans from banks are paid back in a similar manner but without an initial deposit.

An investment annuity (e.g., a regular monthly savings scheme) is often paid at the start of the payment interval. A pension scheme involves two stages, with the first stage involving an investment of regular payments at the start of the payment interval up to retirement and the second stage involving the payment of the retirement annuity. The period of payment of a retirement annuity is usually for the remainder of a person's life (life annuity), or it could be for a period of a fixed number of years.

We could determine the final value of an investment annuity by determining the future value of each payment up to the maturity date and then adding them all together. Alternately, as the future values form a geometric series, we may derive a formula for the value of the investment by using the formula for the sum of a geometric series.

We may determine the present value of an annuity by determining the present value of each payment made and summing these, or we may also develop a formula to calculate the present value.

The repayment of a bank loan is generally with an *amortization annuity*, where the customer borrows a sum of money from a bank (e.g., a personal loan for a car or a mortgage for the purchase of a house). The loan is for a defined period of time, and its repayment is with a regular annuity, where each annuity payment consists of interest and capital repayment. The bulk of the early payments goes on interest due on the outstanding capital with smaller amounts going on capital repayments. However, the bulk of the later payments goes in repaying the capital with smaller amounts going on interest.

An *annuity* is a series of equal cash payments made at regular intervals over a period of time, and they may be used for investment purposes or paying back a loan or mortgage. We first consider the example of an investment annuity.

TABLE 7.2

Calculation of Future Value of Annuity

Year	Amount	Future Value (r = 0.1)
1	10,000	10,000 * 1.1^5 = €16,105
2	10,000	10,000 * 1.1^4 = €14,641
3	10,000	10,000 * 1.1^3 = €13,310
4	10,000	10,000 * 1.1^2 = €12,100
5	10,000	10,000 * 1.1^1 = €11,000
Total		€67,156

Example (Investment Annuity)

Sheila is investing €10,000 a year in a savings scheme that pays 10% interest every year. What will the value of her investment be after 5 years?

Solution (Invested Annuity)

Sheila invests €10,000 at the start of year 1 and so this earns 5 years of compound interest of 10% and so its future value in 5 years is given by $10000 * 1.1^5$ = €16,105. The future value of the payments that she makes is presented in Table 7.2.

Therefore, the value of her investment at the end of 5 years is the sum of the future values of each payment at the end of 5 years = 16,105 + 14,641 + 13,310 + 12,100 + 11,000 = €67,156.

We note that this is the sum of a geometric series and so in general if an investor makes a payment of A at the start of each year for n years at a rate r of interest then the investment value at the end of n years is

$$\begin{aligned} &A(1+r)^n + A(1+r)^{n-1} + \ldots + A(1+r) \\ &= A(1+r)[1 + A(1+r) + \ldots + A(1+r)^{n-1}] \\ &= A(1+r)\frac{(1+r)^n - 1}{(1+r) - 1} \\ &= A(1+r)\frac{(1+r)^n - 1}{r} \end{aligned}$$

We apply the formula to check our calculation.

$$\begin{aligned} &10000\,(1+0.1)\,\frac{(1+0.1)^5 - 1}{0.1} \\ &= 11000\frac{(1.1)^5 - 1}{0.1} \\ &= 11000\frac{(1.61051) - 1}{0.1} \\ &= €67{,}156 \end{aligned}$$

Note 7.1
We assumed that the annual payment was made at the start of the year.

ANNUITY FORMULA

The annuities payment is generally made at the end of the year (or compounding period) and so the formula is slightly different:

$$FV = A\,\frac{(1 + r)^n - 1}{r}$$

The future value formula is adjusted for multiple (m) compounding periods per year, where the interest rate for the period is given by $i = {}^r/_m$, and the number of payment periods n is given by where $n = tm$ (where t is the number of years). The future value of a series of payments of amount A (made at the start of the compounding period) with interest rate i per compounding period, where there are n compounding periods is given by:

$$FV = A\,(1 + i)\frac{(1 + i)^n - 1}{i}$$

ANNUITY FORMULA (MULTIPLE COMPOUNDING PERIODS)

The future value of a series of payments of amount A (made at the end of the compounding period) with interest rate i per compounding period, where there are n compounding periods is given by:

$$FV = A\,\frac{(1 + i)^n - 1}{i}$$

An annuity consists of a series of payments over a period of time, and so it is reasonable to consider its present value with respect to a discount rate r (this is applicable to calculating the present value of the annuity for mortgage repayments discussed in the next section).

The net present value of an annuity is the sum of the present value of each of the payments made over the period, and the method of calculation is clear from Table 7.3.

Example (Present Value Annuities)

Calculate the present value of a series of payments of $1000 with the payments made for 5 years at a discount rate of 10%.

TABLE 7.3

Calculation of Present Value of Annuity

Year	Amount	Present Value (r = 0.1)
1	1000	\$909.91
2	1000	\$826.44
3	1000	\$751.31
4	1000	\$683.01
5	1000	\$620.92
Total		\$3791

Solution (Present Value Annuities)

The regular payment A is 1000, the rate r is 0.1 and $n = 5$. The present value of the first payment received is $1000/1.1 = 909.91$ at the end of year 1; at the end of year 2 it is $1000/(1.1)^2 = 826.45$; and so on. At the end of year 5, its present value is 620.92. The net present value of the annuity is the sum of the present value of all the payments made over the 5 years, and it is given by the sum of the present values from Table 7.3. That is, the present value of the annuity is 909.91 + 826.44 + 751.31+ 683.01 + 620.92 = \$3791.

We may derive a formula for the present value of a series of payments A made over a period of n years at a discount rate of r as follows: clearly, the present value is given by

$$\frac{A}{(1+r)} + \frac{A}{(1+r)^2} + \ldots + \frac{A}{(1+r)^n}$$

This is a geometric series where the constant ratio is $\frac{1}{1+r}$ and the present value of the annuity is given by its sum.

PRESENT VALUE OF ANNUITY

The present value of a series of payments of amount A (made at the end of the compounding period) with interest rate r is given by

$$PV = \frac{A}{r}\left[1 - \frac{1}{(1+r)^n}\right]$$

For the example above, we apply the formula and get

$$
\begin{aligned}
PV &= \tfrac{1000}{0.1}\left[1 - \tfrac{1}{(1.1)^5}\right] \\
&= 10000(0.3791) \\
&= \$3791
\end{aligned}
$$

The annuity formula is adjusted for multiple (m) compounding periods per year, and the interest rate for the period is given by $i = {}^{r}/_{m}$, and the number of payment periods n is given by where $n = tm$ (where t is the number of years). For example, the present value of an annuity of amount A, with interest rate i per compounding period, where there are n compounding periods is given by:

PRESENT VALUE OF ANNUITY (MULTIPLE COMPOUNDING PERIODS)

The present value of a series of payments of amount A (made at the end of the compounding period) with interest rate i per period is given by

$$P = \frac{A}{i}\left[1 - \frac{1}{(1 + i)^n}\right]$$

Example (Retirement Annuity

Bláithín has commenced employment at a company that offers a pension in the form of an annuity that pays 5% interest per annum compounded monthly. She plans to work for 30 years and wishes to accumulate a pension fund that will pay her €2000 per month for 25 years after she retires. How much does she need to save per month to do this?

Solution (Retirement Annuities)

First, we determine the value that the fund must accumulate to pay her €2000 per month, and this is given by the present value of the 25-year annuity of €2000 per month. The interest rate r is 5% and as there are 12 compounding periods per year there are a total of $25 * 12 = 300$ compounding periods, and the interest rate per compounding period is 0.05/12 = 0.004166.

$$
\begin{aligned}
P &= 2000/0.004166[1-(1.004166)^{-300}] \\
&= €342,174.
\end{aligned}
$$

That is, her pension fund at retirement must reach €342,174 and so we need to determine the monthly payments necessary for her to achieve this. The future value is given by the formula:

$$FV = A\frac{(1+i)^{n+1}-1}{i}$$

and so

$$A = FV * i/\ [(1+i)^{n+1} - 1]$$

where $m = 12$, $n = 30 * 12 = 360$ and $i = 0.05/12 = 0.004166$ and $FV = 342174$.

$$\begin{aligned} A &= 342,174 * 0.004166/3.4863 \\ &= €408.87 \end{aligned}$$

That is, Bláithín needs to save €408.87 per month (€4906.44 per year) into her retirement account (sinking fund) for 30 years in order to have an annuity of €2000 per month for 25 years.

7.5 Loans and Mortgages

The purchase of a home or car requires a large sum of money, and so most purchasers need to obtain a loan from the bank to fund the purchase. Once the financial institution has approved the loan, the borrower completes the purchase and pays back the loan to the financial institution over an agreed period of time (the term of the loan). For example, a mortgage is generally paid back over 20–25 years, whereas a car loan is usually paid back in 5 years (Fig. 7.2).

An interest-bearing debt is *amortized* if both the principal and interest are repaid by a series of equal payments (with the exception of possibly the last payment) made at equal intervals of time. That is, the amortization of loans refers to the repayment of interest-bearing debts by a series of equal payments made at equal intervals of time. The debt is repaid by an amortization annuity, where each payment consists of both the repayment of the capital borrowed and the interest due on the capital for that time interval.

Mortgages and many consumer loans are repaid by this method, and the standard problem is to calculate what the annual (or monthly) payment should be to amortize the principal P in n years where the rate of interest is r.

FIGURE 7.2
Loan or Mortgage.

The present value of the annuity is equal to the principal borrowed, i.e., the sum of the present values of the payments must be equal to the original principal borrowed. That is:

$$P = \frac{A}{(1+r)} + \frac{A}{(1+r)^2} + \ldots + \frac{A}{(1+r)^n}$$

We may also use the formula that we previously derived for the present value of the annuity to get

$$P = \frac{A}{r}\left[1 - \frac{1}{(1+r)^n}\right]$$

We may calculate A by manipulating this formula to get:

REPAYMENT AMOUNT OF LOAN

The repayment amount A for principal P (n years at interest rate r) is given by

$$A = \frac{Pr}{\left[1 - \frac{1}{(1+r)^n}\right]}$$

$$A = Pr/[1 - (1+r)^{-n}]$$

Example (Amortization)

Joanne has taken out a €200,000 mortgage over 20 years at 8% per annum. Calculate her repayment amount per annum to amortize the mortgage.

Solution (Amortization)

We apply the formula to calculate her annual repayment:

$$A = \frac{Pr}{\left[1 - \frac{1}{(1+r)^n}\right]}$$

$$A = \frac{200000 * 0.08}{\left[1 - \frac{1}{(1+0.08)^{20}}\right]}$$

$$= \frac{160000}{\left[1 - \frac{1}{4.661}\right]}$$

$$= €20,370$$

We adjust the formula for the more common case where the interest is compounded several times per year (usually monthly), and so n = # years * # compoundings, and the interest rate per compounding period is $i = r/$# compoundings.

REPAYMENT AMOUNT OF LOAN (MULTIPLE COMPOUNDINGS)

The repayment amount A for principal P (n years at interest rate r) is given by

$$A = \frac{Pi}{\left[1 - \frac{1}{(1+i)^n}\right]}$$

$$A = Pi/[1 - (1+i)^{-n}]$$

Example (Amortization)

A mortgage of £150,000 at 6% compounded monthly is amortized over 20 years. Determine the following:

1. Repayment amount per month
2. Total amount paid to amortize the loan
3. The cost of financing.

Solution (Amortization)

The number of payments n = #years * payments per year = 20 * 12= 240.
The interest rate i = 6%/12 = 0.5% = 0.005.

1. We calculate the amount of the repayment A by substituting for n and i and obtain

$$
\begin{aligned}
A &= \frac{150000 * 0.005}{\left[1 - \frac{1}{(1+0.005)^{240}}\right]} \\
&= \frac{750}{\left[1 - \frac{1}{3.3102}\right]} \\
&= £1074.65
\end{aligned}
$$

2. The total amount paid is the number of payments $*$ amount of each payment = $n * A$ = 240 $*$ 1074.65 = £257,916.
3. The total cost of financing = total amount paid – original principal = 257,916 – 150,000 = £107,916.

Example (Amortization)

For the previous example, determine the following at the end of the first period:

1. The amount of interest repaid
2. The amount of the principal repaid
3. The amount of the principal outstanding.

Solution (Amortization)

The amount paid at the end of the first period is £1074.65.

1. The amount paid in interest for the first period is 150000 $*$ 0.005 = £750.
2. The amount of the principal repaid is £1074 – 750 = £324.65.
3. The amount of the principal outstanding at the end of the first interest period is £150,000 – 324.65 = £149,675.35.

The early payments of the mortgage mainly involve repaying interest on the capital borrowed, whereas the later payments mainly involve repaying the capital borrowed with less interest due. We can create an amortization table which shows the interest paid, the amount paid in capital repayment and the outstanding principal balance for each payment interval. Often, the last payment is different from the others due to rounding errors introduced and carried through.

Each entry in the amortization table includes interest, principal repaid and outstanding principal balance. The interest is calculated by the principle balance $*$ periodic interest rate i; the principal repaid is calculated by the payment amount – interest; and the new outstanding principal balance is given by the principal balance – principal repaid.

Example (Amortization)

A borrower is paying €1200 monthly on a 20-year mortgage where the interest rate is 4% compounded monthly. Determine the following:

1. The principal borrowed
2. The principal amount outstanding at the start of year 20.

Solution (Amortization)

The number of payment periods is $20 * 12 = 240$ and the interest rate per period is $4\%/12 = 0.33\% = 0.0033$.

1. The principal borrowed may be determined from the formula for calculating the regular payment to amortize the loan, and so we solve for the unknown value P in the formula.

We manipulate the formula to get that

$$
\begin{aligned}
A &= \frac{Pi}{\left[1 - \frac{1}{(1+i)^n}\right]} \\
P &= A[1-(1+r)^{-n}]/i \\
&= 1200[1-(1.0033)^{-240}]/0.0033 \\
&= €198{,}716.72
\end{aligned}
$$

2. The principal outstanding at the start of year 20 is the present value of the 12 payments made in year 20:

$$
\begin{aligned}
P &= A[1-(1+r)^{-n}]/i \\
&= 1200[1-(1.0033)^{-12}]/0.0033 \\
&= 1200 * 0.03876/0.0033 \\
&= €14{,}095.82
\end{aligned}
$$

7.6 Foreign Exchange

Foreign exchange (*Forex*) is the process of changing one currency for another for a variety of reasons such as tourism or international trade. The forex markets are the largest in terms of average daily volumes, and the daily value of FX activities is several trillion dollars. In another words, the daily activities of the forex market vastly exceed that of the bond or equity markets. The FX (or *forex*) market is a global market place for exchanging national currencies against one another, and currencies trade against each other as exchange rate pairs (e.g., EUR/USD, GBP/JPY).

The largest forex markets are the spot markets, but there are also derivative markets offering forwards, futures, options and currency swaps. The FX market may be used by the market participants for hedging against currency or interest rate risks, and international companies often use the forwards and futures markets to manage their exposure to foreign exchange risks. Many of the participants in the forex market are commercial and investment banks (who are trading on behalf of their clients), as well as individual investors and speculators. There are several popular foreign exchange operations, such as given in Table 7.4.

A company may decide to accept the exchange rate risk (perhaps the currency amount involved in the FX transaction is small), and the company will use the *FX spot rate* for buying/selling, which is the current exchange rate at which a currency may be bought or sold. The spot price is influenced by supply and demand and interest rates, and also the expected future performance of the currencies. These trades take 2 days for settlement.

An alternate approach is to use the *FX forward rate*, which is the exchange rate applicable to a transaction that will take place in the future. Both parties must honour the forward forex rate at the settlement date, irrespective of the movement of currencies during the period.

For *forwards markets*, the contracts are bought and sold over the counter (OTC) between the two parties which determine the terms of their

TABLE 7.4

FX Operations

Type	Description
Spot	The spot market is where currencies are bought and sold according to their current price, and the spot deal is a bilateral transaction where one party delivers the agreed upon currency amount to the other party in return for the agreed amount of a specified currency at an agreed exchange rate.
Forwards/Futures	The forwards and futures markets deal in contracts that represent claims to a specific currency at an agreed price per unit at a future date for settlement.
Swap	This is a simultaneous purchase and sale of identical amounts of one currency for another.
Option	This gives one party the contractual right to buy or sell (there is no legal obligation to exercise the right) a specific amount of currency at a fixed rate on a defined future date.

agreement. In the *futures market*, the contracts are bought and sold based upon a standard size and settlement date on public commodity markets (e.g., the Chicago Mercantile Exchange or the European Eurex Exchange). Both types of contract are legally binding and are typically settled for cash at the exchange in question on the date of expiry of the contract, but they may also be traded before they expire. The forwards and futures markets offer protection against adverse movements in exchanges rates when trading currencies, and they are especially important in international trade where they are used by large corporations for hedging against currency rate fluctuations, as the exchange rate is locked in for the transaction.

Example (Hedging)

Suppose a European company orders $100,000 of components from its US supplier for delivery in 2 months, and the cost of the purchase today when the exchange rate is €1 = $1.1 is €90,909. However, settlement is in 3 months time, and the company decides to settle with the spot rate on the settlement date. However, the exchange rate of the euro to the dollar deteriorates to parity over the 3 months, i.e., €1 = $1, and the cost of purchase is now €100,000 resulting in an FX loss of €9091.

However, if the European company had bought $100,000 on the forwards market, then it could have got a 3-month forward exchange rate of €1 = $1.07 and the cost of purchase would have been €93,458.

Another approach is to use an *FX swap* which is a simultaneous purchase and sale of identical amounts of one currency for another, and it involves a spot and an opposite forward transaction of the same amount that are executed simultaneously. It involves an initial exchange of the two currencies and a later reverse direction exchange, and an FX swap prevents exchange rate risks for either party.

Example (Swap)

A European company in the Euro zone is long on dollars and by selling dollars at the spot rate it can cover its expenses in Europe. However, suppose that it will be paying $50,000 to its US suppliers for various components in 3 months time, and so if it sells its available dollars at the spot rate and buys back the dollars in 3 months time, it is exposed to exchange rate risk. Instead, it performs an FX swap, which involves selling $50,000 at the spot rate and forward buying of $50,000 in 3 months.

An *FX option* gives one party (the option holder) the contractual right to buy or sell a specific amount of currency at a fixed rate on a defined future date. The buyer of an FX option has no obligation to exercise their rights, whereas the seller of the FX option is bound to honour the contract should the option holder decide to exercise their option.

The price of an FX option is dependent on the perceived risks involved in the contract. An FX contact may work like an insurance policy in that if the market moves in an adverse direction, then the option provides protection by removing the risk of unpredictable losses, whereby if the market moves in a positive direction then profits will follow.

7.6.1 Forex and Speculation

The foreign exchange market is subject to volatility caused by several factors such as interest rate changes, the economic climate, geopolitical events, natural catastrophes and so on. That is, financial risk arises due to uncertainty about the financial situation in the future, and a *speculator* is a person or company that aims to profit from changes that may increase or decrease one currency's value with respect to another.

Suppose that a trader forms the view (perhaps due to major differences in interest rates or to a major economic shock) that the exchange rate of the British pound to the US dollar (GBP/USD) will fall, and the trader decides to take a calculated risk and profit from this situation. The exchange rate today is £1 = $2 (i.e., GBP = 2 USD), and as the trader has £100,000 GBP, he sells these and buys dollars and so receives $200,000.

The trader is correct and there are major changes in the pound/dollar exchange rate, where it moves from £1 = $2 to parity £1 = $1. The trader went long on the US dollar (i.e., buys and holds US dollars) and short on British pounds (i.e., sells British pounds), and as a result has made a significant profit (The trader started with £100,000, sold the sterling holding and got $200,000, and if the dollars are now sold for pounds £200,000 will be received meaning a profit of £100,000.

However, if the trader's analysis was incorrect and the exchange rate moved in the opposite direction then significant losses could potentially be incurred. Trading currencies can be difficult and complex, and requires an understanding of economic fundamentals and calculated risk taking.

An investor may profit from the difference in interest rates between two countries by purchasing the currency with the higher interest rate, and shortening the currency with the lower interest rate. This is since higher interest rates in one country will lead to demand for its currency and a subsequent rise in its exchange rates, and so an astute investor who correctly predicts that rates will increase may potentially profit from the situation.

7.6.2 Foreign Exchange Spread

The foreign exchange spread refers to the difference between the bid and ask prices for a given currency pair. The bid price refers to the maximum price that a foreign exchange trader is prepared to pay to buy a certain currency, whereas the ask price is the minimum price that the currency dealer is willing to accept for the currency. It is often expressed as a percentage:

$$\text{Spread\%} = \frac{\text{Ask price} - \text{Bid price}}{\text{Ask price}} * 100$$

The foreign exchange rate spread is influenced by factors such as the trading volumes, economic/political risks and currency volatility.

7.7 Corporate Bonds

A bond is a debt obligation of a corporation to an investor, and investors who buy corporate bonds are, in effect, lending money to the company that is issuing the bond. The investor receives regular interest payments (usually every 6 months) at the defined interest rate (the *coupon rate* of the bond) during the lifetime of the bond, and the investor is repaid the capital in full on the maturity date of the bond. There are also *zero coupon* bonds that do not make interest payments as such, and instead the investor purchases the bond at a discount to the face value, and on the maturity date receives the face value of the bond (the discount is the difference between the face value and the issue price and reflects the interest earned during the lifetime of the bond). Companies use the money raised from a bond issue for different reasons such as growing the business, investing in research and development, refinancing, as well as mergers and acquisitions.

International bonds have an associated credit rating (carried out by credit agencies such as Standard and Poor or Moody's) that indicates the risk associated with the bond issuer (i.e., its creditworthiness). For example, the "triple-A" (AAA) investment grade rating indicates that the risks associated with the bond issuer are low, and that the issuer is highly likely to make the interest payments and repay the capital on time. Similarly, the speculative grade of BB (or *junk bonds*) indicates that there are significant risks associated with the issuer of the bond such as that the risk of default of payment of the interest or capital. Junk bonds often have higher rates of interest to attract investors, but the risk of default is greater than with investment grade bonds.

Corporate bonds are slightly riskier than government bonds (as the latter are backed by the state), and so they generally have a higher interest rate than government debt. Corporate bonds may be traded on the secondary market, and interest rate movements in the market influence the value of a bond.

Bonds are different from equity in a company, in that an investor who buys a bond is lending money to the company, whereas an investor who buys shares in a company purchases an ownership share of the company. The value of the shareholding varies with the share price of the company, and the investor receives dividends from the company based on the profits of the company. There are differences in the rights of bondholders and shareholders when a company defaults and becomes bankrupt, and bondholders and other creditors will have a higher priority claim than shareholders.

A *secured bond* is where the company has pledged specific collateral as security for the bond, and so holders of secured bonds have legal rights on the collateral. *Unsecured* bonds (*debentures*) have no collateral associated with them, and these may be classified into *senior* or *junior debentures*, with seniors having a higher priority claim on assets/cash flows than juniors.

Example (Calculating Present Value of Bond)

Suppose an investor purchases a 5-year €10,000 bond when it is issued on January 1, 2020 at a coupon rate of 6% with interest paid bi-annually, and the market rate of interest is 4%. Determine the present value of the bond on January 1, 2020.

Solution

The present value of the bond is determined from the present value of the interest payments and the present value of the bond maturity amount with respect to the market rate of interest.

$$\text{Value} = \text{Pres Val Int . Payments} + \text{Pres Val. Maturity}$$

We discussed the present value of annuities in section 7.4, and so the present value of the interest payments is determined from the following formula:

$$P = \frac{A}{i}\left[1 - \frac{1}{(1+i)^n}\right]$$

The bond is for 5 years with two payments of interest each year, and so $n = 5 * 2 = 10$. The market rate is 4% and so the interest rate i per interest period is 4%/2 = 2% = 0.02. The interest amount for each interest period is 10,000 * 0.06 * 0.5 = €300.

Therefore, the present value of the interest payments is given by

$$300/0.02(1 - 1/(1.02)^{10}) = 15000(1 - 0.8203) = €2694.77$$

The present value of the maturity amount is given by the present value formula where $n = 5 * 2 = 10$, and the interest rate i per interest period is 4%/2 = 2% = 0.02.

$$
\begin{aligned}
PV &= MV(1+i)^{-n} \\
&= 10000(1.02)^{-10} \\
&= €8202.48
\end{aligned}
$$

Therefore, the present value of the bond on the issue date is given by the sum of the present value of the interest payments and the present value of the maturity amount:

$$
\begin{aligned}
&= 2694.77 + 8202.48 \\
&= €10{,}903.25
\end{aligned}
$$

Note

The reader will note that the present value of the bond (€10,903.25) is greater than the amount of the bond on the issue date. The market rate of interest (4%) is lower than the coupon rate (6%) and so it takes a greater investment to achieve the same return at 4% as the yield from the corporate bond at 6%. If the market rate of interest is higher than the coupon rate then the present value of the bond would be less than the value on the issue date.

Example (Calculating the Value of Bond before Maturity)

Suppose an investor purchases a 5-year €10,000 bond when it is issued on January 1, 2020 at a coupon rate of 6% with interest paid bi-annually, and the market rate of interest is 4%. Determine the value of the bond on July 1, 2023.

Solution

The value of the bond is determined from the present value of the remaining interest payments and the present value of the bond maturity amount with respect to the market rate of interest.

There are three remaining interest payments (December 31, 2023; June 30, 2024; and December 31, 2024) and so $n = 3$, the market rate of interest $i = 0.02$ as before, and the interest amount is €300 as before.

The present value of the interest payments is given by

$$300/0.02(1-1/(1.02)^3) = 15000(0.05767) = €865.16$$

The present value of the maturity amount is given by the present value formula where $n = 3 = 10$, and the interest rate i per interest period is 0.02.

$$
\begin{aligned}
PV &= MV(1+i)^{-n} \\
&= 10000(1.02)^{-3} \\
&= €9423.32
\end{aligned}
$$

Therefore, the value of the bond on July 1, 2023 is given by the sum of the present value of the remaining interest payments and the present value of the maturity amount:

$$= €865.16 + €9423.32$$
$$= €10{,}288.37$$

For more information on banking and financial services, see [RoH:12].

7.8 Review Questions

1. Explain the difference between the central bank, retail banks and commercial banks.
2. Explain the difference between current accounts, savings accounts, loans and mortgages.
3. Explain the difference between bonds, bills and derivatives.
4. Ruth is investing €5000 a year in a savings scheme that pays 8% compound interest every year. What will the value of her investment be after 10 years?
5. Determine the present value of a 20-year annuity of an annual payment of $5000 per year at a discount rate of 5%.
6. Eithne has taken out a €150,000 mortgage over 20 years at 7% per annum. Calculate her repayment amount per annum to amortize the mortgage.
7. Determine the monthly repayment amount of a mortgage of €150,000 at 7% compounded monthly over 20 years.
8. Ashling is paying €1500 monthly on a 20-year mortgage where the interest rate is 5% compounded monthly. Determine the principal borrowed.
9. A borrower is paying €1500 monthly on a 20-year mortgage where the interest rate is 5% compounded monthly. Determine the principal outstanding at the start of year 20.
10. Explain the difference between FX terms spot, forward, swap, hedging and option.

7.9 Summary

Banks are at the heart of the financial system and play a key role in facilitating the flow of money throughout the economy. Banks provide a service where those with excess funds may lend to those who need funds. Banks pay interest to its savers and earn interest from its borrowers, and the spread between the interest paid to savers and the interest charged to borrowers generates revenue for the bank.

There are various types of banks such as the central bank, retail banks and commercial banks. A central bank is responsible for monetary policy and regulating the banks and financial institutions in the state. Retail banks provide banking services to the general public whereas commercial banks provide banking to large corporations and businesses.

An annuity is a sequence of fixed equal payments made over a period of time. For example, in a hire purchase contract, the buyer makes an initial deposit for an item, and then pays an equal amount per month (the payment is generally made at the end of the month) up to a fixed end date.

The purchase of a home or car requires the loan to be repaid to the financial institution with regular monthly payments. An interest-bearing debt is amortized if both principal and interest are repaid by a series of equal payments made at equal intervals of time, and the debt is repaid by an amortization annuity, where each payment consists of repayment of the capital borrowed and the interest due on the capital for that time interval.

Notes

1 We are assuming that the country is not in a cycle of negative interest rates where investors are essentially charged to place their funds on deposit with the bank.

2 The assets of the banks are the loans that it has lent out to borrowers. In the case of a serious recession or depression or property market collapse, many borrowers may be unable to repay their loans, and so some of the bank's assets are toxic (i.e., "toxic assets").

8

Trade Discounts and Pricing

A *trade discount* is a reduction in the list price of a manufactured product, and it is usually stated as a percentage of the list price of the product. Trade discounts are used by manufacturers, distributors and wholesalers as pricing tools for their products, and in communicating changes of prices to their customers. The offer of a trade discount by a supplier is often done as part of an effort to increase the volume of sales made, and they also offer the supplier a way to clear out slow moving or unwanted inventory.

A manufacturer may offer two or more discounts to a customer, where additional discounts are often offered to encourage large volume orders, or early orders for seasonal items. If two or more discounts are applied then these are called a *discount series*, and for every discount series there exists a single equivalent rate of discount.

The *net price factor approach* is often used in calculating the net price of a product following the application of a discount (the standard calculation is to calculate the discount amount and subtract it from the list price).

A *cash discount* may be given to encourage early or prompt payment of an invoice, and the rate of discount and the discount period for when the discount may be applied are specified on the invoice.

Managers need to decide on the prices at which products and services will be offered in the market, and so it is important to consider factors that may affect *pricing policy* and to consider various techniques for setting prices.

8.1 Calculating Discounts

The amount of the discount is given by the rate of discount multiplied by the list price:

$$\text{Amt-Discount} = \text{Rate-Discount} * \text{List-Price}$$
$$A = d * L$$

The list price may be determined from the amount of discount and the rate of discount:

DOI: 10.1201/9781003308140-8

$$\text{List Price} = \frac{\text{Amt-Discount}}{\text{Rate Discount}} = \frac{A}{d}$$

Similarly, the rate of discount may be determined from the amount of discount and the list price:

$$\text{Rate Discount} = \frac{\text{Amt-Discount}}{\text{List Price}} = \frac{A}{L}$$

The net price is given by the list price less the discount:

$$\text{Net Price} = \text{List Price} - \text{Discount}$$
$$N = L - A$$

Example (Trade Discount – Net Price)

A product has a list price of $80 and a trade discount rate of 20%. Find the amount of the discount and the net price.

Solution (Trade Discount – Net Price)

The amount of the discount is given by $A = dL = 0.2 * 80 = \$16$.
The net price is given by $N = L - A = 80 - 16 = \$64$.

Example (Trade Discount – List Price)

The value of a 25% discount is £15. Find the list price and the net price.

Solution (Trade Discount – List Price)

The list price is given by $^{A}/_{d} = 15/0.25 = £60$.
The net price is given by $N = L - A = 60 - 15 = £45$.

Example (Trade Discount – Rate of Discount)

Find the rate of discount for a music system that is listed at €800 less a discount of €120.

Solution (Trade Discount – Rate of Discount)

The rate of discount is given by $^{A}/_{L} = 120/800 = 0.15 = 15\%$.

8.1.1 Net Price Factor Approach

The net price factor (NPF) approach is an alternative approach to the calculation of the net price. The standard calculation of the net price involves first calculating the amount of the discount and then subtracting it from the list price, and so there are two calculations involved. However, the NPF approach employs a single calculation only. The key point to note with the NPF approach is that if you get a discount of say 20% ($d = 0.2$), then you only pay 80% = (0.8 = 1 – d) of the list price. That is, the calculation of the net price using NPF is

$$N = L(1 - d)$$

The NPF is given by the subtraction of the discount from 100%:

$$\begin{aligned} \text{NPF} &= 100\%\text{-rate discount} \\ &= 1 - d \end{aligned}$$

$$\begin{aligned} \text{Net Price} &= L * \text{NPF} \\ N &= L * (1 - d) \end{aligned}$$

Example (Net Factor)

Find the net price of a TV where the list price is \$800 and a discount of 20% is to be applied.

Solution (Net Factor)

The net price is given by N = $L * (1 - d) = 800 * (1 - 0.2) = 800 * 0.8 = \$640$. That is, if you get a discount of 20% off the list price of a TV you end up paying 80% of the list price.

A manufacturer may offer two or more discounts to encourage bulk sales, and the net price in this case is given by

$$\text{Net Price} = L(1 - d_1)\,(1 - d_1) \ldots (1 - d_n)$$

There exists a single equivalent rate of discount for every discount series and it is given by

$$d = 1 - (1 - d_1)\,(1 - d_2) \ldots (1 - d_n)$$

Example (Discount Series)

A mobile phone listed at \$360 is subject to the trade discount series 10%, 25% and 15%. Determine the following:

1. Find the net price
2. Find the amount of the discount
3. Find the single equivalent discount rate.

Solution (Discount Series)

The net price is given by

$$\begin{aligned} N &= L(1-d_1)(1-d_1)\ldots(1-d_n) \\ &= 360 * (1-0.1)(1-0.25)(1-0.15) \\ &= 360 * 0.9 * 0.75 * 0.85 \\ &= \$206.55 \end{aligned}$$

The amount of the discount is given by the difference between the list price and the net price. That is:

$$\begin{aligned} A &= L - N \\ &= 360 - 206.55 \\ &= \$153.45 \end{aligned}$$

The single equivalent rate is given by

$$\begin{aligned} d &= 1-(1-d_1)(1-d_1)\ldots(1-d_n) \\ &= 1-0.9 * 0.75 * 0.85 \\ &= 0.42625 \\ &= 42.63\% \end{aligned}$$

We may check our answer with $d = {}^{A}/_{L} = {}^{153.45}/_{360} = 0.4263 = 42.63\%$.

8.2 Markups and Markdowns

The selling price of an item is made up of the cost of making/buying the item, overhead expenses and the profit for the business. That is:

$$S = C + E + P$$

Markup is used to refer to the sum of expenses and profit and so the selling price of an item is the sum of its cost and the markup of the item. That is:

$$M = E + P$$
$$S = C + M$$

Markup may be stated in two ways: either as a percentage of the cost of the item or as a percentage of its selling price. That is, the rate of markup based on cost is the markup divided by cost, and the rate of markup based on selling price is the markup divided by the selling price. These are then converted to percentages by multiplying by 100:

$$M_C = M/C * 100$$
$$M_S = M/C * 100$$

Example (Markups)

The cost of a table is €600 and it can be sold for €750. Determine the following:

1. The markup.
2. What is the rate of markup based on cost?
3. What is the rate of markup based on price?

Solution (Markups)

1. The markup is given by the difference between the selling price and the cost.

$$M = S - C = 750 - 600 = €150$$

2. The rate of markup based on cost is $^{M}/_{C} * 100 = {}^{150}/_{600} * 100 = 25\%$.
3. The rate of markup based on price is $^{M}/_{S} * 100 = {}^{150}/_{750} * 100 = 20\%$.

Example (Markups)

1. Find the selling price of an item that costs £32 if the markup is 25% of its cost.
2. Find the selling price of an item that costs £32 if the markup is 25% of the selling price.

Solution (Markups)

1. $$\begin{aligned} S &= C + M \\ &= 32 + 32(0.25) \\ &= 32 + 8 \\ &= £40 \end{aligned}$$

2. $$\begin{aligned} S &= C + M \\ &= 32 + S(0.25) \\ &\Rightarrow S - 0.25S = 32 \\ &\Rightarrow 0.75\,S = 32 \\ &\Rightarrow S = £42.67 \end{aligned}$$

The *markdown* of an item refers to the reduction in price of an item sold to a customer, and is stated as a percentage of the price to be reduced. It is computed in a similar manner to a trade discount as discussed earlier. Markdowns are a way to get rid of unwanted inventory and to make space for new items that can be sold more quickly at the full price. They are generally used for clearance at the end of the season or as a way to get rid of merchandise close to the end of its shelf life.

Markdowns are stated either as a percentage of the price to be reduced or the amount of the discount. The new sale price is calculated by subtracting the markdown from the regular selling price:

$$SP = S - MD$$

If the markdown is given by a percentage then the new sales price is given by

$$\begin{aligned} SP &= S - S * MD_r \\ &= S(1 - MD_r) \end{aligned}$$

The markdown rate is given by the markdown divided by the regular selling price:

$$MD_r = MD/S * 100$$

Example (Markdown)

A tablet originally priced at \$240 was sold for \$175. Calculate the following:

1. The markdown amount.
2. The markdown rate.

Solution (Markdown)

1. The markdown is given by MD = S – SP = 240 – 175 = \$65.
2. The markdown rate is given by

$$\begin{aligned} MD_r &= MD/S * 100 \\ &= 65/240 * 100 \\ &= 27.08\% \end{aligned}$$

8.3 Payment Terms and Cash Discounts

The business will issue an invoice to the customer for the goods purchased, and the payment terms are stated on the invoice. The terms may include an incentive to encourage the customer to make the payment promptly, where a cash discount is applied if the payment is made before the end of the discount period.

The payment terms will state the discount rate (e.g., 3%), the discount period for when the discount may be applied (e.g., 10 days), and the credit period during which the invoice must be paid (e.g., 30 days). If payment is made during the discount period then the discount may be applied; if payment is made after the discount period and within the credit period then the full amount of the invoice must be paid; and if payment is made outside the credit period then it may be subject to an interest surcharge for late payment.

Example (Payment Terms)

A company has received an invoice of €1750 from one of its suppliers dated 24 October. The payment terms are 30 days and a discount of 2% is applied if payment is made within 10 days of the invoice date. Determine the following:

1. The amount due if payment is made on 1 November.
2. The amount due if payment is made on 4 November.

Solution (Payment Terms)

The payment date is 24 October and so the discount for early payment is valid up to 3 November.

1. The amount of payment on 1 November is within the discount period, and so the discount to be applied is $1750 * 0.02 =$ €35. The amount to be paid is €1750 – 35 = €1715.
2. The amount of payment on 4 November 4 is outside the discount period, and so the full amount must be paid, i.e., €1750.

8.4 Pricing and Pricing Policy

Every business needs to set the prices of their products and services to achieve profitability, and pricing policy is influenced by several factors such as demand for the product, competition in the market and the life cycle of the product.

Some products are very sensitive to price changes whereas others (especially where quality is the key driver) are less sensitive to changes in price. The competitors will react to any changes in prices, and often prices move in unison (e.g., the sale of petrol in garages) or it may lead to a price war between competitors to maintain market share.

The product lifecycle is the period of time from the initial development of the product, to its introduction into the market, to its growth and maturity in the market, and then to saturation and decline until it is eventually withdrawn from the market. That is, most products do not have a limitless life, and some have a very short lifecycle and disappear soon after they are launched in the market. There may be a different pricing policy at each stage of the product lifecycle.

The business will need to decide on the price that will be charged for a new product, and the pricing needs to take into account the costs in producing the product and the desired profit. The prices will often be influenced by market factors such as what a customer is prepared to pay, and whether there are rival products in the market and the price of these. In general, higher unit prices lead to lower demand, whereas lower unit prices lead to higher demand. There are several types of pricing policies (Table 8.1).

8.4.1 Setting the Price for a Product

The *demand-based* approach to pricing is based on the view that there is a connection between the price, the total quantity and the total revenue and profits. The demand varies with price and so if a realistic estimate of demand can be made at various price levels then it is possible to determine a profit-maximizing price and a revenue-maximizing price. That is, there is a

TABLE 8.1

Pricing Policies

Type	Description
Market penetration pricing	This type of pricing refers to a policy of low pricing when the product is launched with the goal of gaining market share.
Market-skimming pricing	This type of pricing involves charging high prices when the product is launched, and spending heavily on marketing and promotions to obtain sales. The aim of the skimming is to gain high unit profits early in the product's life and to recover the development costs (skimming off the cream), and then progressively lower prices will be charged as competitors enter the market.
Differential pricing	This is where the same product is sold at different prices to different customers, which may happen in different market segments (e.g., prices for cinema for children adults, and old age pensioners), different product versions (e.g., different models of cars have different features), location (e.g., seat in theatre) and time (e.g., rail ticket prices for peak and off-peak).

selling price that maximizes its profit and a selling price that maximizes its sales. That is, if the number of units sold is X then the total revenue is given by TR = SP * X, and the total profit is given by P = TR – TC. Chapter 15 includes a section on cost volume profit analysis, which determines the breakeven point on the number of units that need to be sold to break even.

The *full cost pricing* approach involves calculating the sales price of the product from the full cost of making the product (i.e., including fixed and variable costs), and adding a percentage markup for the profit. The percentage profit markup may be varied to suit the circumstances such as the demand conditions in the market. Chapter 15 includes an example in linear programming where a company is trying to decide how many of each product it should make to maximize profits subject to the constraint of limited resources.

The *variable cost* pricing approach involves adding a profit margin to the variable cost of the product, and the variable cost plus pricing is often called markup pricing (where markups were discussed earlier in the chapter).

The *opportunity cost* approach to pricing (*marginal pricing*) involves taking into account the price for the opportunity costs of the resources consumed in making the product, and so the marginal price is calculated as the incremental cost of producing the product and the opportunity cost of doing so.

The *target cost pricing* approach is where the business decides what the target sales price of the product will be as well as the desired profit margin. The profit is then deducted from the target sales price to determine what the target cost of the product should be. The difference between the current cost and the target cost is termed the cost gap, which needs to be eliminated.

8.5 Review Questions

1. What is a trade discount?
2. Find the net price of a music system where the list price is \$640 and a discount of 15% is to be applied.
3. What is a markup? What is a markdown?
4. A television was sold for \$380 with a rate of markup of 10% on its selling price. Calculate the cost of making the television.
5. A television originally priced at \$480 was sold for \$397. Calculate the rate of markdown.
6. Explain why discounts are often offered for early payments and how this works.

7. A company has received an invoice of €2550 from one of its suppliers dated 15 March. The payment terms are 30 days and a discount of 2% is applied if payment is made within 10 days of the invoice date. Determine the amount due if payment is made on 1 April.
8. Explain the difference between market penetration pricing, market skimming pricing and differential pricing.
9. Explain the difference between demand-based approach to pricing, full cost pricing, variable cost pricing and opportunity cost pricing.

8.6 Summary

A trade discount is a reduction in the list price of a manufactured product, and it is usually stated as a percentage of the list price. Trade discounts are used by manufacturers, distributors and wholesalers as pricing tools for their products, and to increase the volume of sales.

A manufacturer may offer two or more discounts to a customer, where additional discounts are often offered to encourage large volume orders or early orders for seasonal items. If two or more discounts are applied then these are called a discount series, and for every discount series there exists a single equivalent rate of discount.

The markup of a product is the difference between the selling price of the product and the cost of making it, whereas the markdown of an item refers to the reduction in the list price of an item sold to a customer, and is usually stated as a percentage of the price to be reduced.

A cash discount may be given to encourage early or prompt payment of an invoice, and the rate of discount and the discount period for when the discount may be applied are specified on the invoice.

The pricing policy is used to determine the prices at which products and services will be offered in the market, and so several factors such as demand, competition and the product lifecycle will influence pricing policy.

9

Statistics

Statistics is an empirical science that is concerned with the collection, organization, analysis, interpretation and presentation of data. The data collection needs to be planned and this may include surveys and experiments. Statistics are widely used by government and industrial organizations, and they are employed for forecasting as well as for presenting trends. They allow the behaviour of a population to be studied and inferences to be made about the population. These inferences may be tested (*hypothesis testing*) to ensure their validity.

The analysis of statistical data allows an organization to understand its performance in key areas, and to identify problematic areas. Organizations will often examine performance trends over time, and will devise appropriate plans and actions to address problematic areas. The effectiveness of the actions taken will be judged by improvements in performance trends over time.

It is often not possible to study the entire population, and instead a representative subset or sample of the population is chosen. This *random sample* is used to make inferences regarding the entire population, and it is essential that the sample chosen is indeed random and representative of the entire population. Otherwise, the inferences made regarding the entire population will be invalid, as a selection bias has occurred. Clearly, a census where every member of the population is sampled is not subject to this type of bias.

A statistical experiment is a causality study that aims to draw a conclusion between the values of a *predictor variable*(s) and a *response variable*(s). For example, a statistical experiment in the medical field may be conducted to determine if there is a causal relationship between the use of a particular drug and the treatment of a medical condition such as lowering of cholesterol in the population. A statistical experiment involves the following:

- Planning the research
- Designing the experiment
- Performing the experiment
- Analysing the results
- Presenting the results

DOI: 10.1201/9781003308140-9

9.1 Basic Statistics

The field of statistics is concerned with summarizing, digesting and extracting information from large quantities of data. It provides a collection of methods for planning an experiment and analysing data to draw accurate conclusions from the experiment. We distinguish between descriptive statistics and inferential statistics.

Descriptive Statistics

This is concerned with describing the information in a set of data elements in graphical format, or by describing its distribution.

Inferential Statistics

This is concerned with making inferences with respect to the population by using information gathered in the sample.

9.1.1 Abuse of Statistics

Statistics are extremely useful in drawing conclusions about a population. However, it is essential that the random sample chosen is actually random, and that the experiment is properly conducted to ensure that valid conclusions are inferred. Some examples of the abuse of statistics are as follows:

- The sample size may be too small to draw accurate conclusions.
- It may not be a genuine random sample of the population.
- There may be bias introduced from poorly worded questions.
- Graphs may be drawn to exaggerate small differences.
- Area may be misused in representing proportions.
- Misleading percentages may be used.

The quantitative data used in statistics may be discrete or continuous. *Discrete data* is numerical data that has a finite or countable number of possible values, and *continuous data* is numerical data that has an infinite number of possible values.

9.1.2 Statistical Sampling and Data Collection

Statistical sampling is concerned with the methodology for choosing a random sample of a population, and the study of the sample with the goal of drawing valid conclusions about the entire population. If a genuine representative random sample of the population is chosen, then a detailed study of the sample will provide insight into the whole population. This helps to avoid a lengthy expensive (and potentially infeasible) study of the entire population.

The sample chosen must be truly random and the sample size sufficiently large to enable valid conclusions to be made for the entire population. The probability of being chosen for the random sample should be the same for each member of the population.

Random Sample

A *random sample* is a sample of the population such that each member of the population has an equal chance of being chosen.

A large sample size gives more precise information about the population. However, little extra information is gained from increasing the sample size above a certain level, and the sample size chosen will depend on factors such as money and time available, the aims of the survey, the degree of precision required and the number of subsamples required. Table 9.1 summarizes several ways for generating a random sample from the population.

Once the sample is chosen the next step is to obtain the required information from the sample. The data collection may be done by interviewing each member in the sample, conducting a telephone interview with each member, conducting a postal questionnaire survey and so on (Table 9.2).

TABLE 9.1

Sampling Techniques

Sampling Technique	Description
Systematic Sampling	The population is listed and every *k*th member of the population is sampled. For example, to choose a 2% (1 in 50) sample then every 50th member of the population would be sampled.
Stratified Sampling	The population is divided into two or more strata and each subpopulation (stratum) is then sampled. Each element in the subpopulation shares the same characteristics (e.g., age groups, gender). The results from the various strata are then combined.
Multi-Stage Sampling	This approach may be used when the population is spread over a wide geographical area. The area is split up into a number of regions, and a small number of the regions are randomly selected. Each selected region is then sampled. It requires less effort and time but it may introduce bias if a small number of regions are selected, as it is not a truly random sample.
Cluster Sampling	A population is divided into clusters and a few of these clusters are exhaustively sampled (i.e., every element in the cluster is considered). This approach may lead to significant selection bias, as the sampling is not random.
Convenience Sampling	Sampling is done as convenient, and in this case each person selected may decide whether to participate or not in the sample.

TABLE 9.2

Types of Survey

Survey Type	Description
Personal Interview	Interviews are expensive and time consuming, but allow detailed and accurate information to be collected. Questionnaires are often employed and the interviewers need to be trained in interview techniques. Interviews need to be planned and scheduled, and they are useful in dealing with issues that may arise (e.g., a respondent not fully understanding a question).
Phone Survey	This is a reasonably efficient and cost-effective way to gather data. However, refusals or hang-ups may affect the outcome. It also has an in-built bias as only those people with telephones may be contacted and interviewed.
Mail Questionnaire Survey	This involves sending postal questionnaire survey to the participants. The questionnaire needs to be well designed to ensure the respondents understand the questions and answer them correctly. There is a danger of a low response rate that may invalidate the findings.
Direct Measurement	This may involve a direct measurement of all those in the sample (e.g., the height of all students in a class).
Direct Observational Study	An observational study allows individuals to be studied, and the variables of interest to be measured.
Experiment	An experiment imposes some treatment on individuals in order to study their response.

The design of the questionnaire requires careful consideration as a poorly designed questionnaire may lead to invalid results. The questionnaire should be as short as possible, and the questions should be simple and unambiguous. Closed questions where the respondent chooses from simple categories are useful. It is best to pilot the questionnaire prior to carrying out the survey.

9.2 Frequency Distribution and Charts

The data gathered from a statistical study is often raw, and may yield little information as it stands. Therefore, the way the data is presented is important, and it is useful to present the information in pictorial form. The advantage of a pictorial presentation is that it allows the data to be presented in an attractive and colourful way, and the reader is not overwhelmed with excessive detail. This enables analysis and conclusions to be drawn. There are several types of charts or graphs that are often employed in the presentation of the data which are as follows:

- Bar chart
- Histogram
- Pie chart
- Trend graph

A *frequency table* is often used to present and summarize data, where a simple frequency distribution consists of a set of data values and the number of items that have that value (i.e., a set of data values and their frequency). The information is then presented pictorially in a bar chart.

The general frequency distribution is employed when dealing with a larger number of data values (e.g., >20 data values). It involves dividing the data into a set of disjoint data classes, and listing the data classes in one column and the frequency of data values in that category in another column. The information is then presented pictorially in a bar chart or histogram.

Fig. 9.1 presents the raw data of salaries earned by different people in a company, and Table 9.3 presents the raw data in table format using a frequency table of salaries. Fig. 9.2 presents a *bar chart* of the salary data in pictorial form, and it is much easier to read than the raw data presented in Fig 9.1.

A *histogram* is a way of representing data in bar chart format, and it shows the frequency or relative frequency of various data values or ranges of data values. It is usually employed when there are a large number of data values, and it gives a crisp picture of the spread of the data values, and the centring and variance of the data values from the mean.

The data is divided into disjoint intervals where an interval is a certain range of values. The horizontal axis of the histogram contains the intervals

90,000 50,000 50,000 65,000 65,000 45,000 50,000
50,000 50,000 65,000 50,000 50,000 45,000 50,000
65,000

FIGURE 9.1
Raw Salary Data.

TABLE 9.3
Frequency Table of Salary Data

Salary	Frequency
45,000	2
50,000	8
65,000	4
90,000	1

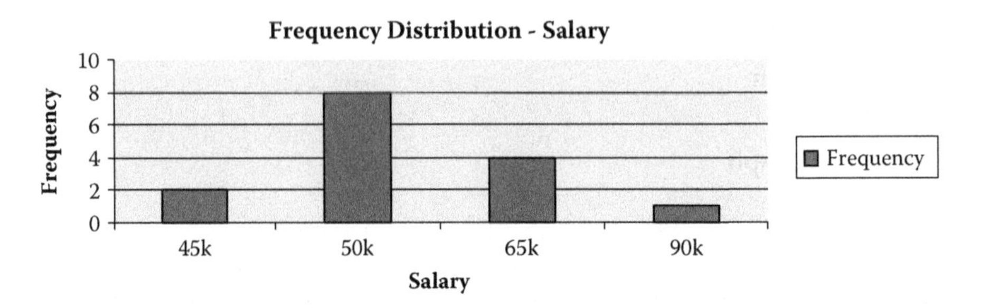

FIGURE 9.2
Bar Chart of Salary Data.

(also known as buckets) and the vertical axis shows the frequency (or relative frequency) of each interval.

The bars represent the frequency and there is no space between the bars. The histogram has an associated shape, e.g., it may be a *normal distribution*, a *bimodal* or *multi-modal distribution*, and it may be positively or negatively skewed. The variation and centring refer to the spread of data, and the spread of the data is important as it may indicate whether the entity under study (e.g., a process) is too variable, or whether it is performing within the requirements.

The histogram is termed process centred if its centre coincides with the customer requirements; otherwise, the process is too high or too low. A histogram allows predictions of future performance to be made, where it can be assumed that the future will resemble the past.

The construction of a histogram first requires that a frequency table be constructed, and this requires that the range of the data values be determined. The number of class intervals is determined, and the class intervals (or data buckets) consist of a particular range of data values. The class intervals are mutually disjoint and span the range of the data values. Each data value belongs to exactly one class interval, and the frequency of each class interval is determined. The frequency (or relative frequency) of each bucket is displayed in bar format.

The results of a class test in mathematics are summarized in Table 9.4. There are 30 students in the class and each student achieves a score somewhere between 0 and 100. There are four data intervals between 0 and 100 employed to summarize the scores, and the result of each student belongs to exactly one interval. Fig. 9.3 is the associated histogram for the frequency data, and it gives a pictorial representation of the marks for the class test.

We may also employ a pie chart as an alternate way to present the class marks. The frequency table is constructed as before, and a visual representation of the percentage in each data class (i.e., the relative frequency) is provided with the pie chart. Each portion of the pie chart represents the percentage of the data values in that interval (Fig. 9.4).

TABLE 9.4

Frequency Table – Test Results

Mark	Frequency
0–24	3
25–49	10
50–74	15
75–100	2

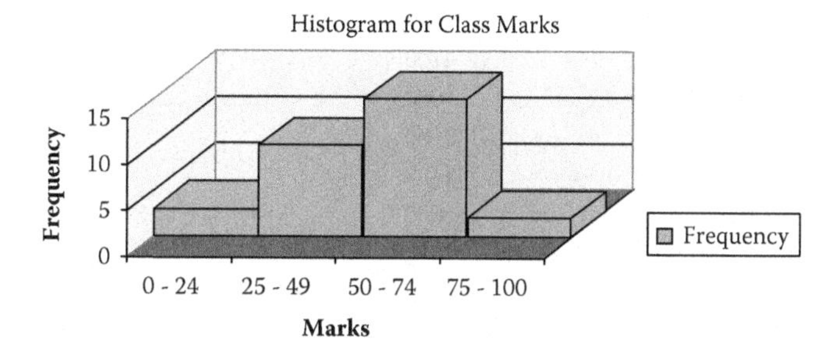

FIGURE 9.3
Histogram Test Results.

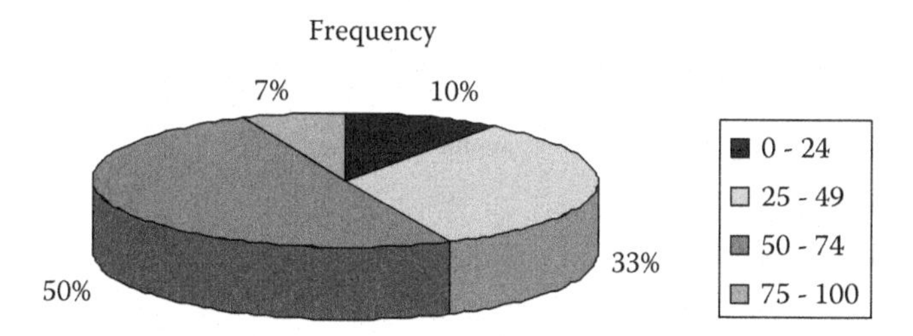

FIGURE 9.4
Pie Chart Test Results.

We present the monthly sales and profit figures for a company in Table 9.5, and Fig. 9.5 gives a pictorial representation of the data in the form of a time series (or trend chart).

9.3 Statistical Measures

Statistical measures are concerned with the basic analysis of the data to determine the average of the data as well as how spread out the data is.

TABLE 9.5
Frequency Table – Test Results

	Sales	Profit
Jan	5500	200
Feb	3000	–400
Mar	3500	200
Apr	3000	600
May	4500	–100
Jun	6200	1200
Jul	7350	3200
Aug	4100	100
Sep	9000	3300
Oct	2000	–500
Nov	1100	–800
Dec	3000	300

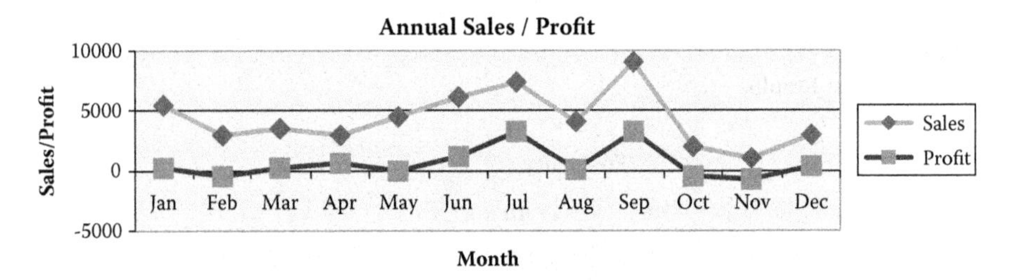

FIGURE 9.5
Monthly Sales and Profit.

The term "average" generally refers to the arithmetic *mean* of a sample, but it may also refer to the statistical *mode* or *median* of the sample. We first discuss the arithmetic mean as it is the mathematical average of the data and is representative of the data. The arithmetic mean is the most widely used average in statistics.

9.3.1 Arithmetic Mean

The *arithmetic mean* (or just mean) of a set of n numbers is defined to be the sum of the data values divided by the number of values. That is, the arithmetic mean of a set of data values $x_1, x_2, \ldots, x_n$ (where the sample size is n) is given by:

ARITHMETIC MEAN

$$\bar{x} = \frac{\sum_{i=1}^{n} x_i}{n}$$

The arithmetic mean is representative of the data as all values are used in its calculation. The mean of the set of values 5, 11, 9, 4, 16, 9 is given by

$$m = {}^{5+11+9+4+16+9}/_{6} = {}^{54}/_{6} = 9.$$

The formula for the arithmetic mean of a set of data values given by a frequency table needs to be adjusted:

x_1	x_2	....	....	x_n
f_1	f_2			f_n

$$\bar{x} = \frac{\sum_{i=1}^{n} f_i x_i}{\sum_{i=1}^{n} f_i}$$

The arithmetic mean for the following frequency distribution is calculated by:

x	2	5	7	10	12
f_x	2	4	7	4	2

The mean is given by

$$m = {}^{(2*2+5*4+7*7+10*4+12*2)}/_{(2+4+7+4+2)} = {}^{(4+20+49+40+24)}/_{19} = {}^{137}/_{19} = 7.2.$$

The actual mean of the population is denoted by μ, and it may differ from the sample mean m.

9.3.2 Mode

The *mode* is the most popular element in the sample, i.e., it is the data element that occurs most frequently in the sample. For example, consider a shop that sells mobile phones, then the mode of the annual sales of phones is the most popular phone sold. The mode of the list [1, 4, 1, 2, 7, 4, 3, 2, 4] is 4, whereas there is no unique mode in the sample [1, 1, 3, 3, 4], and it is said to be bimodal.

The mode of the following frequency distribution is 7 since it occurs the most times in the sample.

x	2	5	7	10	12
f_x	2	4	7	4	2

It is possible that the mode is not unique (i.e., there are at least two elements that occur with the equal highest frequency in the sample), and if this is the case then we are dealing with a bi-modal or possibly a multi-model distribution (where there are more than two elements that occur most frequently in the sample).

9.3.3 Median

The *median* of a set of data is the value of the data item that is exactly half way along the set of items, where the data set is arranged in increasing order of magnitude.

If there are an odd number of elements in the sample, the median is the middle element. Otherwise, the median is the arithmetic mean of the two middle elements.

The median of 34, 21, 38, 11, 74, 90, 7 is determined by first ordering the set as 7, 11, 21, 34, 38, 74, 90, and the median is then given by the value of the 4th item in the list which is 34.

The median of the list 2, 4, 8, 12, 20, 30 is the mean of the middle two items (as there are an even number of elements and the set is ordered), and so it is given by the arithmetic mean of the third and fourth elements, i.e., ${}^{8\,+\,12}/_2 = 10$.

The calculation of the median of a frequency distribution first requires the calculation of the total number of data elements (i.e., this is given by $N = \Sigma f$), and then determining the value of the middle element in the table, which is the ${}^{N+1}/2$ element.

The median for the following frequency distribution is calculated by:

x	2	5	7	10	12
f_x	2	4	7	4	2

The number of elements is given by $N = \Sigma f = 2 + 4 + 7 + 4 + 2 = 19$ and so the middle element is given by the value of the ${}^{N+1}/2$ element, i.e., the ${}^{19+1}/_2$ = the 10th element. From an examination of the table, it is clear

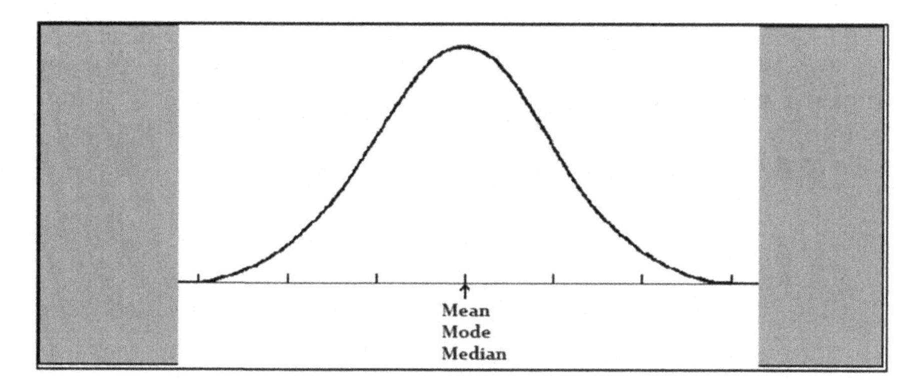

FIGURE 9.6
Symmetric Distribution.

that the value of the 10th element is 7 and so the median of the frequency distribution is 7.

The final average that we consider is the midrange of the data in the sample, and this is given by the arithmetic mean of the highest and lowest data elements in the sample. That is, $m_{mid} = (x_{max} + x_{min})/2$.

The mean, mode and median coincide for symmetric frequency distributions but differ for left- or right-skewed distributions (Fig. 9.6). Skewness describes how non-symmetric the data is.

Dispersion indicates how spread out or scattered the data is, and there are several ways of measuring dispersion including how skewed the distribution is, the range of the data, variance and the standard deviation.

9.4 Variance and Standard Deviation

An important characteristic of a sample is its distribution, and the spread of each element from some measure of central tendency (e.g., the mean). One elementary measure of dispersion is that of the sample *range*, which is defined to be the difference between the maximum and minimum values in the sample. That is, the sample range is defined to be:

$$\text{range} = x_{max} - x_{min}$$

The sample range is not a reliable measure of dispersion as just two elements in the sample are used, and so extreme values in the sample may distort the range and make it very large even if most of the elements are quite close to one another.

The standard deviation is the most common way to measure dispersion, and it gives the average distance of each element in the sample from the arithmetic mean. The *sample standard deviation* of a sample $x_1, x_2, \ldots, x_n$ is denoted by s, and its calculation first requires the calculation of the sample mean. It is defined by:

SAMPLE STANDARD DEVIATION

$$s = \sqrt{\frac{\Sigma(x_i - \bar{x})^2}{n - 1}} = \sqrt{\frac{\Sigma x_i^2 - n\bar{x}^2}{n - 1}}$$

The *population standard deviation* is denoted by σ and is defined by:

POPULATION STANDARD DEVIATION

$$\sigma = \sqrt{\frac{\Sigma(x_i - \mu)^2}{n}} = \sqrt{\frac{\Sigma x_i^2 - n\mu^2}{n}}$$

Variance is another measure of dispersion and it is defined as the square of the standard deviation. The *sample variance* s^2 is given by:

SAMPLE VARIANCE

$$s^2 = \frac{\Sigma(x_i - \bar{x})^2}{n - 1} = \frac{\Sigma x_i^2 - n\bar{x}^2}{n - 1}$$

The *population variance* σ^2 is given by:

POPULATION STANDARD DEVIATION

$$\sigma^2 = \frac{\Sigma(x_i - \mu)^2}{n} = \sqrt{\frac{\Sigma x_i^2 - n\mu^2}{n}}$$

Example (Standard Deviation)

Calculate the standard deviation of the sample 2, 4, 6, 8.

Solution (Standard Deviation)

The sample mean is given by $m = 2 + 4 + 6 + 8/4 = 5$.
The sample variance is given by:

$$\begin{aligned} s^2 &= (2-5)^2 + (4-5)^2 + (6-5)^2 + (8-5)^2/4 - 1 \\ &= 9 + 1 + 1 + 9/3 \\ &= 20/3 \\ &= 6.66 \end{aligned}$$

The sample standard deviation is given by the square root of the variance, and so it is given by:

$$\begin{aligned} s &= \sqrt{6.66} \\ &= 2.58 \end{aligned}$$

The formula for the standard deviation and variance may be adjusted for frequency distributions. The standard deviation and mean often go hand in hand, and for *normal distributions* 68% of the data lie within one standard deviation of the mean; 95% of the data lie within two standard deviations of the mean; and the vast majority (99.7%) of the data lies within three standard deviations of the mean. All data values are used in the calculation of the mean and standard deviation, and so these measures are truly representative of the data.

9.5 Correlation and Regression

The two most common techniques for exploring the relationship between two variables are correlation and linear regression. *Correlation* is concerned with quantifying the strength of the relationship between two variables by measuring the degree of "scatter" of the data values, whereas *regression* expresses the relationship between the variable in the form of an equation (usually a linear equation).

Correlation quantifies the strength and direction of the relationship between two numeric variables X and Y, and the correlation coefficient may be positive or negative and it lies between -1 and $+1$. If the correlation is positive then as the value of one variable increases the value of the other variable increases (i.e., the variables move together in the same direction), whereas if the correlation is negative then as the value of one variable increases the value of the other variable decreases (i.e., the variables move together in the opposite directions). The correlation is zero if there is no relationship between the two variables. The correlation coefficient r is given by the following formula:

CORRELATION (X, Y)

$$\text{Corr}(X.\ Y) = \frac{\overline{XY} - \overline{X}\,\overline{Y}}{\text{Std}(X)\text{Std}(Y)} = \frac{n\Sigma X_i Y_i - \Sigma X_i \Sigma Y_i}{\sqrt{n\Sigma X_i^2 - (\Sigma X_i)^2}\sqrt{n\Sigma Y_i^2 - (\Sigma Y_i)^2}}$$

The sign of the correlation coefficient indicates the direction of the relationship between the two variables. A correlation of $r = +1$ indicates a perfect positive correlation, whereas a correlation of $r = -1$ indicates a perfect negative correlation. A correlation close to 0 indicates no relationship between the two variables; a correlation of $r = -0.3$ indicates a weak negative relationship; whereas a correlation of $r = 0.85$ indicates a strong positive relationship. The extent of the relationship between the two variables may be seen from the following:

- A change in the value of X leads to a change in the value of Y.
- A change in the value of Y leads to a change in the value of X.
- Changes in another variable lead to changes in both X and Y.
- There is no relationship (or correlation) between X and Y.

The relationship (if any) between the two variables can be seen by plotting the values of X and Y in a scatter graph as in Fig. 9.7 and Fig. 9.8. The correlation coefficient identifies linear relationships between X and Y, but it does not detect non-linear relationships. It is possible for correlation to exist between two variables but for no causal relationship to exist, i.e., *correlation is not the same as causation.*

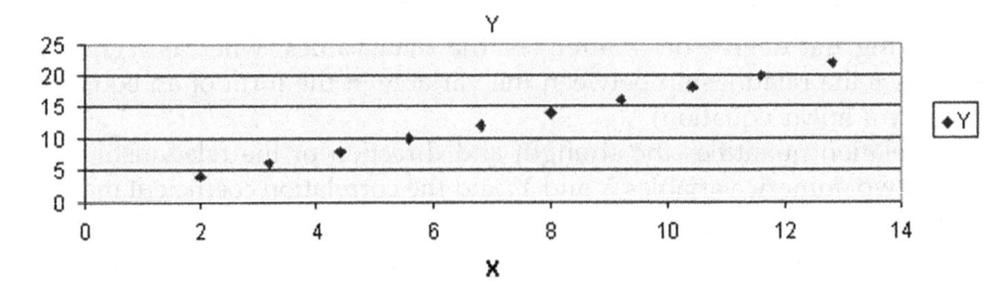

FIGURE 9.7
Strong Positive Correlation.

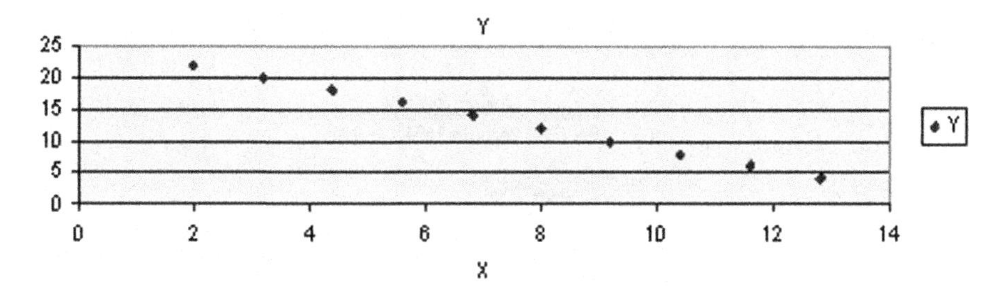

FIGURE 9.8
Strong Negative Correlation.

Example (Correlation)

The data in Table 9.6 is a summary of the cost of maintenance of eight printers, and it is wished to explore the extent to which the age of the machine is related to the cost of maintenance. It is required to calculate the correlation coefficient.

Solution (Correlation)

For this example, $n = 8$ (as there are eight printers) and ΣX_i, ΣY_i, $\Sigma X_i Y_i$, ΣX_i^2, ΣY_i^2 are computed in the last row of the table and so:

$$\begin{aligned} \Sigma X_i &= 76 \\ \Sigma Y_i &= 940 \\ \Sigma X_i Y_i &= 12680 \\ \Sigma X_i^2 &= 978 \\ \Sigma Y_i^2 &= 169450 \end{aligned}$$

TABLE 9.6
Cost of Maintenance of Printers

X (Age)	*Y* (Cost)	*XY*	X^2	Y^2
5	50	250	25	2500
12	135	1620	144	18,225
4	60	240	16	3600
20	300	6000	400	90,000
2	25	50	4	625
10	80	800	100	6400
15	200	3000	225	40,000
8	90	720	64	8100
76	940	12,680	978	169,450

We input these values into the correlation formula and get:

$$
\begin{aligned}
r &= \frac{8 * 12680 - 76 * 940}{\sqrt{8 * 978 - 76^2}\sqrt{8 * 169450 - 940^2}} \\
&= \frac{30000}{\sqrt{2048}\sqrt{472000}} \\
&= \frac{30000}{45.25 * 687.02} \\
&= \frac{30000}{31087} \\
&= 0.96
\end{aligned}
$$

Therefore, $r = 0.96$ and so there is strong correlation between the age of the machine and the cost of maintenance of the machine.

9.5.1 Regression

Regression is used to study the relationship (if any) between dependent and independent variables, and to predict the dependent variable when the independent variable is known. The prediction capability of regression makes it a more powerful tool than correlation, and regression is useful in identifying which factors impact upon a desired outcome variable.

There are several types of regression that may be employed such as linear or polynomial regression, and this section is concerned with linear regression where the relationship between the dependent and independent variables is expressed by a straight line. More advanced statistical analysis may be conducted with multiple regression models, where there are several independent variables that are believed to affect the value of another variable.

Regression analysis first involves data gathering and plotting the data on a scatter graph. The regression line is the line that best fits the data on the scatter graph (Fig. 9.9), and it is usually determined using the method of least squares or one of the methods summarized in Table 9.7. The regression line is a plot of the expected values of the dependant variable for all values of the independent variable, and the formula (or equation) of the regression line is of the form $y = mx + b$, where the coefficients of a and b are determined.

The regression line then acts as a model that describes the relationship between the two variables, and the value of the dependent variable may be predicted from the value of the independent variable using the regression line.

9.5.2 Method of Least Squares

The method of least squares is a procedure to find the best approximating straight line fit ($y = ax + b$) for a set of data points, where the error involved is the sum of the squares of the differences between the y-value on the

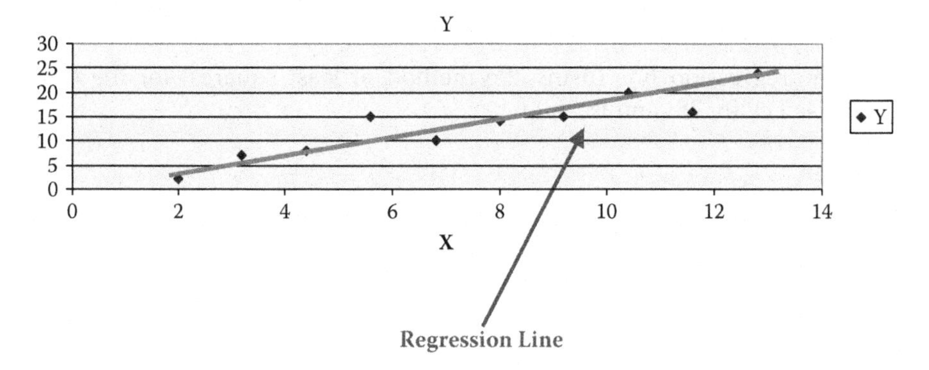

FIGURE 9.9
Regression Line.

TABLE 9.7
Methods to Obtain Regression Line

Methods	Description
Inspection	This is the simplest method and involves plotting the data in a scatter graph and then drawing a line that best suits the data. (This is subjective and so it is best to draw the mean point, and ensure the regression line passes through this point).
Semi-Averages	This involves splitting the data into two equal groups, then finding and drawing the mean point in each group, and joining these points with a straight line (i.e., the regression line).
Least Squares	The method of least squares is a mathematical and involves obtaining the regression line where the sum of the squares of the vertical deviations of all the points from the line is minimal.

approximating line and the given y values, and so the goal is to find the line that minimizes the least-squares error. Suppose there are n data values (x_i, y_i) then the error function $e(a, b)$ for the least-squares line is a function of two variables given by:

$$e(a, b) = \sum_{i=1}^{n} [y_i - (ax_i + b)]^2$$

Further, for a minimum to occur at $e(a, b)$ then the partial derivatives must be 0:

$$\frac{\partial}{\partial a} e(a, b) = \frac{\partial}{\partial b} e(a, b) = 0$$

Example (Regression/Least Squares)[1]

Find the regression line (using the method of least squares) for the data (2, 2), (4, 11), (6, 28) and (8, 40).

Solution

The error function $e(a, b)$ is given by

$$e(a, b) = [2 - (2a + b)^2] + [11 - (4a + b)^2] + [28 - (6a + b)^2] + [40 - (8a + b)^2].$$

The partial derivatives are 0 and so

$$2(2 - 2a - b)(-2) + 2(11 - 4a - b)(-4) + 2(28 - 6a - b)(-6) + 2(40 - 8a - b)(-8) = 0$$

and

$$2(2 - 2a - b)(-1) + 2(11 - 4a - b)(-1) + 2(28 - 6a - b)(-1) + 2(40 - 8a - b)(-1) = 0$$

This simplifies to the two simultaneous equations with two unknowns:

$$240a + 40b = 1072$$
$$40a + 8b = 162$$

The solution to these equations is $a = 6.55$ and $b = -12.5$ and so the regression line using the methods of least squares is (Fig. 9.10)

$$y = 6.55x - 12.5$$

9.6 Statistical Inference and Hypothesis Testing

Inferential statistics is concerned with statistical techniques to infer the properties of a population from samples taken from the population. Often, it is infeasible or inconvenient to study all members of a population, and so the properties of a representative sample are studied and statistical techniques are used to generalize these properties to the population. A statistical experiment is carried out to gain information from

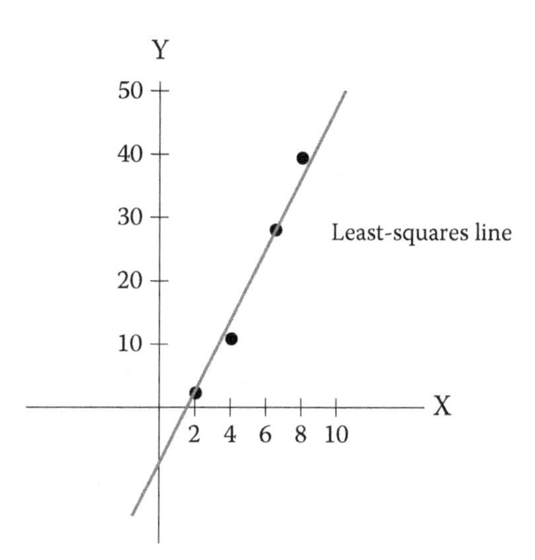

FIGURE 9.10
Least-Squares Regression Line.

the sample, and the experiment may be repeated as many times as required to gain the desired information. Statistical experiments may be simple or complex.

There are two main types of inferential statistics and these are *estimating parameters* and *hypothesis testing*. Estimating parameters is concerned taking a statistic from the sample (e.g., the sample mean or variance) and using it to make a statement about the population parameter (i.e., the population mean or variance). Hypothesis testing is concerned with using the sample data to answer research questions such as whether a new drug is effective in the treatment of a particular disease. A sample is not expected to perfectly represent the population, as sampling errors will naturally occur.

A hypothesis is a statement about a particular population whose truth or falsity is unknown. Hypothesis testing is concerned with determining whether the values of the random sample from the population are consistent with the hypothesis. There are two mutually exclusive hypotheses: one of these is the *null hypothesis* H_0 and the other is the *alternate research hypothesis* H_1. The null hypothesis H_0 is what the researcher is hoping to reject, and the research hypothesis H_1 is what the researcher is hoping to accept.

Statistical testing is employed to test the hypothesis, and the result of the test is that we either *reject the null hypothesis* (and therefore accept the alternative hypothesis), or that we fail to reject it (i.e., we *accept*) *the null hypothesis*. The rejection of the null hypothesis means that the null hypothesis is highly unlikely to be true, and that the research hypothesis should be accepted.

Statistical testing is conducted at a certain level of significance, with the probability of the null hypothesis H_0 being rejected when it is true never greater than α. The value α is called the level of significance of the test, with

TABLE 9.8

Hypothesis Testing

Action	H_0 True, H_1 False	H_0 False, H_1 True
Reject H_1	Correct	False Positive – Type 2 error P (Accept H_0 | H_0 false) = β
Reject H_0	False Negative – Type 1 error P (Reject H_0 | H_0 true) = α	Correct

α usually being 0.1, 0.05 or 0.005. A significance level β may also be applied with respect to accepting the null hypothesis H_0 when H_0 is false.

The objective of a statistical test is not to determine whether or not H_0 is actually true, but rather to determine whether its validity is consistent with the observed data. That is, H_0 should only be rejected if the resultant data is very unlikely if H_0 is true.

The errors that can occur with hypothesis testing include type 1 and type 2 errors. Type 1 errors occur when we reject the null hypothesis when the null hypothesis is actually true. Type 2 errors occur when we accept the null hypothesis when the null hypothesis is false (Table 9.8).

For example, an example of a *false positive* is where the result of a blood test comes back positive to indicate that a person has a particular disease when in fact the person does not have the disease. Similarly, an example of a *false negative* is where a blood test is negative indicating that a person does not have a particular disease when in fact the person does.

Both errors are potentially very serious, with a false positive generating major stress and distress to the recipient, until further tests are done that show that the person does not have the disease. A false negative is potentially even more serious, as early detection of a serious disease is essential to its treatment, and so a false negative means that valuable time is lost in its detection, which could be very serious.

The terms α and β represent the level of significance that will be accepted, and α may or may not be equal to β. In other words, α is the probability that we will reject the null hypothesis when the null hypothesis is true, and β is the probability that we will accept the null hypothesis when the null hypothesis is false.

Testing a hypothesis at the $\alpha = 0.05$ level is equivalent to establishing a 95% confidence interval. For 99% confidence α will be 0.01, and for 99.999% confidence then α will be 0.00001.

The hypothesis may be concerned with testing a specific statement about the value of an unknown parameter θ of the population. This test is to be done at a certain level of significance, and the unknown parameter may, for example, be the mean or variance of the population. An estimator for the unknown parameter is determined, and the hypothesis that this is an accurate estimate is rejected if the random sample is not consistent with it. Otherwise, it is accepted.

The steps involved in hypothesis testing are as follows:

1. Establish the null and alternative hypothesis.
2. Establish error levels (significance).
3. Compute the test statistics (often a *t*-test).
4. Decide on whether to accept or reject the null hypothesis.

The difference between the observed and expected test statistic, and whether the difference could be accounted for by normal sampling fluctuations is the key to the acceptance or rejection of the null hypothesis.

For more detailed information on statistics, see [Dek:10] and [Ros:14].

9.7 Review Questions

1. What is statistics?
2. Explain how statistics may be abused.
3. What is a random sample?
4. Describe the methods available to generate a random sample from a population.
5. Describe the charts available for the presentation of statistical data.
6. Explain how the average of a sample may be determined, and explain the difference between the mean, mode and median of a sample.
7. Explain sample variance and sample standard deviation.
8. Explain the difference between correlation and regression.
9. Explain the methods for obtaining the regression line from data.
10. What is hypothesis testing?

9.8 Summary

Statistics is an empirical science that is concerned with the collection, organization, analysis, interpretation and presentation of data. The data collection needs to be planned, and may include surveys and experiments. Statistics are widely used by government and industrial organizations, and they are employed for forecasting as well as for presenting trends. They allow the behaviour of a population to be studied and inferences to be made about the population.

It is often not possible to study the entire population, and instead a representative subset or sample of the population is chosen. This random sample is used to make inferences regarding the entire population, and it is essential that the sample chosen is indeed random and representative of the entire population.

The data gathered from a statistical study is often raw, and, it is useful to present the information in pictorial form to enable analysis to be done and conclusions to be drawn. There are several types of charts or graphs that are often employed such as bar charts, histograms, pie chart and trend graphs.

Statistical measures are used to determine the average of the data, as well as how spread out the data is. The term "average" generally refers to the arithmetic mean of a sample, but it may also refer to the statistical mode or median of the sample.

An important characteristic of a sample is its distribution, and the spread of each element from some measure of central tendency (e.g., the mean). One elementary measure of dispersion is that of the sample range. The standard deviation is the most common way to measure dispersion, and it gives the average distance of each element in the sample from the arithmetic mean.

Correlation and linear regression are techniques for exploring the relationship between two variables. Correlation is concerned with quantifying the strength of the relationship between two variables, whereas regression expresses the relationship between the variable in the form of an equation (usually a linear equation).

Inferential statistics is concerned with statistical techniques to infer the properties of a population from samples taken from the population. A hypothesis is a statement about a particular population whose truth or falsity is unknown. Hypothesis testing is concerned with determining whether the values of the random sample from the population are consistent with the hypothesis.

Note

1 This example may be skipped as it involves the use of partial differentiation.

10

Probability Theory

Probability is a branch of mathematics that is concerned with measuring uncertainty and random events, and it provides a precise way of expressing the likelihood of a particular event occurring. Probability is also used as part of everyday speech in expressions such as "It is likely to rain in the afternoon," where the corresponding statement expressed mathematically might be "The probability that it will rain in the afternoon is 0.7."

The modern theory of probability theory has its origins in work done on the analysis of games of chance by Cardano in the 16th century, and it was developed further in the 17th century by Fermat and Pascal and refined in the 18th century by Laplace. It led to the classical definition of the probability of an event being:

$$P(\text{Event}) = \frac{\#\text{Favourable Outcomes}}{\#\text{Possible Outcomes}}$$

There are several definitions of probability such as the frequency interpretation and the subjective interpretation of probability. For example, if a geologist states that "there is a 70% chance of finding gas in a certain region" then this statement is usually interpreted in two ways:

- The geologist is of the view that over the long run 70% of the regions whose environmental conditions are very similar to the region under consideration have gas. [*Frequency interpretation*]
- The geologist is of the view that it is likely that the region contains gas, and that 0.7 is a measure of the geologist's belief in this hypothesis. [*Belief interpretation*]

That is, according to the frequency interpretation, the probability of an event is equal to the long-term frequency of the event's occurrence when the same process is repeated many times.

According to the belief, interpretation probability measures the degree of belief about the occurrence of an event or in the truth of a proposition, with a probability of 1 representing the certain belief that something is true and a probability of 0 representing the certain belief that something is false, with a value in between reflecting uncertainty about the belief.

Probabilities may be updated by Bayes' Theorem (section 10.1.2), where the initial belief is the *prior* probability for the event, and this may be

DOI: 10.1201/9781003308140-10

updated to a *posterior* probability with the availability of new information (see section 10.6 for a short account of Bayesian Statistics).

10.1 Basic Probability Theory

Probability theory provides a mathematical indication of the likelihood of an event occurring, and the probability lies between 0 and 1. A probability of 0 indicates that the event cannot occur, whereas a probability of 1 indicates that the event is guaranteed to occur. If the probability of an event is greater than 0.5, then this indicates that the event is more likely to occur than not to occur.

A statistical experiment is conducted to gain certain desired information, and the *sample space* is the set of all possible outcomes of an experiment. The outcomes are all equally likely if no one outcome is more likely to occur than another. An *event* E is a subset of the sample space, and the event is said to have occurred if the outcome of the experiment is in the event E.

For example, the sample space for the experiment of tossing a coin is the set of all possible outcomes of this experiment, i.e., head or tail. The event that the toss results in a tail is a subset of the sample space.

$$S = \{h, t\} \quad E = \{t\}$$

Similarly, the sample space for the gender of a newborn baby is the set of outcomes, i.e., the newborn baby is a boy or a girl. The event that the baby is a girl is a subset of the sample space.

$$S = \{b, g\} \quad E = \{g\}$$

For any two events E and F of a sample space S we can also consider the union and intersection of these events. That is,

- $E \cup F$ consists of all outcomes that are in E or F or both.
- $E \cap F$ (usually written as EF) consists of all outcomes that are in both E and F.
- E^c denotes the complement of E with respect to S, and represents the outcomes of S that are not in E.

If $EF = \emptyset$, then there are no outcomes in both E and F, and so the two events E and F are *mutually exclusive*. Events that are mutually exclusive cannot occur at the same time (i.e., they cannot occur together).

Two events are said to be *independent* if the occurrence (or not) of one of the events does not affect the occurrence (or not) of the other. Two mutually exclusive events cannot be independent, since the occurrence of one excludes the occurrence of the other.

The union and intersection of two events can be extended to the union and intersection of a family of events $E_1, E_2, \ldots, E_n$ (i.e., $\cup^n_{i=1} E_i$ and $\cap^n_{i=1} E_i$).

10.1.1 Laws of Probability

The probability of an event E occurring is given by

$$P(E) = \frac{\#\text{Outcomes in Event E}}{\#\text{Total Outcomes (in S)}}$$

The laws of probability essentially state that the probability of an event is between 0 and 1, and that the probability of the union of a mutually disjoint set of events is the sum of their individual probabilities. The probability of an event E is 0 if E is an impossible event, and the probability of an event E is one if it is a certain event (Table 10.1).

The probability of the union of two events (not necessarily disjoint) is given by

$$P(E \cup F) = P(E) + P(F) - P(EF)$$

The complement of an event E is denoted by E^c and denotes that event E does not occur. Clearly, $S = E \cup E^c$ and E and E^c are disjoint and so:

$$P(S) = P(E \cup E^c) = P(E) + P(E^c) = 1$$
$$\Rightarrow P(E^c) = 1 - P(E)$$

TABLE 10.1

Axioms of Probability

Axiom	Description
1	$P(S) = 1$
2	$P(\varnothing) = 0$
3	$0 \leq P(E) \leq 1$
4	For any sequence of mutually exclusive events $E_1, E_2, \ldots, E_n$ (i.e., $E_i E_j = \varnothing$ where $i \neq j$) the probability of the union of these events is the sum of their individual probabilities, i.e., $P(\cup^n_{i=1} E_i) = \Sigma^n_{i=1} P(E_i)$

The probability of an event E occurring given that an event F has occurred is termed the *conditional probability* (denoted by $P(E \mid F)$) and is given by:

CONDITIONAL PROBABILITY

$$P(E \mid F) = \frac{P(EF)}{P(F)} \quad \text{where } P(F) > 0$$

This formula allows us to deduce that

$$P(EF) = P(E \mid F)P(F)$$

Example (Conditional Probability)

A family has two children. Find the probability that they are both girls given that they have at least one girl.

Solution (Conditional Probability)

The sample space for a family of two children is $S = \{(g, g), (g, b), (b, g), (b, b)\}$. The event E where is at least one girl in the family is given by $E = \{(g, g), (g, b), (b, g)\}$, and the event G that both are girls is $G = \{(g, g)\}$, so we will determine the conditional probability $P(G \mid E))$ that both children are girls given that there is at least one girl in the family:

$$P(EG) = P(G) = P(g, g) = {}^{1}/_{4}$$
$$P(E) = {}^{3}/_{4}$$
$$P(G \mid E) = \frac{P(EG)}{P(E)} = \frac{{}^{1}/_{4}}{{}^{3}/_{4}} = {}^{1}/_{3}$$

Two events E and F are independent if the knowledge that F has occurred does not change the probability that E has occurred. That is, $P(E \mid F) = P(E)$ and since $P(E \mid F) = P(EF)/P(F)$ we have that two events E and F are independent if:

$$P(EF) = P(E)P(F)$$

Two events E and F that are not independent are said to be *dependent*.

10.1.2 Bayes' Formula

Bayes' formula enables the probability of an event E to be determined by a weighted average of the conditional probability of E given that the event F occurred and the conditional probability of E given that F has not occurred:

$$\begin{aligned} E &= E \cap S = E \cap (F \cup F^c) \\ &= EF \cup EF^c \end{aligned}$$

BAYES' FORMULA

$$\begin{aligned} P(E) &= P(EF) + P(EF^c) \quad (\text{since } EF \cap EF^c = \emptyset) \\ &= P(E \mid F)P(F) + P(E \mid F^c)P(F^c) \\ &= P(E \mid F)P(F) + P(E \mid F^c)(1 - P(F)) \end{aligned}$$

We may also get another expression of Bayes' formula by noting that

$$P(F \mid E) = \frac{P(FE)}{P(E)} = \frac{P(EF)}{P(E)}$$

Therefore, $P(EF) = P(F \mid E)P(E) = P(E \mid F)P(F)$.

$$P(E \mid F) = \frac{P(F \mid E)P(E)}{P(F)}$$

This version of Bayes' formula[1] allows the probability to be updated where the initial or preconceived belief (i.e., P(E)) is the *prior* probability for the event, and this may be updated to a *posterior* probability (i.e., $P(E \mid F)$ is the updated probability), with the new information or evidence (i.e., P(F) and the likelihood that the new information leads to the event.

Bayes' theorem may be used to determine the actual probability that you have a certain rare disease following a positive test (as medical tests are not perfectly accurate and yield false positives and false negatives).

Example 1 (Bayes' Formula)

A medical lab is 99% effective in detecting a certain disease when it is actually present, and it yields a false positive for 1% of healthy people tested. If 0.25% of the population actually have the disease, what is the probability that a person has the disease if the patient's blood test is positive?

Solution (Bayes' Formula)

Let T be the event that the patient's test result is positive, and D the event that the tested person has the disease. Then the desired probability is P $(D \mid T)$ and is given by

$$\begin{aligned} P(D|T) &= \frac{P(DT)}{P(T)} = \frac{P(T|D)P(D)}{P(T|D)P(D) + P(T|D^c)P(D^c)} \\ &= \frac{0.99 * 0.0025}{0.99 * 0.0025 + 0.01 * 0.9975} = 0.1988 \end{aligned}$$

The reason that only approximately 20% of the population whose test results are positive actually have the disease may seem surprising, but it is explained by the low incidence of the disease (just one person out of every 400 tested will have the disease and the test will correctly confirm that 0.99 have the disease, but the test will also state that 399 $*$ 0.01 = 3.99 have the disease and so the proportion of time that the test is correct is $^{0.99}/_{0.99+3.99}$ = 0.1988.

Example 2 (Bayes' Formula)

Approximately 1% of women have breast cancer (i.e., 99% do not) with 80% of mammogram tests detecting breast cancer when it is there (i.e., 20% miss it), and 9.6% of mammograms detect breast cancer when it is not there (false positive) with 90.4% correctly returning a negative). What is the probability that a woman has breast cancer given a positive mammogram test?

Solution (Bayes' Formula)

Let T be the event that the patient's mammogram test result is positive, and D the event that the tested person has breast cancer. Then the desired probability is P(D | T) and is given by

$$\begin{aligned} P(D|T) &= \frac{P(DT)}{P(T)} = \frac{P(T|D)P(D)}{P(T|D)P(D) + P(T|D^c)P(D^c)} \\ &= \frac{0.8 * 0.01}{0.8 * 0.01 + 0.096 * 0.99} \\ &= 7.8\% \end{aligned}$$

That is, the probability that she has breast cancer following a positive mammogram test is 0.078.

10.2 Random Variables

Often, some numerical quantity determined by the result of the experiment is of interest rather than the result of the experiment itself. These numerical quantities are termed *random variables*. A random variable is termed *discrete* if it can take on a finite or countable number of values, otherwise it is termed *continuous*.

The *probability distribution function* (denoted by $F(x)$) of a random variable is the probability that the random variable X takes on a value less than or equal to x. It is given by:

PROBABILITY DISTRIBUTION FUNCTION (DISCRETE RANDOM VARIABLE)

$$F(x) = P\{X \le x\}$$

All probability questions about X can be answered in terms of its distribution function F. For example, the computation of P $\{a < X < b\}$ is given by

$$\begin{aligned} P\{a < X < b\} &= P\{X \le b\} - P\{X \le a\} \\ &= F(b) - F(a) \end{aligned}$$

The *probability mass function* for a discrete random variable X (denoted by $p(a)$) is the probability that the random variable is a certain value. It is given by

$$p(a) = P\{X = a\}$$

Further, $F(a)$ can also be expressed in terms of the probability mass function:

$$F(a) = P\{X \le a\} = \sum_{\forall x \le a} p(x)$$

X is a continuous random variable if there exists a non-negative function $f(x)$ (termed the *probability density function*) defined for all $x \in (-\infty, \infty)$ such that:

PROBABILITY DENSITY FUNCTION (CONTINUOUS RANDOM VARIABLE)

$$P\{X \in B\} = \int_B f(x)dx$$

All probability statements about X can be answered in terms of its density function $f(x)$. For example:

$$P\{a \le X \le b\} = \int_a^b f(x)dx$$

$$P\{X \in (-\infty, \infty)\} = 1 = \int_{-\infty}^{\infty} f(x)dx$$

The function $f(x)$ is termed the probability density function, and the probability distribution function $F(a)$ is defined by:

PROBABILITY DISTRIBUTION FUNCTION

$$F(a) = P\{X \leq a\} = \int_{-\infty}^{a} f(x)dx$$

Further, the first derivative of the probability distribution function yields the probability density function. That is:

$$d/_{da}\, F(a) = f(a).$$

The expected value (i.e., the *mean*) of a discrete random variable X (denoted E[X]) is given by the weighted average of the possible values of X:

EXPECTED VALUES

$$E[X] = \begin{cases} \sum_i x_i P\{X = x_i\} & \text{Discrete Random variable} \\ \int_{-\infty}^{\infty} xf(x)dx & \text{Continuous Random variable} \end{cases}$$

Further, the expected value of a function of a random variable is given by E[g(X)] and is defined for the discrete and continuous case respectively.

$$E[g(X)] = \begin{cases} \sum_i g(x_i) P\{X = x_i\} & \text{Discrete Random variable} \\ \int_{-\infty}^{\infty} g(x)f(x)dx & \text{Continuous Random variable} \end{cases}$$

The *variance* of a random variable is a measure of the spread of values from the mean, and it is defined by:

VARIANCE OF A RANDOM VARIABLE

$$Var(X) = E[X^2] - (E[X])^2$$

The standard deviation σ is given by the square root of the variance. That is:

$$\sigma = \sqrt{\mathrm{Var}(X)}$$

The *covariance* of two random variables is a measure of the relationship between two random variables X and Y, and indicates the extent to which they both change (in either similar or opposite ways) together. It is defined by:

COVARIANCE OF A RANDOM VARIABLE

$$\mathrm{Cov}(X, Y) = E[XY] - E[X].E[Y].$$

It follows that the covariance of two independent random variables is 0. Variance is a special case of covariance (when the two random variables are identical). This follows since Cov(X, X) = E[X.X] – (E[X])(E[X]) = E[X2] – $(E[X])^2$ = Var(X).

A positive covariance (Cov(X, Y) ≥ 0) indicates that Y tends to increase as X does, whereas a negative covariance indicates that Y tends to decrease as X increases.

The *correlation* of two random variables is an indication of the relationship between two variables X and Y (we discussed correlation and regression in section 9.5). If the correlation is negative and close to –1 then Y tends to decrease as X increases, and if it is positive and close to 1 then Y tends to increase as X increases. A correlation close to 0 indicates no relationship between the two variables; a correlation of $r = -0.4$ indicates a weak negative relationship; whereas a correlation of $r = 0.8$ indicates a strong positive relationship. The correlation coefficient is between ±1 and is defined by:

CORRELATION

$$\mathrm{Corr}(X, Y) = \frac{\mathrm{Cov}(X, Y)}{\sqrt{\mathrm{Var}(X)\mathrm{Var}(Y)}}$$

Once the correlation between two variables has been calculated, the probability that the observed correlation was due to chance can be computed. This is to ensure that the observed correlation is a real one and not due to a chance occurrence.

10.3 Binomial and Poisson Distributions

The Binomial and Poisson distributions are two important distributions in probability theory, and the Poisson distribution may be used as an approximation for the binomial. The binomial distribution was first used in games of chance, and it has the following characteristics:

- The existence of a trial of an experiment is defined in terms of two states namely success or failure.
- The identical trials may be repeated a number of times yielding several successes and failures.
- The probability of success (or failure) is the same for each trial.

BERNOULLI TRIAL

A *Bernoulli trial* is where there are just two possible outcomes of an experiment, i.e., success or failure. The probability of success and failure is given by

$$P\{X = 1\} = p$$
$$P\{X = 0\} = 1 - p$$

The mean of the Bernoulli distribution is given by p (since $E[X] = 1.p + 0.(1 - p) = p$), and the variance is given by $p(1 - p)$ (since $E[X^2] - E[X]^2 = p - p^2 = p(1 - p)$).

The *Binomial distribution* involves n Bernoulli trials, where each trial is independent and results in either success (with probability p) or failure (with probability $1 - p$). The binomial random variable X with parameters n and p represents the number of successes in n independent trials, where X_i is the result of the ith trial and X is represented as:

BINOMIAL DISTRIBUTION

$$X = \sum_{i=1}^{n} X_i \quad X_i = \begin{cases} 1 & \text{if the } i^{\text{th}} \text{ trial is a success} \\ 0 & \text{otherwise} \end{cases}$$

The probability of i successes from n independent trials which is then given by the binomial theorem:

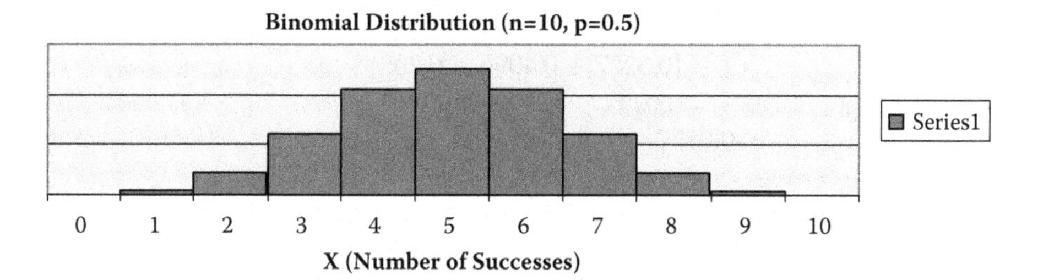

FIGURE 10.1
Binomial Distribution.

$$P\{X = i\} = \binom{n}{i} p^i (1 - p)^{n-i} \quad i = 0, 1, \ldots n$$

Clearly, $E[X_i] = p$ and $Var(X_i) = p(1 - p)$ (since X_i is an independent Bernoulli random variable). The mean of the Binomial distribution E[X] is the sum of the mean of the $E[X_i]$, i.e., $\Sigma_1^n E[X_i] = np$, and the variance Var(X) is the sum of the $Var(X_i)$ (since the X_i are independent random variables) and so $Var(X) = np(1 - p)$. The binomial distribution is symmetric when $p = 0.5$, and the distribution is skewed to the left or right when $p \neq 0.5$ (Fig. 10.1).

Example (Binomial Distribution)

The probability that a printer will need correcting adjustments during a day is 0.2. If there are five printers running on a particular day determine the probability of:

1. No printers need correcting
2. One printer needs correcting
3. Two printers require correcting
4. More than two printers require adjusting.

Solution (Binomial Distribution)

There are five trials (with $n = 5$, $p = 0.2$, and the success of a trial is a printer needing adjustments). And so:

1. This is given by $P(X = 0) = (^5{}_0)\ 0.2^0 * 0.8^5 = 0.3277$
2. This is given by $P(X = 1) = (^5{}_1)\ 0.2^1 * 0.8^4 = 0.4096$
3. This is given by $P(X = 2) = (^5{}_2)\ 0.2^2 * 0.8^3 = 0.205$
4. This is given by 1 – P(2 or fewer printers need correcting)

$$= 1 - [P(X = 0) + P(X = 1) + P(X = 2)$$
$$= 1 - [0.3277 + 0.4096 + 0.205]$$
$$= 1 - 0.9423$$
$$= 0.0577$$

The *Poisson distribution* may be used as an approximation to the binomial distribution when n is large (e.g., $n > 30$) and p is small (e.g., $p < 0.1$). The characteristics of the Poisson distribution are as follows:

- The existence of events that occur at random and may be rare (e.g., road accidents).
- An interval of time is defined in which events may occur.

The probability of i successes (where $i = 0, 1, 2, \ldots$) is given by:

POISSON DISTRIBUTION

$$P(X = i) = \frac{e^{-\lambda} \lambda^i}{i!}$$

The mean and variance of the Poisson distribution is given by λ.

Example (Poisson Distribution)

Customers arrive randomly at a supermarket at an average rate of 2.5 customers per minute, where the customer arrivals form a Poisson distribution. Determine the probability that:

1. No customers arrive in any particular minute
2. Exactly one customer arrives in any particular minute
3. Two or more customers arrive in any particular minute
4. One or more customers arrive in any 30-second period.

Solution (Poisson Distribution)

The mean λ is 2.5 per minute for parts 1–3 and λ is 1.25 for part 4.

1. $P(X = 0) = e^{-2.5} * 2.5^0/0! = 0.0821$
2. $P(X = 1) = e^{-2.5} * 2.5^1/1! = 0.2052$
3. $P(2 \text{ or more}) = 1 - P(X = 0 \text{ or } X = 1)] = 1 - [P(X = 0) + P(X = 1)] = 0.7127$
4. $P(1 \text{ or more}) = 1 - P(X = 0) = 1 - e^{-1.25} * 1.25^0/0! = 1 - 0.2865 = 0.7134.$

FIGURE 10.2
Carl Friedrich Gauss.

10.4 The Normal Distribution

The *normal distribution* is the most important distribution in statistics and it occurs frequently in practice. It is shaped like a bell and it is popularly known as the bell-shaped distribution, and the curve is symmetric about the mean of the distribution. The empirical frequencies of many natural populations exhibit a bell-shaped (normal) curve, such as the frequencies of the height and weight of people. The largest frequencies cluster around the mean and taper away symmetrically on either side of the mean. The German mathematician, Gauss (Fig. 10.2), originally studied the normal distribution, and it is also known as the Gaussian distribution.

The normal distribution is a continuous distribution, and it has two parameters namely the mean μ and the standard deviation σ. It is a continuous distribution, and so it is not possible to find the probability of individual values, and thus it is only possible to find the probabilities of ranges of values. The normal distribution has the important properties that 68.2% of the values lie within one standard deviation of the mean, with 95% of the values within two standard deviations; and 99.7% of values are within three standard deviations of the mean. In another words, the vast majority of the values are within three standard deviations of the mean, with minimal values more than three standard deviations from the mean. The shaded area under the curve in Fig. 10.3 represents two standard deviations of the mean and comprises 95% of the population.

The *normal distribution* N has mean μ, and standard deviation σ. Its density function $f(x)$ (where $-\infty < x < \infty$) is given by:

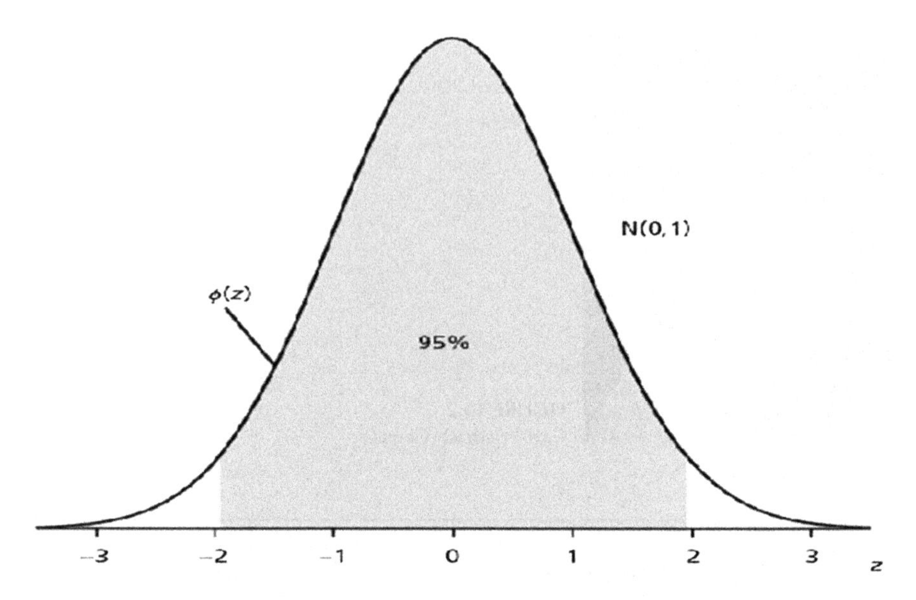

FIGURE 10.3
Standard Normal Bell Curve (Gaussian Distribution).

NORMAL DISTRIBUTION

$$f(x) = \frac{1}{\sqrt{2\pi}\,\sigma} e^{-(x-\mu)^2/2\sigma^2}$$

10.4.1 Unit Normal Distribution

The *unit* (or *standard*) normal distribution Z(0, 1) has mean of 0 and standard deviation of 1. Every normal distribution may be converted to the unit normal distribution by $Z = (X - \mu)/\sigma$, and every probability statement about X has an equivalent probability statement about Z. The unit normal density function is given by:

UNIT NORMAL DISTRIBUTION

$$f(y) = \frac{1}{\sqrt{2\pi}} e^{-\frac{1}{2}y^2}$$

The standard normal distribution is symmetric about 0 (as the mean is 0), and the process of converting a normal distribution with mean μ and standard deviation σ is termed standardizing the *x*-value. There are tables of values that give the probability of a Z score between 0 and the one specified.

Example (Normal Distribution)

Weights of bags of oranges are normally distributed with mean 3 lbs and standard deviation 0.2 lb. The delivery to a supermarket is 350 bags at a time. Determine the following:

1. Standardize to a unit normal distribution.
2. What is the probability that a standard bag will weigh more than 3.5 lbs?
3. How many bags from a single delivery would be expected to weigh more than 3.5 lbs?

Solution (Normal Distribution)

1. $Z = X - \mu/\sigma = X - 3/0.2$.
2. Therefore, when $X = 3.5$, we have $Z = 3.5 - 3/0.2 = 2.5$.
   For $Z = 2.5$, we have from the unit normal tables that

   $$P(Z \le 2.50) = 0.9938 = P(X \le 3.5)$$
   $$\text{Therefore, } P(X > 3.5) = 1 - P(X \le 3.5) = 1 - 0.9938 = 0.0062$$

3. The proportion of all bags that have a weight greater than 3.5 lbs is 0.0062, and so it would be expected that there are $350 * 0.0062 = 2.17$ bags with a weight > 3.5, and so in practical terms we would expect 2 bags to weigh more than 3.5 lbs.

The normal distribution may be used as an approximation to the binomial when n is large (e.g., $n > 30$), and when p is not too small or large. This is discussed in the next section, where the mean of the normal distribution is np and the standard deviation is $\sqrt{np(1-p)}$.

10.4.2 Confidence Intervals and Tests of Significance

The study of normal distributions helps in the process of estimating or specifying a range of values, where certain population parameters (such as the mean) lie from the results of small samples. Further, the estimate may be stated with a certain degree of confidence, such as there is 95% or 99% confidence that the mean value lies between 4.5 and 5.5. That is, *confidence intervals* (also known as confidence limits) specify a range of values within which some unknown population parameter lies with a stated degree of confidence, and it is based on the results of the sample.

The confidence interval for an unknown population mean where the sample mean, sample variance, the sample size and desired confidence level are known is given by

$$\bar{x} \pm z\frac{s}{\sqrt{n}}$$

where $\bar{x}$ is the sample mean, s is the sample standard deviation, n is the sample size and z is the confidence factor (for a 90% confidence interval $z = 1.64$, for the more common 95% confidence interval $z = 1.96$, and $z = 2.58$ for the 99% confidence interval).

Example (Confidence Intervals)

Suppose a new motor fuel has been tested on 30 similar cars, and the fuel consumption was 44.1 mpg with a standard deviation of 2.9 mpg. Calculate a 95% confidence interval for the fuel consumption of this model of car.

Solution (Confidence Intervals)

The sample mean is 44.1, the sample standard deviation is 2.9, the sample size is 30 and the confidence factor is 1.96; so the 95% confidence interval is

$$\begin{aligned} \bar{x} \pm z\frac{s}{\sqrt{n}} &= 44.1 \pm 1.96 * \frac{2.9}{\sqrt{30}} \\ &= 44.1 \pm 1.96 * 0.5295 \\ &= 44.1 \pm 1.0378 \\ &= (43.0622, 45.1378) \end{aligned}$$

That is, we can say with 95% confidence that the fuel consumption for this model of car is between 43.0 and 45.1 mpg.

The confidence interval for an unknown population mean where the sample proportion and sample size are known is given by

$$p \pm z\sqrt{\frac{p(1-p)}{n}}$$

where p is the sample proportion, n is the sample size and z is the confidence factor.

Example (Confidence Intervals)

Suppose three faulty components are identified in a random sample of 20 products taken from a production line. What statement can be made about the defect rate of all finished products?

Solution (Confidence Intervals)

The proportion of defective products in the sample is $p = {}^{3}/_{20} = 0.15$, and the sample size is $n = 20$. Therefore, the 95% confidence interval for the population mean is given by

$$
\begin{aligned}
& p \pm z\sqrt{\frac{p(1-p)}{n}} = 0.15 \pm 1.96\sqrt{\frac{0.15(1-0.15)}{20}} \\
& = 0.15 \pm 1.96 * 0.0798 \\
& = 0.15 \pm 0.1565 \\
& = (-0.0065,\ 0.3065)
\end{aligned}
$$

That is, we can say with 95% confidence that the defective rate of finished products lies between 0 and 0.3065.

Tests of Significance for the Mean

Tests of significance are related to confidence intervals and use the concepts from the normal distribution. To test whether a sample of size n with sample mean $\bar{x}$ and sample standard deviation s could be considered as having been drawn from a population with mean μ, the test statistic must lie in the range −1.96 to 1.96.

$$z = \frac{\bar{x} - \mu}{\left[\frac{s}{\sqrt{n}}\right]}$$

That is, the test is looking for evidence of a significant difference between the sample mean $\bar{x}$ and the population mean μ, and evidence is found if z lies outside of the stated limits, whereas if z lies within the limits then there is no evidence that the sample mean is different from the population mean.

Example (Tests of Significance)

A new machine has been introduced and management is questioning whether it is more productive than the previous one. Management takes 15 samples of this week's hourly output to test whether it is less productive, and the average production per hour is 1250 items with a standard deviation of 50. The output per hour of the previous machine was 1275 items per hour. Determine with a test of significance whether the new machine is less productive.

Solution (Tests of Significance)

The sample mean is 1250, the population mean is 1275, the sample standard deviation is 50 and the sample size is 15.

$$z = \frac{\bar{x} - \mu}{\left[\frac{s}{\sqrt{n}}\right]} = \frac{1250 - 1275}{\left[\frac{50}{\sqrt{15}}\right]} = \frac{-25}{12.91} = -0.1936$$

This lies within the range −1.96 to 1.96 and so there is no evidence of any significant difference between the sample mean and the population mean, and so management are unable to make any statement on differences in productivity.

10.4.3 The Central Limit Theorem

A fundamental result in probability theory is the *Central Limit Theorem*, which essentially states that the sum of a large number of independent and identically distributed random variables has a distribution that is approximately normal. That is, suppose $X_1, X_2, \ldots, X_n$ is a sequence of independent random variables each with mean μ and variance σ^2. Then for large n the distribution of

$$\frac{X_1 + X_2 + \ldots + X_n - n\mu}{\sigma\sqrt{n}}$$

is approximately that of a unit normal variable Z. One application of the central limit theorem is in relation to the binomial random variables, where a binomial random variable with parameters (n, p) represents the number of successes of n independent trials, where each trial has a probability of p of success. This may be expressed as:

$$X = X_1 + X_2 + \ldots + X_n$$

where $X_i = 1$ if the ith trial is a success and is 0 otherwise. The mean of the Bernoulli trial $E(X_i) = p$, and its variance is $Var(X_i) = p(1 - p)$. (The mean of the Binomial distribution with n Bernoulli trials is np and the variance is $np(1-p)$). By applying the central limit theorem it follows that for large n

$$\frac{X - np}{\sqrt{np(1 - p)}}$$

will be approximately a unit normal variable (which becomes more normal as n becomes larger).

The sum of independent normal random variables is normally distributed, and it can be shown that the sample average of $X_1, X_2, \ldots, X_n$ is normal, with a mean equal to the population mean but with a variance reduced by a factor of $1/n$:

$$\mathrm{Var}(\overline{X}) = \frac{1}{n^2}\sum_{i=1}^{n}\mathrm{Var}(X_i) = \frac{\sigma^2}{n}$$

$$\mathrm{E}(\overline{X}) = \sum_{i=1}^{n}\frac{E(X_i)}{n} = \mu$$

It follows from this that the following is a unit normal random variable:

$$\sqrt{n}\frac{(X-\mu)}{\sigma}$$

The term *six-sigma* (6σ) is a methodology concerned with continuous process improvement to improve business performance, and it aims to develop very high quality close to perfection. It was developed by Bill Smith at Motorola in the early 1980s, and it was later used by leading companies such as General Electric. A 6σ process is one in which 99.9996% of the products are expected to be free from defects (3.4 defects per million) [ORg:14].

There are many other well-known distributions such as the *hypergeometric distribution* that describes the probability of i successes in n draws from a finite population without replacement, the *uniform distribution*, the *exponential distribution* and the *gamma* distribution. The mean and variance of these distributions are summarized in Table 10.2.

10.5 Bayesianism

Bayesian thinking is named after Thomas Bayes who was an 18th-century English theologian and statistician, and it differs from the frequency

TABLE 10.2

Probability Distributions

Distribution Name	Density Function	Mean/Variance
Hypergeometric	$P\{X=i\}=\binom{N}{i}\binom{M}{n-i}/\binom{N+M}{n}$	$nN/N+M$, $np(1-p)[1-(n-1)/N+M-1]$
Uniform	$f(x)=1/(\beta-\alpha)\ \alpha \le x \le \beta, 0$	$(\alpha+\beta)/2$, $(\beta-\alpha)^2/12$
Exponential	$f(x)=\lambda e^{-\lambda x}$	$1/\lambda$, $1/\lambda^2$
Gamma	$f(x)=\lambda e^{-\lambda x}(\lambda x)^{\alpha-1}/\Gamma(\alpha)$	α/λ, α/λ^2

interpretation of probability in that it considers the probability of an event to be a measure of one's personal belief in the event. According to the frequentist approach only repeatable events such as the result from flipping a coin have probabilities, where the probability of an event is the long-term frequency of occurrence of the particular event. Bayesians view probability in a more general way and probabilities may be used to represent the uncertainty of an event or hypothesis. It is perfectly acceptable in the Bayesian view of the world to assign probabilities to non-repeatable events, whereas a strict frequentist would claim that such probabilities do not make sense as they are not repeatable.

Bayesianism provides a way of dealing rationally with randomness and risk in daily life, and it is very useful when the more common frequency interpretation is unavailable or has limited information. It interprets probability as a measure of one's personal belief in a proposition or outcome, and it is essential to first use all your available prior knowledge to form an initial estimate of the probability of the event or hypothesis. Further, when reliable frequency data becomes available the measure of personal belief would be updated accordingly to equal the probability calculated by the frequency calculation. Further, the probabilities must be updated in the light of new information that becomes available, as probabilities may change significantly from new information and knowledge. Finally, no matter how much the odds move in your favour there is eventually one final outcome (which may or may not be the desired event).

Often, in an unreliable and uncertain world, we base our decision-making on a mixture of reflection and our gut instinct (which can be wrong). Often we encounter several constantly changing random events and so it is natural to wonder on the extent to which rational methods may be applied to risk assessment and decision-making in an uncertain world.

An initial estimate is made of the belief in the proposition, and if you always rely on the most reliable and objective probability estimates while keeping track of possible uncertainties and updating probabilities in line with new data then the final probability number computed will be the best possible. We discussed how probabilities are updated in the light of new information in section 10.1.2.

We illustrate the idea of probabilities being updated with an adapted excerpt from a children's story called "Fortunately," which was written by Remy Charlip in the 1960s [Cha:93]:

- A lady went on a hot air balloon trip
- Unfortunately she fell out
- Fortunately she had a parachute on
- Unfortunately the parachute did not open
- Fortunately there was a haystack directly below

TABLE 10.3
Probability of Survival

Step	Prob. Survival
A lady went on a hot air balloon trip	$p = 0.999998$
Unfortunately she fell out	$p = 0.000001$
Fortunately she had a parachute on	$p = 0.999999$
Unfortunately the parachute did not open	$p = 0.000001$
Fortunately there was a haystack directly below	$p = 0.5$
Unfortunately there was a pitchfork sticking out at the top of haystack	$p = 0.000001$
Fortunately she missed the pitchforks	$p = 0.5$
Unfortunately she missed the haystack	$p = 0.000001$

- Unfortunately there was a pitchfork sticking out at the top of haystack
- Fortunately she missed the pitchforks
- Unfortunately she missed the haystack.

The story illustrates how probabilities can change dramatically based on new information, and despite all the changes to the probabilities during the fall the final outcome is a single result (i.e., either life or death). Let p be the probability of survival then the value of p changes as she falls through sky based on new information at each step. Table 10.3 illustrates an estimate of what the probabilities might be:

However, even if probability calculations become irrelevant after the event they still give the best chances over the long term. Over our lives we make many thousands of decisions about where and how to travel, what diet we should have and so on, and though the impact of each of these decisions on our life expectancy is very small, their combined effects is potentially significant. Clearly, careful analysis is needed for major decisions rather than just making a decision based on gut instinct.

For the example above, we could estimate probabilities for the various steps based on the expectation of probability of survival on falling without a parachute, the expectation of probability of survival on falling onto a haystack without a parachute and we would see wildly changing probabilities from the changing circumstances.

We discussed Bayes' formula in section 10.1.2, which allows the probability to be updated where the initial or preconceived belief (i.e., P(E) is the *prior* probability for the event), and this may be updated to a *posterior* probability (i.e., P(E | F) is the updated probability), with the new information or evidence (i.e., P(F)) and the likelihood that the new information leads to the event.

The reader is referred to [Ros:14, Dek:10] for a more detailed account of probability and statistics.

10.6 Review Questions

1. What is probability?
2. Explain the laws of probability.
3. What is a sample space? What is an event?
4. Prove Boole's inequality $P(\cup^{n}_{i=1}E_i) \leq \Sigma^{n}_{i=1}P(E_i)$ where E_i are not necessarily disjoint.
5. A couple has two children. What is the probability that both are girls if the eldest is a girl?
6. What is a random variable?
7. Explain the difference between the probability mass function and the probability density function (for both discrete and continuous random variables).
8. Explain variance, covariance and correlation.
9. What is the binomial distribution and what is its mean and variance?
10. What is the Poisson distribution and what is its mean and variance?
11. What is the normal distribution and what is its mean and variance?
12. What is the unit normal distribution and what is its mean and variance?
13. Explain the significance of the central limit theorem.
14. What is Bayes' theorem? Explain the importance of Bayesian thinking.

10.7 Summary

Probability is a branch of mathematics that is concerned with measuring uncertainty and random events, and it provides a precise way of expressing the likelihood of a particular event occurring, and the probability is a numerical value between 0 and 1. A probability of 0 indicates that the event cannot occur, whereas a probability of 1 indicates that the event is guaranteed to occur. If the probability of an event is greater than 0.5, then this indicates that the event is more likely to occur than not to occur.

A sample space is the set of all possible outcomes of an experiment, and an event E is a subset of the sample space, and the event is said to have occurred if the outcome of the experiment is in the event E. Bayes' formula enables the probability of an event E to be determined by a weighted average of the conditional probability of E given that the event F occurred and the conditional probability of E given that F has not occurred.

Often, some numerical quantity determined by the result of the experiment is of interest rather than the result of the experiment itself. These numerical quantities are termed random variables. The distribution function of a random variable is the probability that the random variable X takes on a value less than or equal to x.

The binomial and Poisson distributions are important distributions in statistics, and the Poisson distribution may be used as an approximation for the binomial. The binomial distribution involves n Bernoulli trials, where each trial is independent and results in either success or failure. The mean of the Bernoulli distribution is given by p and the variance by $p(1 - p)$.

The normal distribution is the most important distribution in statistics, and it is shaped like a bell. It is a continuous distribution, and the curve is symmetric about the mean of the distribution. It has two parameters namely the mean μ and the standard deviation σ. Every normal distribution may be converted to the unit normal distribution by $Z = (X - \mu)/\sigma$, and every probability statement about X has an equivalent probability statement about the unit distribution Z.

The Central Limit Theorem essentially states that the sum of a large number of independent and identically distributed random variables has a distribution that is approximately normal.

Bayesianism provides a way of dealing rationally with randomness and risk in daily life, and it interprets probability as a measure of one's personal belief in a proposition or outcome. It involves first using all your available prior knowledge to form an initial estimate of the probability of the event or hypothesis. Further, when reliable frequency data becomes available the measure of personal belief would be updated accordingly to equal the probability calculated by the frequency calculation. Further, the probabilities must be updated in the light of new information that becomes available, as probabilities may change significantly from new information and knowledge. Finally, no matter how much the odds move in your favour, there is eventually one final outcome (which may or may not be the desired event).

Note

1 Bayes' formula is named after the Reverend Thomas Bayes, an 18th-century English statistician and Presbyterian minister. It allows probabilities to be updated in light of relevant evidence.

11

The Insurance Industry

Insurance is a way to manage risk and especially a way to manage the occurrence of undesirable events such as damage or loss to one's home due to fire, the theft of personal belongings from one's home, or damage or loss of a vehicle and injury to others in an automobile accident. Insurance provides a level of protection to the policyholder should such an undesired event occur, as the cost of such incidents prior to the introduction of insurance was potentially very high. The concept of cooperative (pooling) where the risk is shared with others is a fundamental principle underlying the insurance field, and it means that the risk is pooled with all other policyholders. There are several types of insurance which are described in Table 11.1.

Insurance is a form of risk management that is used by the policyholder to avoid or manage the risk of financial loss. The policyholder pays a premium to the insurance company, in return for the insurance company's commitment to compensate the insured party in the event of a loss that is covered under the policy. The insurance contract details the conditions of the agreement between the insured party and the insurance company, and the circumstances in which the insurance company will compensate the insured party for their loss. There are a number of key principles underlying insurance:

- Utmost good faith
- Insurable interest
- Indemnity
- Contribution
- Subrogation
- Proximate cause.

The principle of utmost good faith (*uberrimae fides*) is binding on both parties, and demands the disclosure of all material facts that would affect the judgement of the underwriter in accepting or declining the risk, or accepting the risk at an increased premium. The proposer is expected to disclose all relevant facts whenever the policy is issued, renewed or whenever alterations are made to the policy during the insurance period. Non-disclosure may be innocent or it could be a fraudulent misrepresentation of the facts.

DOI: 10.1201/9781003308140-11

TABLE 11.1

Types of Insurance

Type	Description
Car insurance	Automobile insurance is mandatory in most countries, and it provides protection to the policyholder should an accident happen.
Home insurance	This covers the insured should any negative incidents arise in the home such as damage to the house and its contents caused by fire or flooding, or the theft of cash and personal effects from the home.
Health insurance	The cost of medical care is quite high (especially in private hospitals), and as there are often lengthy delays in public hospitals medical insurance has become essential in the western world.
Life insurance	This type of insurance provides protection to the family should the policyholder die, and it generally covers the funeral and mortgage expenses and gives financial protection to the family.
Life annuity	A life annuity may be purchased at retirement that pays the holder a regular amount for the remainder of their life. That is, a life annuity manages the risk of the holder living longer than their available savings, and the risk is transferred to the insurer.
Travel insurance	Travel insurance protects the policyholder should accidents occur on a trip, should theft occur, should flights be delayed or cancelled, should a medial incident arise and so on.
Disability insurance	This covers the insured in the event of being unable to work in the short/long term due to a disability, and this type of insurance provides partial/total income protection.
Public liability insurance	This type of cover provides protection to the policyholder where a member of the public suffers injury or damage to their property while on the policyholder's property.
Pet insurance	Veterinary medicine has become expensive, and pet insurance provides partial/total cover for the treatment of the insured person's pet.

The principle of *insurable interest* essentially states that there is a legal right to insure what is proposed and that it has monetary value, and that the party seeking insurance has some legal ownership of the item (such as sole ownership or joint ownership of a property).

The principle of *indemnity* essentially states that the insurance company will attempt to put the insured party in the same financial position after a loss as before. The insured party cannot claim more than the insurable interest (the insured party cannot profit from the loss). The insurance company has the option of choosing the method of indemnity, which may be provided by repair, replacement or cash.

The principle of *contribution* deals with the situation where the insured has two or more policies in place (covering the same risk) at the time of a loss, and it essentially states that the insurance company has the right to call upon the other insurance companies to contribute to the payment.

The principle of *subrogation* essentially states that the insurance company has the right, after indemnifying the insured party, to stand in place of the insured and to take over all of the insured party's rights and to seek appropriate remedies to diminish their loss.

The principle of *proximate cause* essentially states that the onus is on the insured party to prove that the loss was caused by a peril covered by the insurance policy, and not by an uninsured or expected peril. That is, if several events have contributed to the loss, the original or proximate cause must be determined before liability can be decided.

Normally, only a small percentage of policyholders suffer losses, and these losses are paid out of the pool of premiums collected from the pool of policyholders, where the pool compensates the few who have suffered loss. Each policyholder therefore exchanges an unknown loss for the payment of a known (much lower) premium. That is, the risk of unanticipated losses is transferred from the policyholder to the insurance company.

11.1 Basic Mathematics of Motor Insurance

Motor insurance protects the policyholder against financial loss in the event of an accident involving a vehicle that he or she owns (Fig. 11.1). The total amount of losses depends on the number of losses that occur during a specified period, and the amount of each loss. For example, a mature policyholder with motor insurance will generally have no claims most years, but in some years he/she could have one or more accidents requiring hospitalization. The frequency distribution and the severity distribution are combined to yield the overall loss distribution. The material in this section has been adapted from [Rsa:05].

Example

Suppose that a car owner has an 80% probability of zero accidents in the year, a 20% probability of one accident and a zero probability of being in more than one accident in a year. If an accident does occur, then there is a 50% probability that the repairs will cost €500, a 40% probability that the repairs will cost €5000 and a 10% probability that the car will need to be replaced at a cost of €15,000.

Combine the frequency and severity distributions to determine the car owner's expected loss in the year, and determine the variability of the loss by calculating the standard deviation.

FIGURE 11.1
Driving to Summit of Mt. Washington, 1899.

Solution

The overall loss distribution $L(x)$ is determined from the frequency distribution $F(x)$ of the frequency of loss random variable, and the severity distribution $S(x)$ of the amount of loss random variable. The car owner's loss is then calculated as the expected value (i.e., mean) of this distribution. The car owner's loss is zero where there are no accidents, and when an accident has occurred the loss may be €500, €5000 or €15,000 (depending on the severity of the accident). The calculation of the overall loss function $L(x)$ is given in Table 11.2.

Therefore, the expected value of the overall loss function is given by

$$L(x) = \begin{cases} 0.8, & x = 0 \\ 0.1, & x = 500 \\ 0.08, & x = 5000 \\ 0.02, & x = 15{,}000 \end{cases}$$

TABLE 11.2
Loss Function

$F(x)$	$S(x)$	$L(x)$
0.8	–	0.8, 0
0.2	0.5	0.1, 500
	0.4	0.08, 5000
	0.1	0.02, 15000

$$\begin{aligned}\Sigma xL(x) &= 0.8 * 0 + 0.1 * 500 + 0.08 * 5000 + 0.02 * 15000 \\ &= 0 + 50 + 400 + 300 \\ &= €750\end{aligned}$$

That is, the average car owner spends €750 on repairs after accidents, and we determine the variability from the standard deviation:

$$\begin{aligned}\sigma^2 &= \Sigma(\mathrm{X} - \mu)^2 L(x) = \Sigma(\mathrm{X} - \mathrm{E}(\mathrm{X}))^2 L(x) \\ &= (0 - 750)^2 * 0.8 + (500 - 750)^2 * 0.1 + (5000 - 750)^2 * 0.08 \\ &\quad + (15{,}000 - 750)^2 * 0.02 \\ &= 0.8 * 750^2 + 0.1 * (-250)^2 + 0.08 * 4250^2 + 0.02 * 14250^2 \\ &= 450{,}000 + 6250 + 1{,}445{,}000 + 4{,}061{,}250 \\ &= 5{,}962{,}500\end{aligned}$$

Therefore, $\sigma = \sqrt{5{,}962{,}500} = €2442$, and so there is quite a large variation in outcomes.

It might be thought that if an insurance company sells n policies to n individuals it assumes the total risk of the sum of the risk of the individuals. However, the risk is actually smaller than the sum of the risks associated with each policyholder, and this follows from the well-known Central Limit Theorem (see section 10.4.3 in Chapter 10).

This theorem states that the mean of the sum of n independent random variables (each with mean μ and variance σ^2) is $n\mu$, and the variance is $n\sigma^2$ and the standard deviation is $\sqrt{n}\sigma$ (which is less than the sum of the standard deviations for each policy). Further, the coefficient of variation (i.e., the ratio of the standard deviation to the mean) is smaller for the sum than the coefficient of variation of each individual policy which is $^{\sigma}/_{\mu}$.

$$\frac{\sqrt{n}\sigma}{n\mu} = \frac{\sigma}{\sqrt{n}\,\mu}$$

The coefficient of variation is a measure of the insurer's risk, and as n becomes larger and larger the coefficient of variation gets closer and closer to zero, and so the insurer's risk (the coefficient of variation) tends to zero (Fig. 11.2).

Each car owner has an expected loss of €750 and a standard deviation of €2442, so the coefficient of variation is 3.26. The expected repair loss for 100 car owners is 100 * 750 = €75,000 and the standard deviation is 24,418, so the coefficient of variation is 0.326.

FIGURE 11.2
A Car Crash. Public Domain.

It would be reasonable to expect that the payment the policyholder makes to the insurer would include the expected claim payments, plus an amount to cover the insurer's cost in selling and servicing the policy. The net premium is the expected amount of the claim payments, whereas the gross premium is the net premium plus an amount to cover the insurer's expenses and profit.

It is quite common for an insurance policy to be subject to certain limitations such as a *policy excess* and a maximum payout limit. The policy excess specifies that the losses that will be reimbursed are only those that are in excess of a certain threshold value (stated by the excess value on the policy), and the maximum payout limit sets an upper bound on what will be paid out by the insurance company for any loss. That is, the insurance company does not cover the entire loss, and the policyholder is required to cover part of the loss.

The advantage of the policy excess is that it limits the number of claims that will be made, as small losses will not result in a claim. This leads to saving on the administration and processing of claims, and it also provides a financial incentive to the policyholder to take care to avoid losses that would lead to claims as the policyholder is at risk of financial loss. The claim payment is reduced by the amount of the excess, so the claim payment is always less than the loss incurred, and where a small loss is less than the excess there will be no claim payment. This means that it is necessary to distinguish between losses and claim payments as to both frequency and severity, and we illustrate this by adjusting the previous example to include a policy excess and showing how this affects the distributions and claim payments.

Example (ctd)

Suppose the policy for the car pool of 100 cars that was discussed has a €500 excess. Calculate the expected claims payment and the insurer's risk.

Solution

The claim payment distribution is determined from Table 11.3.

$$C(y) = \begin{cases} \text{loss} = 0 \text{ or } 500, \ 0.9 \ y = 0 \\ \text{loss} = 5000, \ 0.08 \ y = 4500 \\ \text{loss} = 15{,}000, \ 0.02 \ y = 14{,}500 \end{cases}$$

The expected claims payment and standard deviation for a single policy are:

$$\begin{aligned} E[Y] &= 0.9 * 0 + 0.08 * 4500 + 0.02 * 14500 \\ &= 360 + 290 \\ &= €650 \end{aligned}$$

TABLE 11.3

Claim Payment Function (Excess)

F(x)	*S(x)*	*L(x)*	*C(y)*
0.8	–	0.8, 0	0.9, 0
0.2	0.5	0.1, 500	0.08, 4500
	0.4	0.08, 5000	0.02, 14500
	0.1	0.02, 15000	

$$\begin{aligned}\sigma_Y^2 &= 0.9 * (0 - 650)^2 + 0.08 * (4500 - 650)^2 + 0.02 * (14500 - 650)^2 \\ &= 5,402,500\end{aligned}$$

$$\sigma_Y = \sqrt{5402500} = 2324$$

That is, the expected claim payments for the 100 policies would be €65,000 and the variance would be 540,250,000 and the standard deviation would be 23,243.

This example shows that the policy excess eliminated the need to process small claims below €500, and the probability of a claim has fallen from 20% to 10% of policies. Further, the excess reduces the amount of the expected claim payments from €75,000 to €65,000, and the standard deviation has fallen from 24,418 to 23,243.

We mentioned that it is quite common for a maximum benefit limit to be set in the insurance policy, and this reduces the risk taken on by the insurance company as well as ensures that the insurer has the reserves to meet all claims. There may be a choice of benefit limits that the policyholder may choose with a corresponding reduction in premium for choosing a lower limit. There are various ways in which benefit limits may be set in a policy. Next, we continue our example and show how the maximum payout affects the claim payments.

Example (ctd)

Suppose the policy for the car pool of 100 cars that was discussed has a €500 excess and a maximum payout of €12,500. Calculate the expected claims payment and the insurer's risk.

Solution

The claim payment distribution is determined from Table 11.4.

TABLE 11.4

Claim Payment Function (Excess/Limit)

$F(x)$	$S(x)$	$L(x)$	$C(y)$
0.8	–	0.8, 0	0.9, 0
0.2	0.5	0.1, 500	0.08, 4500
	0.4	0.08, 5000	0.02, 12500
	0.1	0.02, 15000	

$$C(y) = \begin{cases} \text{loss} = 0 \text{ or } 500, & 0.9\ y = 0 \\ \text{loss} = 5000, & 0.08\ y = 4500 \\ \text{loss} = 15{,}000, & 0.02\ y = 12{,}500 \end{cases}$$

The expected claims payment and standard deviation for a single policy are:

$$\begin{aligned} E[Y] &= 0.9 * 0 + 0.08 * 4500 + 0.02 * 12500 \\ &= 360 + 250 \\ &= €610 \end{aligned}$$

$$\begin{aligned} \sigma_Y^2 &= 0.9 * (0 - 610)^2 + 0.08 * (4500 - 610)^2 + 0.02 * (12500 - 610)^2 \\ &= 4,372,900 \end{aligned}$$

$$\sigma_Y = \sqrt{4372900} = 2091$$

The expected claim payment for the 100 policies is then €61,000 with the variance 437,290,000 and the standard deviation 20,911. That is, by employing a policy excess and maximum limit the insurer's expected claim payments has fallen from €75,000 to €61,000 and the standard deviation has fallen from 24,418 to 20,917.

The insurance company would need to take inflation into account as the cost of repairs will increase over a period, so there will be a need to adjust the excess and benefit limit to reflect inflation.

11.2 Basic Mathematics of Health Insurance

The cost of medical care is quite high in the western world and medical insurance has become essential. Suppose an individual takes out an annual health insurance policy that covers the insured for hospitalization during the year (Fig. 11.3).

Example

The probability of an individual being hospitalized during the year is $P(H) = 0.15 = 15\%$, and once hospitalized the charges X have a continuous probability density function $f_X(x \mid H = 1) = 0.1e^{-0.1x}$ for $x > 0$.

FIGURE 11.3
Florence Nightingale. An Angel of Mercy. Scutari Hospital. 1855.

Determine the expected value, standard deviation and the ratio of the standard deviation to the mean of hospital charges.

Solution

The expected value of hospital charges is given by

$$
\begin{aligned}
E[X] &= P(\mathrm{H} \neq 1)E[X|H \neq 1] + P(H = 1)E[X|H = 1] \\
&= 0.85 * 0 + 0.15 \int_0^{\infty} x 0.1 e^{-0.1x} dx \\
&= 0.15 \int_0^{\infty} 0.1x\, e^{-0.1x} dx \\
&= 0.15(-xe^{-0.1x}|_0^{\infty} + \int_0^{\infty} e^{-0.1x} dx) \\
&= 0.15(0 - 10e^{-0.1x}|_0^{\infty}) \\
&= 1.5
\end{aligned}
$$

Similarly, we may calculate $E[X2]$ and then compute the variance from the formula:

$$\mathrm{Var}(X) = \sigma_X^2 = E[X^2] - (E[X])^2$$

$$
\begin{aligned}
E[X]^2 &= P(H \neq 1)E[X^2|H \neq 1] + P(H = 1)E[X^2|H = 1] \\
&= 0.85 * 0 + 0.15 \int_0^{\infty} x^2 0.1 e^{-0.1x} dx \\
&= 0.15 \int_0^{\infty} 0.1x^2 e^{-0.1x} dx \\
&= 0.15\left(-x^2 e^{-0.1x}|_0^{\infty} + 2\int_0^{\infty} x\, e^{-0.1x} dx\right) \\
&= 0.15\left(2 * 10 \int_0^{\infty} 0.1x\, e^{-0.1x} dx\right) \\
&= 2 * 10 * 0.15\left(\int_0^{\infty} 0.1x\, e^{-0.1x} dx\right) \\
&= 20 * 1.5 \\
&= 30
\end{aligned}
$$

$$
\begin{aligned}
\mathrm{Var}(X) &= \sigma_X^2 = E[X^2] - (E[X])^2 \\
&= 30 - 1.5^2 \\
&= 27.75
\end{aligned}
$$

Therefore, as the standard deviation σ_X is given by the square root of the variance $\sigma^2{}_X = 27.75$, we have $\sigma_X = \sqrt{27.75} = 5.27$.

The ratio of the standard deviation to the mean (the coefficient of variation) is given by

$$\sigma_X / E[X] = 5.27/1.5 = 3.51$$

Example (ctd)

Suppose there is an insurance pool that reimburses hospital charges for 200 policyholders. Determine the expected claims payments, standard deviation and coefficient of variation where the claims for each individual are independent of others.

Solution

Let X_i represent the charges for policyholder X_i. The expected value of the hospital charges for the pool is given by

Let $S = \sum_1^{200} X_i$. Then

$$E[S] = 200 * E[X] = 200 * 1.5 = 300$$
$$\sigma_S^2 = 200 * \sigma_X^2 = 200 * 27.75 = 5500$$
$$\sigma_S = \sqrt{5500} = 74.50(=\sqrt{200} * \sigma_X)$$

The coefficient of variation is given by $\sigma_S/E[S] = 74.50/300 = 0.25$.

The insurance company will generally include a deductible (policy excess) that specifies the losses that will be reimbursed are only those that are in excess of a certain threshold value (stated by the excess value on the policy). This affects the expected claims payment and standard deviation for the insurance pool.

Example (ctd)

Suppose the insurance company includes a deductible of 5 on annual claim payments for each member. Determine the expected claims payments and standard deviation for the pool.

Solution

There are three cases to consider:

a. No hospitalization and so no claim payments
b. Hospitalization but charges less than the deductible
c. Hospitalization but charges more than the deductible

$$E[Y] = P(H \neq 1)E[Y|H \neq 1] + P(X \leq 5, H = 1)E[Y|X \leq 5, H = 1]$$
$$+ P(X > 5, H = 1)E[Y|X > 5, H = 1].$$

The calculation results in $E[Y] = 0.91$ and $\sigma^2_Y = 17.37$, we have $\sigma_X = \sqrt{17.37} = 4.17$.

The details are in [Rsa:05].

11.3 Basic Mathematics of Pensions

Financial planning for the future has become important as improvements to the health and well being of the world's population has led to significant increases in life expectancy. This means that a person who retires at the age of 65 could live for many years after retirement, and so they will need to fund their retirement through a mixture of the state pension (where the state provides such a service), their savings and investments, and their company or personal pension.

Financial planning for retirement is usually not a priority when someone starts his or her first job, but it is an important matter that needs to be considered early in one's career. Large companies often provide company pension schemes (*defined benefit* schemes or the more common *defined contribution* schemes), and mandate their employees to join the scheme. If the employer does not provide such a scheme then the employee could set up their own personal pension plan with an insurance company that they could use throughout their career. As employees may change jobs several times during their career, they will need to transfer their company pension fund and contributions from one company to another.

At retirement age, the pension holder may have accumulated a large pension pot. For example, suppose the value of the pension fund is €500,000 and under the rules of the pension scheme the holder is allowed to take a 25% tax-free sum of €125,000 from the capital sum, with the balance of €375,000 used to purchase an annuity. For example, this annuity might pay the annuitant a regular annual income of €20,000 for life.

A *life annuity* is a financial product provided by insurance companies that pays out a regular income to an individual for the rest of their life. An annuity may be purchased from the proceeds of a personal pension or a company pension, and people generally invest in annuities at retirement. The key attraction of annuities is that they are designed to provide financial security to the annuitant for the rest of their life, as they are provided with a regular consistent income from the date of their retirement. Some annuities provide a flat fixed amount for life, others provide a fixed rate of increase per year, and others provide inflation protection by offering a rate of increase that is equal to the rate of inflation during the year (this may be capped at a maximum rate). An annuity that pays a fixed amount per year is subject to the perils of inflation, for example, for the annuity that is paying out €20,000 for life the present value of receiving €20,000 in year 20 where the inflation rate has been 2% per annum over the period is given by $€20{,}000(1 + 0.02)^{-20}$ or roughly €13,500.

An annuity dies with its holder (although some annuities pay out for a minimum period of 5–10 years, and provide a death benefit to the estate of the deceased). The amount remaining following the death of the annuitant is shared among the pool of the other annuitants. It has become more expensive to purchase annuities in recent years due to the low yields on government bonds and the increase in the life expectancy of the population. The cost of an annuity for a female is generally higher than a male due to the difference in life expectancy between the genders.

Early work on calculating the price of an annuity was done by Johan Hudde in the Netherlands in the 17th century. He used statistics from a group of 1495 people who had purchased annuities in the period from 1586 to 1590, and who were now all deceased. He then constructed a mortality table based on the death rates of people who had bought the annuities, rather than basing the table from the death rate of the general population. He assumed a 4% growth rate ($r = 0.04$) and used it to calculate the present value of the payments made to a purchaser of an annuity. For example, suppose the regular annual payment is A then the present value of n annual payments of amount A is given by

$$A(1+r)^{-1} + A(1+r)^{-2} + \ldots + A(1+r)^{-n}$$
$$= A/r * [1-(1+r)^{-n}]$$

Hudde recommended a purchase price of 17.2 times the annual payment for a six-year-old boy, and we can calculate the life expectancy of a six years old in the Netherlands in the 17th century from this:

$$17.2A = A/r * [1-(1+r)^{-n}]$$
$$17.2 = 1/0.04 * [1-(1.04)^{-n}]$$
$$17.2 * 0.04 = [1-(1.04)^{-n}]$$
$$0.688 = 1-(1.04)^{-n}$$
$$-0.312 = -(1.04)^{-n}$$
$$0.312 = (1.04)^{-n}$$
$$\log(0.312) = \log(1.04)^{-n}$$
$$-0.5058 = -n * 0.0170$$
$$0.5058 = n * 0.0170$$
$$0.5058/0.0170 = n$$
$$n = 29.75$$

That is, Huddes mortality table predicted that a six-year-old boy would live an average of another 29.75 years (i.e., to 35.75 years), and so it provides information on life expectancy and the high infant mortality rate in the Netherlands (and in other countries) in the 17th century.

Edmond Halley (of Halley's comet fame) constructed the world's first modern mortality table in 1693 leading to the scientific foundation of life insurance [Hal:93]. He analysed the records of deaths in the city of Breslau, which was then in Germany (it is now called Wroclaw and is in Poland). He started with a population of 1000 at age 1 and determined the numbers surviving at each age up to 84. The figures for the first 8 years are given in Table 11.5.

TABLE 11.5

Table of Deaths for Breslau in 17th Century

Age	1	2	3	4	5	6	7	8
Number	1000	855	798	760	732	710	692	680

Suppose that one wants to calculate a life annuity of an amount A for a child who is five years old assuming a 4% growth rate. Then we compute the present value of the expected payments to the child. The present value of the first payment is

$$A(1.04)^{-1} = 0.9615A$$

It is clear from Table 11.5 that the probability of a child surviving from year 5 to year 6 is given by $^{710}/_{732}$, so the net present value is given by

$$A(1.04)^{-1} * {}^{710}/_{732} = 0.9326A$$

Similarly, the net present value of the second payment is

$$A(1.04)^{-2} * {}^{692}/_{732} = 0.8740A$$

Similarly, the net present value of the third payment is

$$A(1.04)^{-3} * {}^{680}/_{732} = 0.8258A$$

The calculation is continued until the end of the mortality table and then all the net present values are added together. The sum gives the price of the annuity and it involves a long calculation (in the age before computers or spreadsheets) as the mortality table numbers do not conform to a simple formula. If they did correspond to a simple formula the calculations could be shortened through the use of algebra.

Abraham de Moivre (of De Moivre's Theorem fame in complex analysis fame) introduced an assumption on Halley's mortality tables with the goal of simplifying the calculation. He assumed that the number of people dying decreased in an arithmetic progression. Another words, if the initial population is N and d people die every year then the population after k years is $N - kd$. It turned out that de Moivre's results were sufficiently close to those of Halley to be of practical use, although De Moivre' simplification is no longer used in mortality calculations today as the introduction of modern computer technology and software (e.g., spreadsheets) has made it easy to perform these calculations.

11.3.1 Mortality Tables

A mortality table (also call actuarial table) shows the rate of deaths in a population during a particular time interval. A mortality table shows the probability of a person's death before their next birthday based on their current age. Actuarial tables may include several other characteristics such as whether the person is a smoker, their occupation, the person's socio-economic background and health condition (e.g., whether the person is obese).

A life insurance policy provides a benefit to the beneficiary (or beneficiaries) on the death of the policyholder, and it allows the outstanding mortgage and other debts to be paid, as well as providing a level of financial security to the beneficiary. The life insurance industry relies heavily on mortality tables to determine the risk associated with a person wishing to take out a life insurance policy, where the data in the table is used to calculate the appropriate premium for the policy.

Mortality tables generally go from birth through age 100 in 1-year increments, and the table gives the probability of death for someone at any age before their next birthday and may include other data such as life expectancy at that age. The first step in using a mortality table is to identify the person's age, and then to read the corresponding row in the table for that age. There will be entries in the row for the probability of death at that age. The probability of death increases with age (e.g., the mortality rate of an 80-year-old male is hundreds of time that of a 10-year-old boy).

The two main types of mortality table are the *period life table* and the *cohort life table*. The period life table shows the current probability of death for people of different ages in the current year, whereas the cohort life table shows the probability of death of people from a given cohort (especially their birth year) over the course of their lifetime.

There are usually separate mortality tables for males and females due to the differences in life expectancy between the genders.

11.3.2 Life Insurance

Life insurance provides insurance against the possibility of death on guaranteed terms and the insurance period is long term (e.g., 20 years or longer), and the modern life insurance industry commenced in London in the mid-18th century. A life insurance policy pays a death benefit in the case of the demise of the policyholder during the policy term (i.e., the period of insurance), and the benefit is usually a lump sum that is paid to the beneficiaries of the policy or the estate of the policyholder.

The premium for life insurance is determined from the level of cover required and the mortality risk of the individual to be insured. The amount that the policyholder insures their life is based on several factors such as their age, their salary, the number of years remaining to retirement, their occupation, their family size, their net worth, their medical history and so on. For example, a person aged 35 with an annual salary of £35,000 and 30 years to retirement may wish to insure their life for 30 * £35,000 = £1,050,000.

The actuary will calculate the probability of death and how much it costs the life insurance company to insure the proposer. It is essential that the premiums charged by the insurance company are adequate to cover future claims, and the insurance company needs to be prudent in assessing risk. A mortality table is required to calculate the premium, and due to the exceptionally long-term nature of life insurance it may be prudent for the actuary to assume that the mortality may be worse than is thought likely. The actuary needs an understanding of the accumulation of invested funds, and it may be prudent to assume that the investment return may be lower than expected and that the operating expenses will be higher than expected. The actuary will need to check periodically that the investment assets are sufficient to cover the liabilities.

11.4 The Actuary

An actuary has the mathematical and statistical background as well as business acumen to determine the expected costs and risks of any situation where there is financial uncertainty, and the actuary uses the data to create models of the risks. It involves determining the premiums to be paid as well as the amount that the insurer needs to have available to ensure that benefits and expenses can be paid as they arise.

The first task for the actuary is to estimate the frequency and severity distribution for a certain insurance pool, and this involves an analysis of past experience and historical data from the insurance pool or from a similar group to the pool. There is a need for care with historical data as it may not be a reliable predictor of the future, as there may be fundamental changes that will alter the underlying distributions. However, assuming that the historical data is representative then it will be a good predictor of the underlying probability distributions. The frequency and severity functions are developed from an analysis of the past experience and are combined to form the loss distribution. The claims distribution may then be derived, which involves adjusting the loss distribution to reflect provisions made for the policy excess and a maximum limit to the benefits that will be paid out.

The actuary may need to take inflation into account and predict future inflation based on past expertise, where the claim payments could be affected by inflation. Further, as premiums are often invested to cover future payments the actuary will need to calculate the expected investment returns. Actuaries will also carry out many other projections of the insurer's future financial position under different circumstances, and to identify potential negative situations before they become significant.

For more detailed information on the mathematics employed in the insurance field, see [Oli:11].

11.5 Review Questions

1. Explain the various types of insurance that are available.
2. Explain the key principles of insurance?
3. Explain the difference between a loss and a claim.
4. Explain how insurance is used to manage risk.
5. What is an insurance annuity? How does it differ from a standard annuity in the banking sector?
6. Explain the role of the actuary in the insurance sector.
7. Explain the concept of pooling that is used in the insurance field.
8. Explain how the total risk from several policies is smaller than the sum of the risks associated with each policyholder.
9. Explain how mortality tables are used in the calculation of life annuities.
10. What is the purpose of the policy excess and the maximum benefits limit?

11.6 Summary

Insurance is a form of risk management that is used by the policyholder to avoid or manage the risk of financial loss. The policyholder pays a premium to the insurance company, in return for the insurance company's commitment to compensate the insured party in the event of a loss that is covered under the policy. Insurance is a way to manage the occurrence of undesirable events such as the loss of one's home due to fire, or damage to a vehicle in an automobile accident. Insurance provides a level of protection to the policyholder should such an undesired event occur.

The concept of cooperative (pooling) where the risk is shared with others is a fundamental principle underlying the insurance field, and it means that the risk is pooled with all other policyholders.

There are a number of key principles underlying insurance such as utmost good faith, insurable interest, indemnity, contribution, subrogation and proximate cause. These essentially state that the proposer should disclose all material facts to the insurance company, as these could affect the judgement of the underwriter in accepting or declining the risk. Further, the party seeking insurance must have appropriate legal ownership of the item. The insurance company will attempt to put the insured party in the same financial position after a loss as before. If the insured has two or more policies in place (covering the same risk) at the time of a loss, then the insurance company has the right to call upon the other insurance companies to contribute to the payment. Further, the insurance company has the right, after indemnifying the insured party, to stand in place of the insured. Finally, the onus is on the insured party to prove that the loss was caused by a peril covered by the insurance policy, and not by an uninsured or expected peril.

Motor insurance protects the policyholder against financial loss in the event of an accident involving a vehicle he or she owns, whereas medical insurance protects the policyholder against financial loss in the event of hospitalization. A life annuity is a financial product provided by insurance companies that pays out a regular income to an individual for the rest of their life.

An actuary has the mathematical background to determine the expected costs and risks of any situation where there is financial uncertainty, and to create models of the risks. The role involves determining the premiums to be paid as well as the amount that the insurer needs to have available to ensure that benefits and expenses can be paid.

12

Data Science and Data Analytics

Information is power in the digital age, and the collection, processing and use of information need to be regulated. Data science involves the extraction of knowledge from data sets that consist of structured and unstructured data, and data scientists have a responsibility to ensure that this knowledge is used wisely and not abused. Data science may be regarded as a branch of statistics as it uses many concepts from the field, and in order to prevent errors occurring during data analysis it is essential that both the data and models are valid.

The question of ownership of the data is important; as if, for example, I take a picture of another individual does the picture belong to me (as owner of the camera and the collector of the data)? Or does it belong to the individual who is the subject of the image? Most reasonable people would say that the image is my property, and if so what responsibilities or obligations do I have (if any) to the other individual? That is, although I may technically be the owner of the image, the fact that it contains the personal data (or image) of another should indicate that I have an ethical responsibility or obligation to ensure that the image (or personal data) is not misused in any way to harm that individual. Further, if I misuse the image in any way then I may be open to a lawsuit from the individual.

Personal data is collected about individuals from their use of computer resources such as their use of email, their Google searches, their Internet and social media use to build up revealing profiles of the user that may be targeted to advertisers. Modern technology has allowed governments to conduct mass surveillance on its citizens, with face recognition software allowing citizens to be recognized at demonstrations or other mass assemblies.

Further, smartphones provide location data that allows the location of the user to be tracked, and may be used for mass surveillance. Many online service providers give customer data to the security and intelligence agencies (e.g., NSA and CIA), and these agencies often have the ability to hack into electronic devices. It is important that such surveillance technologies are regulated and not abused by the state. Privacy has become more important in the information age, and it is the way in which we separate ourselves from other people, and is the right to be left alone. The European GDPR law has become an important protector of privacy and personal data, and both European and other countries have adapted it.

Companies collect lots of personal data about individuals, and so the question is: How should a company respond to a request for personal

DOI: 10.1201/9781003308140-12

information on particular users? Does it have a policy to deal with that situation? What happens to the personal data that a bankrupt company has gathered? Is the personal data part of the assets of the bankrupt company and sold on with the remainder of the company? How does this affect privacy information agreements and compliance to them or does the agreement cease on termination of business activities?

The consequence of an error in data collection or processing could result in harm to an individual, and so the data collection and processing need to be accurate. Decisions may be made on the basis of public and private data, and often individuals are unaware as to what data was collected about them, whether the data is accurate, and whether it is possible to correct errors in the data.

Further, the conclusions from the analysis may be invalid due to errors in incorrect or biased algorithms, and so a reasonable question is how to keep algorithmically driven systems from harming people? Data scientists have a responsibility to ensure that the algorithm is fit for purpose and uses the right training data, and as far as practical to detect and eliminate unintentional discrimination in algorithms against individuals or groups.

That is, problems may arise when the algorithm uses criteria tuned to fit the majority, as this may be unfair to minorities. Another words, the results are correct, but presented in an over-simplistic manner. This could involve presenting the correct aggregate outcome but ignoring the differences within the population, and so leading to the suppression of diversity, and discriminating against the minority group. Another example is where the data may be correct but presented in a misleading way (e.g., the scales of the axis may be used to present the results visually in an exaggerated way).

12.1 What Is Data Science?

There has been a phenomenal growth in the use of digital data in information technology, with vast amounts of data collected, processed and used, so the ethics of data science has become important. There are social consequences to the use of data, and the ethics of data science aims to investigate what is fair and ethical in data science, and what should or should not be done with data.

A fundamental principle of ethics in data science refers to *informed consent*, and this has its origins in the ethics of medical experiments on individuals. The concept of informed consent in medical ethics is where the individual is informed about the experiment, and gives their *consent voluntarily*. The individual has the right to withdraw consent at any time during the experiment. Such experiments are generally conducted to benefit society, and often there is a board that approves the study and oversees it to ensure that all participants have given their informed consent, and attempts to balance the benefits to society with any potential harm to individuals. Once individuals have given their informed consent data may be collected about them.

The principle of informed consent is part of information technology, in the sense that individuals accept the terms and conditions before they may use software applications, and these terms state that data may be collected, processed and shared. However, it is important to note that generally users do not give informed consent in the sense of medical experiments, as the details of the data collection and processing are hidden in the small print of the terms and condition, and this is generally a long and largely unreadable document. Further, the consent is not given voluntarily, in the sense that if a user wishes to use the software, then he or she has no choice but to click acceptance of the terms and conditions of use for the site. Otherwise, they are unable to access the site, and so for many software applications (apps) consent is essentially coerced rather than freely given.

There was some early research done on user behaviour by Facebook in 2012, where they conducted an experiment to determine if they could influence the mood of users by posing happy or sad stories to their news feed. The experiment was done without the consent of the users, and while the study indicated that happy or sad stories did influence the user's mood and postings, it led to controversy and major dissatisfaction with Facebook when users became aware that they were the *subject of a psychological experiment without their consent.*

The dating site OkCupid uses an algorithm to find compatibility matches for its users based on their profiles, and two people are assigned a match rating based on the extent to which the algorithm judges them to be compatible. OkCupid conducted psychological experiments on its users without their knowledge, with the first experiment being a "love is blind" day where all images were removed from the site, and so compatibilities were determined without the use of images.

Another experiment that was conducted by the site was controversial and unethical, as the users were deceived on their match ratings (e.g., two people with a compatibility rating of 90% were given a rating of 30%, and vice versa). The site was trying to determine the extent that two people would get along irrespective of the rating that they were given, and it showed that two people talked more when falsely told that the algorithm matched them, and vice versa. The controversy arose once users became aware of the deception by the company, and it provides a case study on the *socially unacceptable manipulation of user data* by an Internet company.

Data collection is not a new phenomenon as devices such as cameras and telephones have been around for some time. People have reasonable expectations on privacy, and do not expect their phone calls to be monitored and eavesdropped by others, or they do not expect to be recorded in a changing room or in their home. Individuals will wish to avoid the harm that could occur due to data about them being collected, processed and shared. The question is whether reasonable rules can be defined and agreed, and whether tradeoffs may be made to balance the conflicting rights and to protect the individual as far as is possible. Some relevant questions about data collection and ownership are considered in Table 12.1.

TABLE 12.1

Some Reasons for Data Collection

Question	Answers
Who owns the data?	A user's personal information may legally belong to the data collector, but the data subject may have some control as the data is about him/her. • The author of the biography of an individual owns the copyright not the individual. • The photographer of a (legally taken) photo owns the image not the subject. • Recording of audio/video is similar • May be a need to acknowledge copyright (if applicable) • May be limits in rights as to how data is collected and used (e.g., privacy of phone calls) • The data subject may have some control of the data collected
What are the expected responsibilities of the collector?	The collector of the data is expected to: • Collect only required data • Collect legal and ethical data only • Preserve confidentiality/integrity of collected personal data • Not misuse the data (e.g., alter image) • Use data only for purpose gathered • Share data only with user consent
What is the purpose of the data collection?	The purpose may be to: • Carry out service for a user • Improve user experience • Understand users • Build up profile of user behaviour • Exploit user data for commercial purposes
How is user consent to data collection given?	User consent may be given in various ways: • User informed of purpose of data collection • User consents to use of data • May be hidden in terms and conditions of site
User control	This refers to the ability of the user to control the way that their personal data is being collected/used: • Ability of user to modify their personal data • Ability of user to delete their personal data

12.1.1 Data Science and Data Scientists

Data science is a multi-disciplinary field that extracts knowledge from data sets that consist of structured and unstructured data, and large data sets (*big data*[1]) may be analysed to extract useful information. The field has great power to help and to harm, and data scientists have a responsibility to use this power wisely. Data science may be regarded as a branch of statistics as it uses many concepts from the field, and in order to prevent errors occurring during data analysis it is essential that both the data and models are valid.

The consequence of an error in the data analysis or with the analysis method could result in harm to an individual. There are many sources of error such as the sample chosen, which may not be representative of the entire population. Other problems arise with knowledge acquisition by machine learning, where the learning algorithm has used incomplete training data for pattern (or other knowledge) recognition. Training data may also be incomplete if the future population differs from the past population.

The data that needs to be collected as well as the desired attributes of the data needs to be identified, and often the attributes chosen are limited to what is available, and the data scientist will also need to decide what to do with missing attributes. Often errors arise in data processing tasks such as analysing text information or recognizing faces from photos. There may be human errors in the data (e.g., spelling errors or where the data field was misunderstood), and errors may lead to poor results and possible harm to the user. The problem with such errors is that often decisions are made on the basis of public and private data, and often individuals are unaware as to what data was collected and whether there is a method to correct it.

Even with perfect data the conclusions from the analysis may be invalid due to errors in the model, and there are many ways in which the model may be incorrect. Many machine-learning algorithms just estimate parameters to fit a pre-determined model, without knowing whether the model is appropriate or not (e.g., the model may be attempting to fit a linear model to a non-linear reality). This becomes problematic when estimating (or extrapolating) values outside of the given data unless there is confidence in the correctness of the model.

Further, care is required before assigning results to an individual from an analysis of group data, as there may be other explanations (e.g., Simpson's paradox in probability/statistics is where a trend that appears in several groups of data disappears or reverses when these groups are combined). It is important to think about the population that you are studying, and to make sure that you are collecting data on the right population, and whether to segment it into population groups, as well as how best to do the segmentation.

It may seem reasonable to assume that data-driven analysis is fair and neutral, but unfortunately the problem is that humans may unintentionally introduce bias, as they set the boundary conditions. The bias may be through their choice of the model, the use of training data that may not be representative of the population, or the past population may not be representative of the future population and so on. This may potentially lead to algorithmic decisions that are unfair (e.g., the case of the Amazon hiring algorithm that was biased towards the hiring of males), and so the question is how to be confident that the algorithms are fair and unbiased. Data scientists have a responsibility to ensure that the algorithm is fit for purpose and uses the right training data, and as far as practical to detect and eliminate unintentional discrimination (individual or target group).

Another problem that may arise is data that is correct but presented in a misleading way. One simple way to do this is to manipulate the scales of the axis to present the results visually in an exaggerated way. Another example is where the results are correct, but presented in an over-simplistic manner (e.g., there may be two or more groups in the population with distinct behaviour where one group is the dominant), where the correct aggregate outcome is presented but this is misleading due to the differences within the population, and by suppressing diversity there may be discrimination against the minority group. In other words, the algorithm may use criteria tuned to fit the majority and may be unfair to minorities.

Exploration is the first phase in data analysis, and a hypothesis may be devised to fit the observed data (this is the opposite of traditional approaches where the starting point is the hypothesis, and the data is used to confirm or reject the hypothesis based on the data from the control and target groups, so this approach needs to be used carefully to ensure the validity of the results).

12.1.2 Data Science and Society

Data science has consequences for society with one problem being that algorithms tend to learn and codify the current state of the world, and is therefore harder to change the algorithm to reflect the reality of a changing world. The impact of innovative technologies affects the different cohorts and social groups in society in different ways, and there may also be differences between how different groups view privacy. Data scientists are often focused on getting the algorithm to perform correctly and to do the right processing, and often they may not consider the wide societal impacts of the technology.

Algorithms may be unfair to individuals in that an individual may be classified as being a member of a group in view of the value of a particular attribute, and so the individual could be typecast due to their perceived membership of the group. Another words, the individual may be assigned opinions or properties of the group, and this means that there is a danger of developing a stereotype view of the individual. Further, it may be difficult for individuals to break out of these stereotypes, as these biases become embedded within the algorithm thereby helping to maintain the status quo.

There are further dangers when predictions are made, as predictions are probabilistic and may be wrong, and only suggest a greater likelihood of occurrence of an event. Predictive techniques have been applied to predictive policing and to the prediction of uprisings, but there are dangers of false positives and false negatives (see type I and type II errors in probability/statistics in Chapter 9).

It is important that the societal consequences of algorithms are fully considered by companies, in order to ensure that the benefits of data science are achieved, and harm to individuals is avoided.

12.2 What Is Data Analytics?

Data analytics is the science of handling data collection by computer-driven systems, where the goal is to generate insights that will improve decision-making. It involves the overlap of several disciplines such as statistics, information technology and domain knowledge. It is widely used in social media, e-commerce, the Internet of Things (IoT), recommendation engines, gaming, and may potentially be applied to other fields such as information security, logistics and so on.

Data analytics involves the analysis of data to create structure, order, meaning and patterns from the data. It uses the collected data to produce information as well as generate insights from the data for decision-makers. This is essential in making informed decisions to meet current and future business needs. Data analytics may involve machine learning, or it could be quick and simple if the data set is ready, and the goal is to perform just simple descriptive analysis. There are four types of data analytics (Table 12.2).

Descriptive analysis is a data analysis method that is used to give a summary of what is going on and nothing more. It provides information as to

TABLE 12.2

Types of Data Analytics

Type	Description
Descriptive	These metrics describe what happened in the past and give a summary of what is going on.
Diagnostic	These are concerned with why it happened and involve analysis to determine why something has happened.
Predictive	These are concerned with what is likely to happen in the future.
Prescriptive	These are concerned with analysis to make better decisions, and it may involve considering several factors to suggest a course of action for the business. It may involve the use of AI techniques such as machine learning, and the goal is to make progress and avoid problems in the future.

what happened, and it allows the data collected by the system to be used to identify what went wrong. This type of data is often used to summarize large data sets, and to describe a summary of the outcomes to the stakeholders. The most relevant metrics produced include the key performance indicators (KPIs).

Diagnostic analysis is concerned with the analysis of the descriptive metrics to solve problems, and to identify where the issue could potentially be, and to understand why something has happened.

Predictive analysis involves predicting what is likely to happen in the future based on data from the past, i.e., it is attempting to predict the future based on actions in the past, and it may involve the use of statistics and modelling to predict future performance, based on current and historical data. Other techniques employed include neural networks, regression and decision trees.

Prescriptive analysis is used to help business to make better decisions through the analysis of data, and is effective when the organization knows the right questions to ask and responds appropriately to the answers. It often uses AI techniques such as machine learning to process a vast amount of data, to find patterns and to recommend a course of action that will resolve or improve the situation. The recommended course of action is based on past events and outcomes, and the use of machine learning strategies builds upon the predictive analysis of what is likely to happen to recommend a future course of action.

Prescriptive analytics may be used to automate prices based on several factors such as demand, weather and commodity prices. These algorithms may automatically raise or lower prices at a much faster rate than human intervention.

Companies may use data analytics to create and sell useful products by drilling down into customer data to determine what they are looking for. This includes understanding the features desired of the product and the price that they are willing to pay, and so data analytics has a role to play in new product design. They may be used by the business to improve customer loyalty and retention, and this may be done by gathering data (e.g., the opinions of customers from social media, email and phone calls) to ensure that the voice of the customer is heard and acted upon appropriately.

Marketing groups often use data analytics to determine how successful their marketing campaign has been, and to make changes where required. The marketing team may use the analytics to run targeted marketing and advertisement campaigns to segmented audiences (i.e., subsets of the population based on their unique characteristics such as demographics, interests, needs and location). Market segmentation is useful in getting to know the customers, and determining what is needed in their market segment, and to determine how best to meet their needs.

Big data analytics may be used for targeted advertisements. For example, Netflix collects data on its customers including their searches and viewing history, and this data provides an insight into the specific interests of the customer, which is then used to send suggestions to the customer on the next movie that they should watch.

Big data analytics involves examining large amounts of data to identify the hidden patterns and correlations, and to give insights to enable the right business decisions to be made. Big data analytics is often done with sophisticated software systems that provide fast analytic procedures, where the use of big data allows the business to identify patterns and trends. It enables the business to collect as much data it requires to understand the customers and to derive critical insights to maintain customers.

12.2.1 Business Analytics and Business Intelligence

Business analytics involves converting business data into useful business information through the use of statistical techniques and advanced software. It includes a set of analytical methods for solving problems and assisting decision-making, especially in the context of vast quantities of data. The combination of analysis with intuition provides useful insights into business organizations, and helps them to achieve their objectives. Many organizations use the principles and practice of business analytics.

Business intelligence (BI) processes all of the data generated by a business and uses it to generate clear reports (e.g., a dashboard report of the key metrics), as well as the key trends and performance measures that are used by management in decision-making. That is, BI is data analytics with insight that allows managers to make informed decisions, and so it is focused on the decision-making part of the process. It may employ data mining, performance benchmarking, process analysis and descriptive analytics. That is, business analytics allows management issues to be explored with data and solved.

The effectiveness of management decision-making is influenced by the accuracy and completeness of the information that managers have, with inaccurate or incomplete information leading to poor decisions. Companies often have data that is unstructured or in diverse formats, and such data is generally more difficult to gather and analyse. This has led software firms to offer BI solutions to organizations that wish to make better use of their data and to optimize the information gathered from the data. There are several software applications designed to unify a company's data and analytics.

12.2.2 Big Data and Data Mining

The term "Big data" refers to the large, diverse sets of data that arrives at ever-increasing rates and volumes. It encompasses the volume of data, the velocity or speed at which it is created and collected, and the variety or scope of the data points being covered (these are generally referred to as the three V's of big data). There has been an explosion in the volume of big data with approximately 40 zettabytes[2] (ZB) of data employed globally.

Big data often comes from *data mining*, where data mining involves exploring and analysing large blocks of data to gather meaningful patterns and trends. Data is gathered and loaded into data warehouses by organizations (i.e., the data is centralized into a single database or program), and then stored either on in-house servers or on the cloud. The user decides how to organize the data, and application software sorts the data accordingly, and the data is presented in an easy-to-read format such as a graph or report.

The data may be internal or external. It may be structured or unstructured, where structured data is often already managed in the organization's databases or spreadsheets, and may be numeric and easily formatted. Unstructured data uses data that may be unformatted, so it does not fall into a predetermined format (i.e., it is free form), and it may come from search engines or from forum discussions on social media.

Big data may be collected in various ways such as from publicly shared comments on social media, or gathered from personal electronics or apps, through questionnaires, product purchases and so on. Big data is generally stored in databases, and is analysed with software that is designed to handle large and complex data sets (often software as a service SaaS).

12.2.3 Data Analytics for Social Media

Data analytics provides a quantitative insight into human behaviour on a social media website, and is a way to understand users and how to communicate with them better. It enables the business to understand its audience better, to improve the user experience and to create content that will be of interest to them. Data analytics consists of a collection of data that says something about the social media conversation, and it involves the collection, monitoring, analysis, summarization and a graph to visualize insight into the behaviour of users.

Another words, *data analytics* involves learning to read a social media community through data, and the interpretations of the quantifiable data (or metrics) give information on the activities, events and conversations. This includes what users like when they are online, but other important information such as their opinions and emotions need to be gathered through *social listening*. Social listening involves monitoring keywords and mentions in social media conversations in the target audience and industry, to understand and analyse what the audience is saying about the business and allows the business to engage with its audience.

Social media companies use data analytics to gain an insight into customers, and elementary data such as the number of likes, the number of followers, the number of times a video is played on YouTube and so on are gathered to obtain a quantified understanding of a conversation. This data is valuable in judging the effectiveness of a social media campaign, where the focus is to determine how effective the campaign has been in meeting its goals. The goals may be to increase the number of users or to build a brand, and data analytics combined with social listening help in understanding how people are interacting, as well as what they are interacting about and how successful the interactions have been.

Facebook and Twitter maintain a comprehensive set of measurements for data analytics, with Facebook maintaining several metrics such as the number of page views and the number of likes and the reach of posts (i.e., the number of people who saw posts at least once). Twitter includes a dashboard view to summarize how successful tweet activity has been, as well as the interests and locations of the user's followers. Social listening considers user opinions, emotions, views, evaluations and attitude, and social media data contains a rich collection of human emotions.

The design of a social media campaign is often an iterative process, with the first step being to determine the objective of the campaign and designing the campaign to meet the requirements. The effectiveness of a campaign is judged by a combination of social media analytics and social listening, with the campaign refined appropriately to meet its goals and the cycle repeating. The KPIs may include increased followers/subscribers or an increase in the content shared, and so on.

12.3 Mathematics in Data Science and Analytics

The main mathematics used in data science and analytics are given in Table 12.3.

Other areas of mathematics that may arise in data analytics include discrete mathematics and graph theory [ORg:21], and operations research (Chapter 15).

TABLE 12.3

Mathematics in Data Analytics

Type	Description
Probability	An introduction to some of the concepts in probability theory such as basic probability, expectation, conditional probability, Bayes' Theorem and probability density functions are discussed in Chapter 10.
Statistics	Statistics is a vast area and an introduction to some of the important concepts in the field including descriptive statistics; measures of central tendency such as the mean, mode and median; variance and covariance; and correlation are discussed in Chapter 9.
Linear Algebra	This includes topics such as matrix theory and Gaussian elimination as discussed in Chapter 14, as well as basic algebra as discussed in Chapter 2.
Calculus	This includes the study of differentiation and integration and includes topics such as limits, continuity, rules of differentiation, Taylor's series and area and volume as discussed in Chapters 17 and 18.

12.4 Privacy

Privacy is a fundamental concept in modern society and it has become an important area in the computing field and especially with the rise of social media, the IoT and artificial intelligence. There are various definitions of privacy such as the right to be left alone, for secret or intimate information to be kept secure from others, and for *control over personal information* where individuals are able to decide what information will be shared, when it will be shared and how it will be communicated and shared with others. Privacy concerns are not a new phenomenon, and they initially grew out of the development of early technologies such as the first cameras, microphones and telephones, where indiscreet or unauthorized images or recordings could be made.

There is traditionally a difference between rural and urban living, where in a small town people know everything about every other person in the town, and there is essentially very little privacy from all the gossip (*pueblo pequeño infierno grande*). In a larger city, people are anonymous and nobody knows or cares about what others are doing, so there is a greater sense of privacy.

There are some parallels of the Internet being like the small village, except that the relationship is asymmetric. Another words, in a small town, each individual knows as much about another as vice versa (i.e., it is a symmetric relationship), whereas the relationship is asymmetric for the Internet. This makes it a very unequal relationship, with one party gathering lots of information and building up a profile about all other parties, and using that information for commercial purposes. The other parties are not actively gathering information, and have a very limited picture of what is going on with all the data that is gathered (Fig. 12.1).

Further, while events and information may be forgotten in a village over time this does not happen with the Internet, i.e., it is very difficult to forget things on the Internet with web pages surviving forever in some archives even if taken down. That is, there is no way of really deleting something once it has been published on the web. This could create major problems for individuals who pose indiscreet content online, as that content may be there in perpetuity.

People need an understanding of how their personal information and data is collected, shared and used across the many computer platforms that they use, and the extent to which they have control over their personal information. New technology has led to major changes in which privacy is experienced by society, and it is important to understand the nature of privacy, and to consider the problems and risks that exist. The main sources of personal data are given in Table 12.4.

The collected data is commercially valuable, especially when data about individuals are linked from several sources. *Data brokers* are companies that

FIGURE 12.1
Cardinals Eavesdropping in the Vatican.

TABLE 12.4
Sources of Personal Data

Source	Answers
Data collected by merchants and service providers	This includes personal data entered for the purchase of products and services such as name, address, date of birth, products and services purchased, etc.
Activity tracking	This involves monitoring the user's activity on the site (or app), and recording the user's searches, and the products browsed and purchased. It may involve recording the user's interests, their activities and their interactions and communications with others on the site.
Search profile	The history of a person's searches over a period of time on a search engine such as Google reveals information about the individual and their interests.
Sensors from devices	There are many sensors in the world around us such as personal devices as part of the Internet of Things that may record information such as health data or what the individual is eating. Third party devices such as security cameras may be conducting public or private surveillance. GPS technology on smartphones may be tracking the user's location.

aggregate and link information from multiple sources to create more complete and valuable information products (i.e., profiles of individuals) that may then be sold on to interested parties. Meta data (i.e., data about the data such as the time of a phone call or who the call is made to) also provides useful information that may be collected and shared.

For example, suppose the probability of an individual buying a pair of hiking books is very low (say 1 in 5000 probability). Next, that individual starts scanning a website (say Amazon) for boots then that individual is now viewed as being more likely to buy a pair of hiking boots (say a 1 in 100 probability). This large increase in probability will mean that the individual is now of interest to advertisers and sellers, and various targeted (popup) advertisements will appear to the individual advertising different hiking boots. This may become quite tedious and annoying to the individual, who may have been just browsing, and is now subject to an invasion of advertisements, but many apps are free and often the source of their revenue is from advertisements, so they gather data about the user that is then sold on to advertisers.

Users should be in control of how their data is used, and most user agreements are "all-or-nothing" in the sense that a user must give up control of their data to use the application, so essentially the user has no control. That is, a user must click acceptance of the terms and conditions in order to use the services of a web application. Clearly, users would be happier and feel that they are in control if they were offered graduated choices by the vendor, to allow them to make tradeoffs, and to choose a level of privacy that they are comfortable with.

The importance of privacy in the information technology field became apparent in the early 1970s with the introduction of databases. These could hold private information about individuals, and there was a need for a set of rules to protect how information should be collected and used. This led to the development of a set of fair information processing principles (FIPs) in the United States that mandated that the organization collecting personal data is doing so openly (i.e., it is not secretly or covertly collecting data), and an individual must be able to access any data that the organization has about her.

Computing technology has evolved in a major way from the mainframes and databases of the early 1970s, and includes a plethora of leading edge technologies such as smartphones, social media, the IoT and artificial intelligence. It is reasonable to ask what privacy means in the modern digital world and whether there is privacy anymore. Users of social media share large parts of their lives with a massive online audience as well as with large corporations, and social media companies gather lots of data about its users that may be used to determine patterns, and to generate profiles that may be targeted to advertisers. So much data is being collected about individuals, and the question is: Where does it go? Who controls it? Are companies adequately managing risks of data breaches? What happens when data privacy is breached or data is not secured properly? Is there transparency? Is user data encrypted? Is confidentiality and authenticity maintained?

The IoT is not a single technology as such; instead, it is a collection of devices, sensors and services that capture data to monitor and control the world around them. It refers to interconnected technology that is now an integral part of modern society, where computation and data communication are embedded in the environment. It allows everyday devices to connect to other devices or people over the Internet, and this may include smartphone to smartphone communication, vehicle to vehicle communication, connected cameras, GPS tracking, the smart grid and so on. It allows a vast amount of data to be gathered and transmitted to and processed by companies. It means that information processing is now an integral part of people's lives, and IoT connects many devices to the Internet.

The level of interconnectivity and data gathered with IoT means that security and privacy have become important concerns, and it is essential to control both the devices and the data. For example, control could be lost if someone hacks into the smartphone, as the smartphone often links to bank accounts, email accounts and even household appliances. A lot of user data is potentially gathered painting a profile of individual users through their online activities as well as their searches, and the data gathered is used to improve the user experience, and the profile of users may be sold on to advertisers. Data should only be gathered with user consent, and there are risks of hacking or eavesdropping.

There has been major growth in AI technology in recent years, and AI has been applied to self-driving cars, facial recognition, machine translation and so on. Facial recognition technology may be used to unlock phones to authenticate identity, and it has also been applied to read facial expression during job interviews, as well as following the movement of individuals. A vehicle may contain several on-board computers for processing various vehicle controls as well as for entertainment systems. Vehicles that connect to the Internet are potentially at risk of being hacked, where a hacker could potentially commandeer vehicle controls such as steering and the brakes.

It is often unclear who is collecting personal information, the type of information that they are collecting, what is being done with the data, and who the data is being shared with. Information privacy refers to control over information, and is a value that in a sense protects from certain kinds of harm. For example, if others have information about a particular individual they may be able to use it against the individual. For example, if the individual has been the victim of phishing or identity theft where their personal financial information such as credit cards are stolen, then the perpetrators have power over the individual since they have personal and sensitive information about the individual.

12.4.1 Privacy and Social Media

Social media involves the use of computer technology for the creation and exchange of user-generated content. These web-based technologies allow users to discuss and modify the created content, and it has led to major changes in communication between individuals, communities and organizations.

Social media is designed to have the individual share as much information as possible, and to continue to do so while they are on the site, and with every disclosure (or post) the individual reveals a little bit more about himself or herself. It is very easy to post photos and information on social media sites such as Facebook or Twitter, and social media is designed in such a way that it is addictive and poses risks to the privacy of an individual (Fig. 12.2).

There is a danger that both social media companies and other users could harm the individual's privacy. The harm from other users may arise when a piece of the user's information is shared with the wrong audience, and this later leads to problems for the user. There are two distinct audiences for the individual's information namely other users and the platform itself. The social media platform maintains a vast quantity of electronic information consisting of immense databases, which can collect a vast amount of data on the individual and other users.

There is a power imbalance between the platform and the user, with the platform designed to have the individual share as much as possible, and people may potentially pose risks in social interaction. An individual's

FIGURE 12.2
Young People on Smartphones and Social Media. Public Domain.

information may be viewed by friends, family, employer, work colleagues and nameless others, so everyone in the individual's network as well as others could be an unwanted audience.

Users often may not realize the full extent of their audience when they post, and the people who are authorized in an individual's network may not be the desired recipients of certain posts (disclosures). It is difficult to delete online messages, and destructive posts may last long after an incident. Therefore, it is very much in the interests of users to keep their social media posts discreet, as both friends and outsiders of their social media network pose risks to their privacy.

12.4.2 Privacy and the Law

Data collection laws focus on how data is collected, used and shared, and data protection includes the right to information self-determination. The web is full of privacy policies that specify what type of personal data will be collected, how it will be processed and used, how it is shared and what can be done about it. Further, individuals may take a lawsuit against another for a tort, for example, when someone pries or stalks them, or publishes a defamatory article, or violates their privacy. There are three main areas that impact an individual's privacy:

- The media
- Surveillance
- Personal data

Media laws protect an individual against intrusion, where another party may be held liable for the invasion of the individual's privacy (e.g., phone tapping, snooping, examining a person's bank account and so on). The tort of the public disclosure of private facts prevents others from widely spreading private facts such as the individual's face or identity for their own benefit, and there are slander and libel laws to protect an individual's good name and reputation, and to prevent defamation of character.

There are laws and rights to regulate surveillance with search warrants required in most countries to search the home of a private individual, as well as the right to seize personal property. Warrants are generally required to obtain personal electronic records held by telecommunication companies (e.g., the calls made and received as well as metadata such as geo-location data), and warrants may be required to obtain records held by Internet technology companies (e.g., emails, web sites visited, searches and other electronic messages).

Countries vary in their laws for the protection of security and privacy, but many countries recognize that the security and privacy commitments made by a company in their policies should be fully implemented. Further, companies should be held accountable for any security breaches that occur that lead to data security or privacy being compromised, and the company may be liable for any losses suffered by individuals as a result of the breach.

Further, people must not be misled about the functionality of a website or mobile app that places their security or privacy at risk, and users must give their consent to any changes to the privacy policy that would allow for the collection of additional personal data, and users must be informed about the extensiveness of tracking and data collection.

12.4.3 EU GDPR Privacy Law

Europe has been active in the development of data protection regulations, and the European General Data Protection Regulation (EU GDPR 2016/679) is a comprehensive data protection framework that became operational in 2018. Privacy and data protection are regarded as fundamental human rights in the EU, and GDPR aims to give individuals control over their personal data. It has had a huge impact on privacy laws of other countries around the world, and it also protects the transfer of personal data outside of the EU, as it prohibits its transfer to countries that do not provide an equivalent or adequate data protection framework as GDPR.

GDPR consists of a data governance framework that attempts to place privacy on a par with other laws. It creates protections that follow the data, and it places responsibilities on companies in managing privacy and information. GDPR applies whenever personal data is processed, and it starts from the presumption that the processing of the personal data is illegitimate. This means that companies carry the burden of legitimizing their actions, and they must be able to show that they have a legitimate basis for processing data. That is, they must be able to show that they have the consent of the data subject, or that the processing is necessary as a result of the contract that exists between them and the data subject, or where they have a legitimate interest, and where the interest of the data controller prevails over that of the data subject. The company must be able to demonstrate adherence to fair information practice.

This means that data must be obtained legitimately and is used in the manner of the purpose for which it was acquired, and there must be openness and transparency so that individuals will know how their data will be used. There should be special protections for sensitive data with the ability to opt in for consent (e.g., race, sexual orientation, political beliefs), and there must be standards for enforcement to ensure compliance to the standards. The *Data Privacy Impact Assessment* (DPIA) is mentioned in GDPR, and it is needed if the processing of personal information is likely to result in a high risk to the rights and freedoms of individuals. This assessment helps to ensure that companies are complying with privacy requirements.

The standard for informed consent is very high which means that it is freely given and informed. GDPR also gives very strong data subject rights, including the right to access data, data portability, the right to rectify data, the right to erase data, the right to object to processing and the right to restrict processing. These provide solid rights for the data subjects to exercise control over their personal data.

12.5 Security

The privacy of personal information may be easily violated if the security of the system is easily breached, and the introduction of the Internet led to a major growth in attacks on businesses and individuals. The Internet was developed based on trust with security features added as a response to different types of attacks. Today, good cybersecurity is essential, and is a prerequisite for the protection of personal data, and security is an essential part of modern electronic systems. The consequences of poor security may include the theft of personal information and credit cards of customers, resulting in a loss of trust and damage to the reputation and credibility of the company. The security of the system refers to its ability to protect itself from attacks, and the characteristics of security are as follows:

- Confidentiality
- Integrity
- Availability

Confidentiality means that the information may be viewed and accessed only by those authorized, and encryption may be employed to ensure that the unauthorized access of information is meaningless to anyone other than the intended parties. Other approaches include access controls where only those with the appropriate access privileges may access the data. *Integrity* means that the data may only be modified by those authorized to do so, and *availability* refers to the fact that the system and its data are available for use at all times (i.e., it is not subject to a denial of service attack).

Security is holistic and it is essential to identify any security vulnerabilities and to correct them. It is important to be able to limit the access that malicious software may have within the company's network. There may be controls that detect and repel attacks that shut down parts of the system or restrict access thereby preventing the malicious software from moving around the network. Hackers may use phishing emails to install malicious software to get credentials and steal passwords. They are often motivated by personal gain, and they exploit security and system vulnerabilities to steal, exploit or sell data or intellectual property. The attacks may lead to:

- Unauthorized data access and usage
- Unauthorized data theft and deletion
- Unauthorized data manipulation

The system needs to be designed for security, as it is difficult to add security after the system has been implemented. Security loopholes may be introduced in the development of the system, and so care needs to be taken to prevent these as well as prevent hackers from exploiting them. Early risk analysis is conducted to determine what needs to be protected, and the threats and vulnerabilities of the current system are analysed to produce a risk profile. This leads to the security requirements including the required security measures and supporting technologies. There is a tradeoff between security risks and the cost of security measures, and this is a continuous process due to continued changes in technology. A comprehensive security system requires the following:

- Preventive measures
- Detective measures
- Administrative measures

Preventive measures are used to stop unauthorized attacks from occurring before they succeed and do any harm, *detective measures* are used to discover any unauthorized attacks that may be ongoing or completed and *administrative measures* which are used to clarify processes, procedures, rules and responsibilities and standards within an organization. Organizational and administrative measures are as important as technical measures in securing a system.

Preventive measures often involve the use of encryption for communication and stored data, to ensure that the data is meaningless to those who are not authorized to view it. The encryption and decryption are performed with keys, where the same secret key is used for symmetric cryptography and different keys (public and private) are used for asymmetric cryptography with digital certificates used to verify the authenticity of the key owner (see chapter 10 of [ORg:21]).

Preventive measures also include access control mechanisms for *authentication* to verify if the person is who she claims to be, and a range of measures such as user id and password, smart card and biometric data may be used for verification. The next step is to ensure that those authenticated have the appropriate level of *authorization* to access the data, and an authorization matrix that includes roles and the level of access for each role may be used for this.

The goal of detective measures is to monitor whether a system is actually secure, and to detect attacks. Security audits may be conducted to verify that the planned and recommended security measures have been implemented.

Penetration testing is a way to find and remove security weaknesses, and it involves experts (e.g., white hat hackers) playing the role of attackers trying to find vulnerabilities in a system, and improvement actions are then taken to improve any identified weakness in the system.

12.6 Review Questions

1. What is data science?
2. What is the role of the data scientist?
3. What is privacy? Why is it important?
4. What are the main sources of personal data collected online?
5. What are the main risks to an individual using social media?
6. What are the main risks to an individual using a fitness device (as part of the Internet of Things)?
7. What are the main risks with AI facial recognition technology?
8. Explain the importance of the EU GDPR law.
9. What is a digital privacy impact assessment?

12.7 Summary

Companies collect lots of personal data about individuals from their use of computer resources such as email, search engines and Internet and social media use, and the data is processed to build up revealing profiles of the user that may be targeted to advertisers. Data science involves the extraction of knowledge from data sets that consist of structured and unstructured data, and data science may be regarded as a branch of statistics. It is essential that both the data and models are valid, as otherwise errors will be made during data analysis.

Data analytics involves the analysis of data to create structure, order, meaning and patterns from the data. It uses the collected data to produce information as well as generate insights from the data for decision-makers. This is essential in making informed decisions to meet current and future business needs. Data analytics may involve machine learning, or it could be quick and simple if the data set is ready, and the goal is to perform just simple descriptive analysis. It is widely used in fields such as social media, e-commerce, the IoT and recommendation engines.

The field has great power to help and to harm, and data scientists have a responsibility to use this power wisely. The consequence of an error in the data analysis or with the analysis method could result in harm to an individual. There are many sources of error such as the sample chosen, which may not be representative of the entire population. Other problems arise with knowledge acquisition by machine learning, where the learning algorithm has used incomplete training data for pattern (or other knowledge) recognition. Training data may also be incomplete if the future population differs from the past population.

Notes

1 Big data involves combining data from lots of sources such as bar codes, CCTV, shopping data, drivers license and so on.
2 A zettabyte is 1 sextillion bytes = 2^{70} bytes (approximately a billion terabytes or 1000 exabytes or a trillion gigabytes).

13

Business Metrics and Problem Solving

Measurement is an essential part of mathematics and the physical sciences, and measurement programmes have been successfully applied in industry. The purpose of a measurement programme is to establish and use quantitative techniques to manage an organization, and this chapter presents a selection of business metrics to give an insight into a business and to understand its current performance. We show how measurement may be applied to decision-making, and in providing a quantitative understanding of its capability in delivering projects on time, on budget and with the right quality.

Business metrics are used to track all areas of business and provide visibility into the various functional areas in an organization, with the measurement data allowing trends to be seen over time. The analysis of the measurements allows action plans to be produced for continuous improvement. *Key performance indicators* (KPIs) are business metrics that target critical areas of performance, and often companies will maintain a *dashboard* of KPIs to manage performance in key areas.

Measurements may be employed to track the quality, timeliness, cost, schedule and effort of software projects. The term "metric" and "measurement" are used interchangeably in this book. The formal definition of measurement given by Fenton [Fen:95] is:

> *Measurement is the process by which numbers or symbols are assigned to attributes or entities in the real world in such a way as to describe them according to clearly defined rules.*

Measurement provides a more precise understanding of the entity under study, and measures are used in everyday life to calculate the distance to the planets and stars, determining the mass of objects, computing the speed of mechanical vehicles, calculating the electric current flowing through a wire, computing the rate of inflation, estimating the unemployment rate and so on.

Often several measurements are used to provide a detailed understanding of the entity under study. For example, the cockpit of an airplane contains measurements of altitude, speed, temperature, fuel, latitude, longitude and various devices essential to modern navigation and flight, and clearly an airline offering to fly passengers using just the altitude measurement would not be taken seriously.

Metrics play a key role in problem solving, and various problem-solving techniques will be discussed later in this chapter. Measurement data are essential

DOI: 10.1201/9781003308140-13

in quantifying how serious a particular problem is, and they provide a precise quantitative measure of the extent of the problem. For example, a telecommunications outage is measured as the elapsed time between the down time and the subsequent up time, and the longer the outage lasts the more serious it is. It is essential to minimize outages and their impact should one occur, and measurement data are invaluable in proving an objective account of the extent of the problem. Measurement data may be used to perform analysis on the root cause of a particular problem, and to verify that the actions taken to correct the problem have been effective.

Metrics provide an internal view of the quality of a product, but care is needed before deducing the behaviour that a product will exhibit externally from the various internal measurements of the product. A *leading measure* is a measure that usually precedes the attribute that is under examination, for example, the arrival rate of software problems is a leading indicator of the maintenance effort. Leading measures provide an indication of the likely behaviour of the product in the field and need to be examined closely. A *lagging indicator* is a software measure that is likely to follow the attribute being studied, for example, escaped customer defects are an indicator of the quality and reliability of the software. It is important to learn from lagging indicators even if the data can have little impact on the current project.

13.1 The Goal Question Metric Paradigm

Many metrics programs have failed because they had poorly defined, or non-existent goals and objectives, with the metrics defined unrelated to the achievement of the business goals. The *Goal Question Metric* (GQM) paradigm was developed by Victor Basili and others of the University of Maryland [Bas:88]. It is a rigorous goal-oriented approach to measurement, in which goals, questions and measurements are closely integrated.

The business goals are first defined, and then questions that relate to the achievement of the goal are identified. For each question, a metric that gives an objective answer to the particular question is defined. The statement of the business goal is precise, and it is related to individuals or groups. It involves managers and engineers proceeding according to the following three stages:

- Set goals specific to needs in terms of purpose, perspective and environment.
- Refine the goals into quantifiable questions.
- Deduce the metrics and data to be collected (and the means for collecting them) to answer the questions.

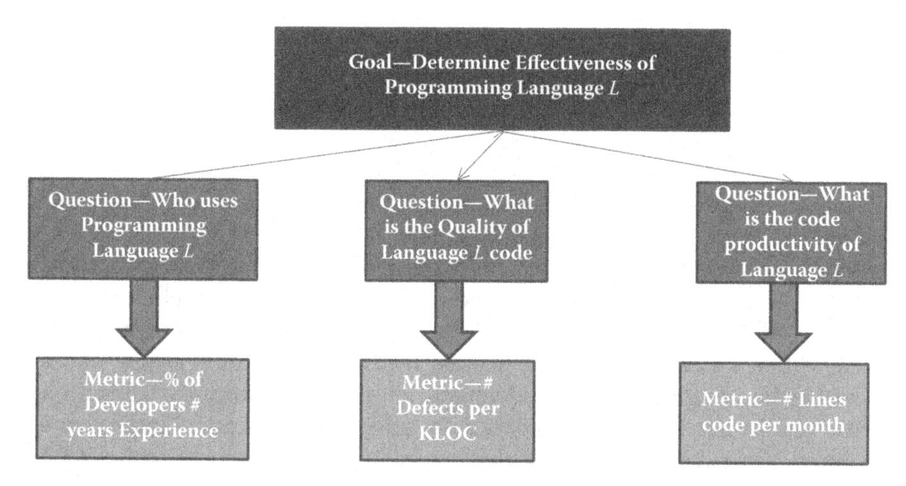

FIGURE 13.1
GQM Example.

GQM has been applied to several domains, and we consider an example from the software field. Consider the goal of determining the effectiveness of a new programming language L. There are several valid questions that may be asked at this stage: Who are the programmers that use L? What is their level of experience? What is the quality of software code produced with language L? What is the productivity of language L? This leads naturally to the quality and productivity metrics as detailed in Fig. 13.1.

13.1.1 Goal

The focus of the improvements should be closely related to the business goals, and the first step is to identify the key goals that are essential for business success. The business goals are related to the strategic direction of the organization and the problems that it is currently facing. There is little sense in directing improvement activities to areas that do not require improvement, or for which there is no business need to improve, or from which there will be a minimal return to the organization.

13.1.2 Question

These are the key questions that determine the extent to which the goal is being satisfied, and for each business goal the set of pertinent questions needs to be identified. The information that is required to determine the current status of the goal is determined, and this naturally leads to the set of questions that must be answered to provide this information. Each question is analysed to determine the best approach to obtain an objective answer, and to define the metrics that are needed, and the data that needs to be gathered to answer the question objectively.

13.1.3 Metrics

These are measurements that give a quantitative answer to the key questions, and they are closely related to the achievement of the goals. They provide an objective picture of the extent to which the goal is currently satisfied. The GQM approach leads to measurements that are closely related to the goal, rather than measurement for the sake of measurement.

GQM helps to ensure that the defined measurements are relevant, and are used by the organizations to understand its current performance, and to improve and satisfy the business goals more effectively. Successful improvement requires clear improvement goals that are related to the business goals. GQM is a rigorous approach to software measurement, and the measures are from various viewpoints, e.g., manager viewpoint, project team viewpoint, etc. The idea is always first to identify the goals, and once the goals have been defined appropriate questions and measurements are determined.

13.2 The Balanced Scorecard

The balanced scorecard (BSC) (Fig. 13.2) is a management tool that is used to clarify and translate the organization vision and strategy into action. It was developed by Kaplan and Norton [KpN:96], and has been applied to many organizations. The European Software Institute (ESI) developed a tailored version of the BSC for the IT sector (the IT BSC).

The BSC assists in selecting appropriate measurements to indicate the success or failure of the organization's strategy. There are four perspectives

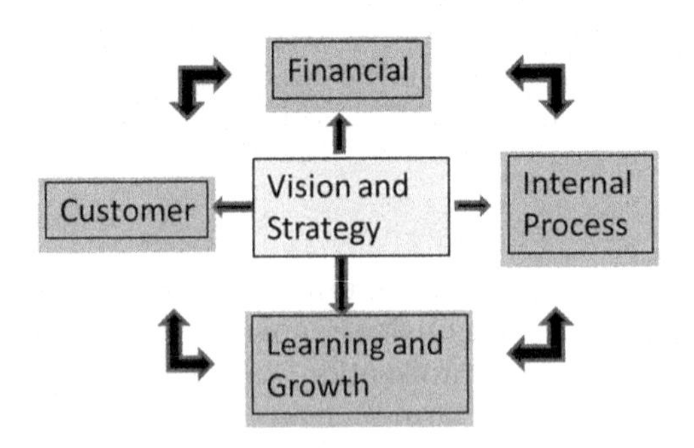

FIGURE 13.2
The Balanced Scorecard.

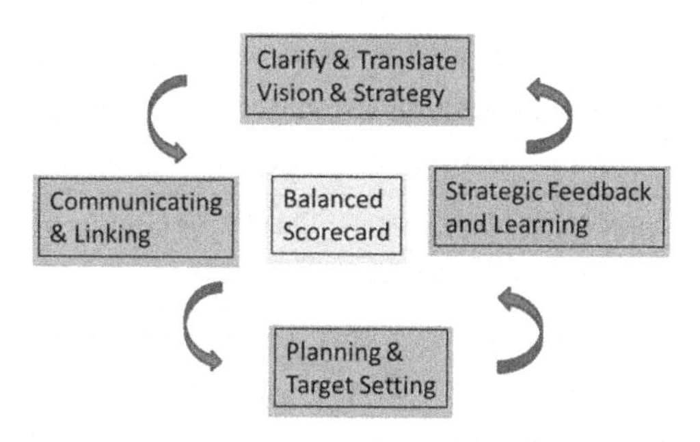

FIGURE 13.3
Balanced Scorecard and Implementing Strategy.

in the scorecard: *customer, financial, internal process* and *learning and growth*. Each perspective includes objectives to be accomplished for the strategy to succeed, measures to indicate the extent to which the objectives are being met, targets to be achieved in the perspective, and initiatives to achieve the targets. The BSC includes financial and non-financial measures.

The BSC is useful in selecting the key processes that the organization should focus its process improvement efforts on in order to achieve its strategy (Fig. 13.3). Traditional improvement is based on improving quality, reducing costs, and improving productivity, whereas the BSC takes the future needs of the organization into account and identifies the processes that the organization needs to excel at in the future to achieve its strategy. This results in focused process improvement, where the intention is to yield the greatest business benefit from the improvement program.

The starting point is for the organization to define its *vision* and *strategy* for the future. This often involves strategy meetings with the senior management to clarify the vision and to achieve consensus on the strategic direction for the organization. The vision and strategy are then translated into *objectives* for the organization or business unit. The next step is communication, and the vision and strategy and objectives are communicated to all employees. These critical objectives must be achieved in order for the strategy to succeed, and so all employees (with management support) will need to determine their own local objectives to support the organization's strategy. Goals are set and rewards are linked to performance measures.

The financial and customer objectives are first determined from the strategy, and the key business processes to be improved are then identified. These are the key processes that will lead to a breakthrough in performance for customers and shareholders of the company. It may require new processes with employees trained on the new processes. The BSC is effective in driving organization change.

The financial objectives require targets to be set for customer, internal business process, and the learning and growth perspective. The learning

TABLE 13.1

BSC Objectives and Measures for IT Service Organization

Financial	Customer
Increase revenue	Quality Service
Reduce costs	Reliability of solution
Cost of provision of services	Rapid response time
Cost of hardware/software	Accurate information
Timeliness of Solution	Timeliness of Solution
99.999% network availability	99.999% network availability
24/7 customer support	24/7 customer support
Internal Business Process	**Learning and Growth**
Requirements definition	Expertise of staff
Design	Leadership
Implementation	Employee satisfaction
Testing	Software development capability
Maintenance	Project management
Customer support	Customer support
Security/proprietary information	Staff development career structure
Disaster prevention and recovery	Objectives for staff

and growth perspective will examine the competencies and capabilities of employees and the level of employee satisfaction. Fig. 13.3 describes how the BSC may be used for implementing the organization vision and strategy.

Table 13.1 presents sample objectives and measures for the four perspectives in the BSC for an IT service organization.

13.3 Metrics for an Organization

The objective of this section is to present a case study of how metrics may be used to provide visibility into various areas of an organization, and to facilitate improvement. The metrics presented may be applied or tailored to other organizations, and the goal is to show how metrics may be employed to manage performance. Many organizations have monthly quality or operation reviews in which metrics are presented and performance trends monitored.

We present sample metrics for the various functional areas in a software development organization, including sales and marketing, financial, human resources, customer satisfaction, supplier quality, internal audit, project management, requirements and development, testing and process improvement.

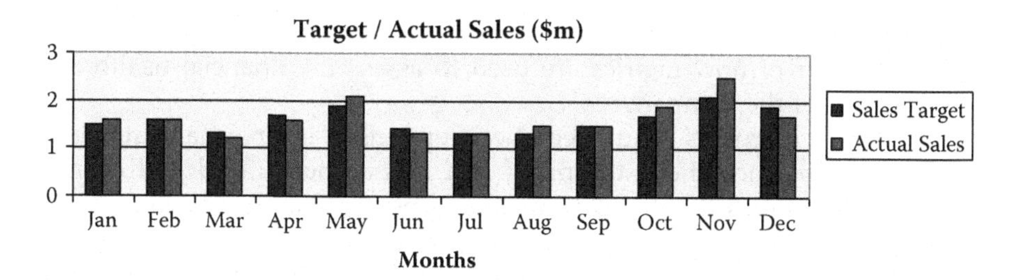

FIGURE 13.4
Target and Actual Sales per Month.

These metrics are often presented at a monthly management review, and performance trends observed. The main output from a management review is a series of improvement actions, and the effectiveness of the improvement actions is judged by improvements in performance trends.

13.3.1 Sales and Marketing Metrics

There are many sales and marketing metrics to give visibility into the effectiveness of marketing and to allow analysis of sales performance. We consider just two metrics in this section including sales revenue per month and customer churn. The sales revenue per month is given in Fig. 13.4.

The customer churn rate is determined from the number of new customers per month and the number of customers that leave per month (Fig. 13.5).

13.3.2 Financial Metrics

Every company needs to be fully aware of its financial performance, and a business will often create a KPI dashboard for its key financial metrics. There are many financial metrics used to assess the health of a company,

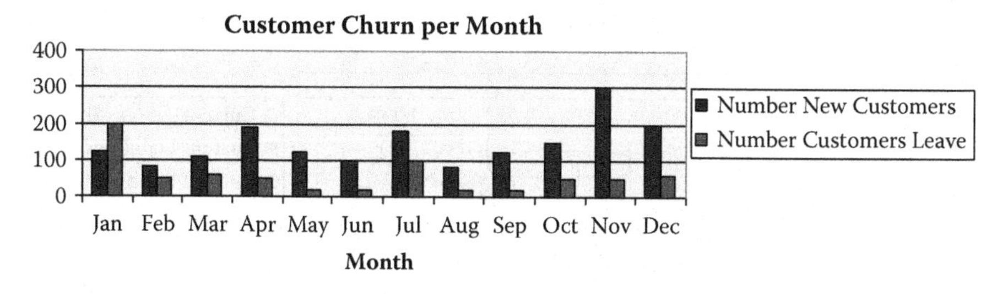

FIGURE 13.5
Customer Churn per Month.

and we consider a small selection of the available metrics in this section to give a flavour of how metrics are used to assess the financial health and profitability of the company.

Chapter 16 discusses fundamental concepts used in financial statements such as the balance sheet, the profit and loss account, fixed and current assets and liabilities. The financial metrics considered in this section include metrics for the total revenue and net income of the company per annum (or per quarter), the current ratio of the current assets to current liabilities (a measure of the financial health of the company through its working capital), and the amounts of the a/c payable and receivable per month. The first metric that we consider is a metric for the total revenue and net income of a company (Fig. 13.6).

The current ratio is a measure of the working capital and it reflects the company's abilities to repay all of its financial obligations in the year, and it is calculated as the ratio of current assets to current liabilities (Fig. 13.7). A ratio less than one indicates that the company may be unable to fulfil all of its financial obligations or unable to grow, and a healthy current ratio is usually between 1.5 and 3.

The current ratio may be refined to yield a quick ratio metric that determines if the company's short-term current assets allow it to repay its short-term future liabilities (Fig. 13.7).

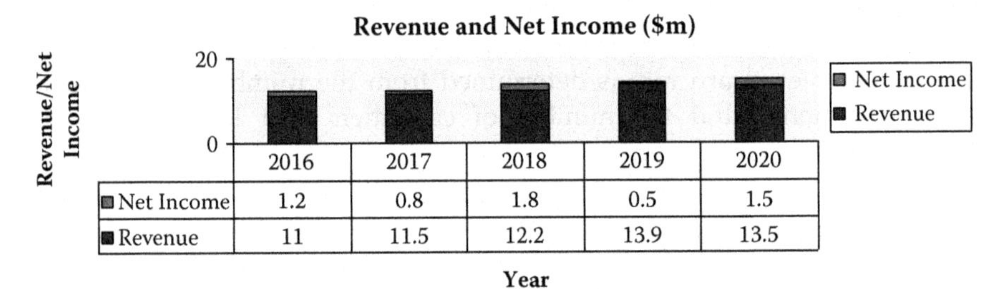

	2016	2017	2018	2019	2020
Net Income	1.2	0.8	1.8	0.5	1.5
Revenue	11	11.5	12.2	13.9	13.5

FIGURE 13.6
Revenue and Net Income per Annum.

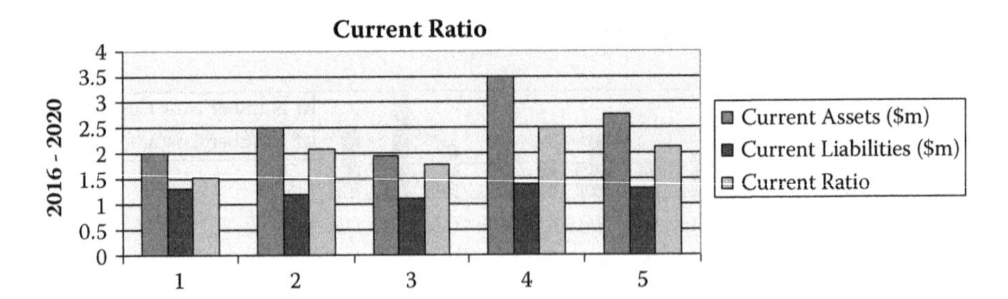

FIGURE 13.7
Current Ratio.

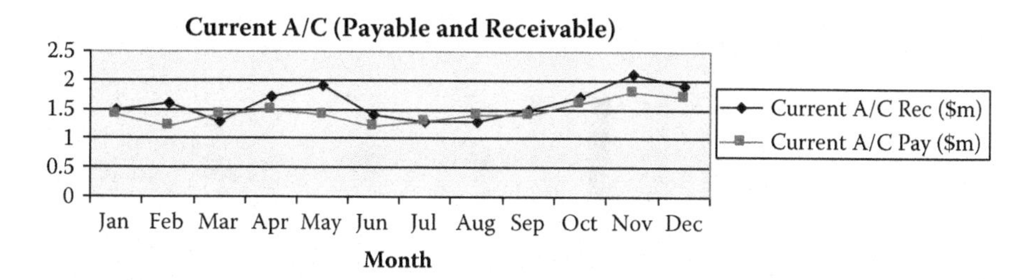

FIGURE 13.8
Current A/C Receivables and Payable.

The current a/c receivable indicates the money owed to a business by its debtors, and the current a/c payable metric indicates the money that a business owes to its suppliers and creditors per month (Fig. 13.8).

13.3.3 Customer Satisfaction Metrics

Fig. 13.9 shows the customer survey arrival rate per customer per month, and it indicates that there is a customer satisfaction process in place, that customers are surveyed, and the extent to which they are surveyed. It does not provide any information as to whether the customers are satisfied, whether any follow-up activity from the survey is required, or whether the frequency of surveys is sufficient (or excessive) for the organization.

Fig. 13.10 gives the customer satisfaction measurements in several categories including quality, the ability of the company to meet the committed

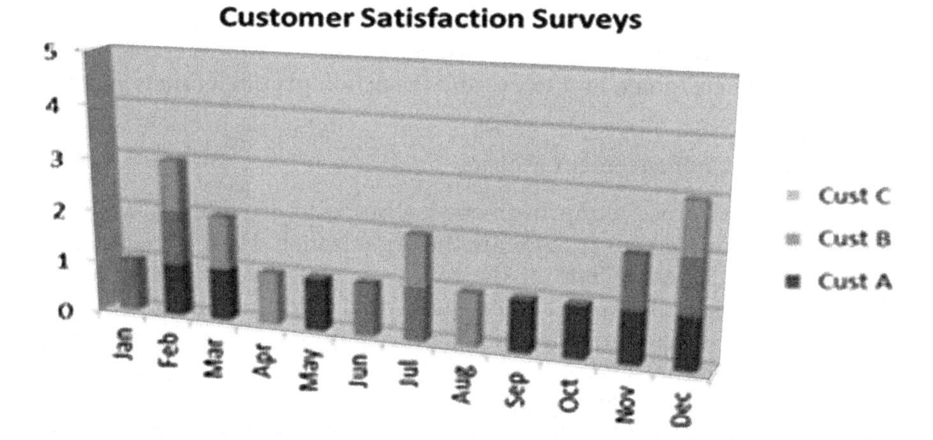

FIGURE 13.9
Customer Survey Arrivals.

FIGURE 13.10
Customer Satisfaction Measurements.

dates and to deliver the agreed content, the ease of use of the software, the expertise of the staff and the value for money. Fig. 13.5 is interpreted as follows:

8–10	Exceeds expectations
7	Meets expectations
5–6	Fair
0–4	Below expectations

Another words, a score of 8 for quality indicates that the customers consider the software to be of high quality, and a score of 9 for value for money indicates that the customers consider the solution to be excellent value. It is essential that the customer feedback is analysed (with follow-up meetings with the customer where appropriate). There may be a need to produce an action plan to deal with customer issues, and to communicate the plan to the customer, and to execute the action plan in a timely manner.

13.3.4 Process Improvement Metrics

The objective of process improvement metrics is to provide visibility into the process improvement program in the organization. Fig. 13.11 shows the arrival rate and closure rate of improvement suggestions. The chart indicates that initially the arrival rate is high and the closure rate low, which is consistent with the commencement of a process improvement program. The closure rate then improves which indicates that the improvement team is active and acting upon the improvement suggestions. The closure rate is low during July and August, which may be explained by the traditional holiday period in the northern hemisphere.

The chart does not indicate the effectiveness of the process improvement suggestions and their overall impact on quality and productivity. There are

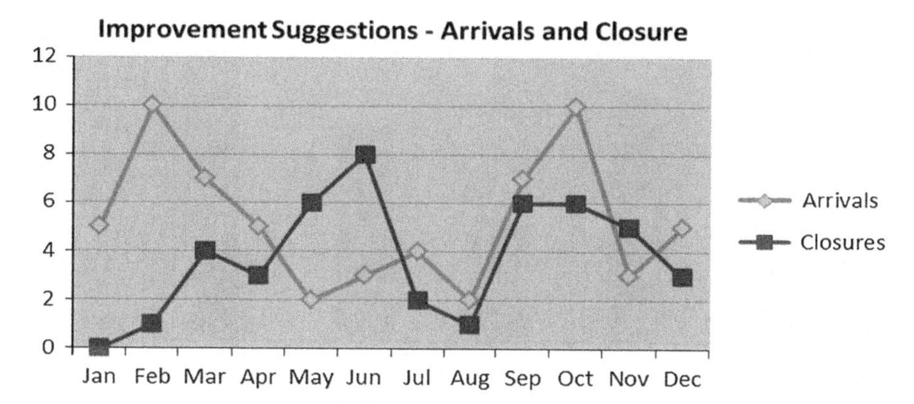

FIGURE 13.11
Process Improvement Measurements.

no measurements of the cost of performing improvements, and this is important for a cost-benefit analysis of the benefits of the improvements obtained versus the cost of the improvements.

Fig. 13.12 provides visibility into the status of the improvement suggestions, and the number of raised, open and closed suggestions per month. The chart indicates that gradual progress has been made in the improvement program with a gradual increase in the number of suggestions that are closed.

Fig. 13.13 provides visibility into the age of the improvement suggestions, and this is a measure of the productivity of the improvement team and its ability to do its assigned work.

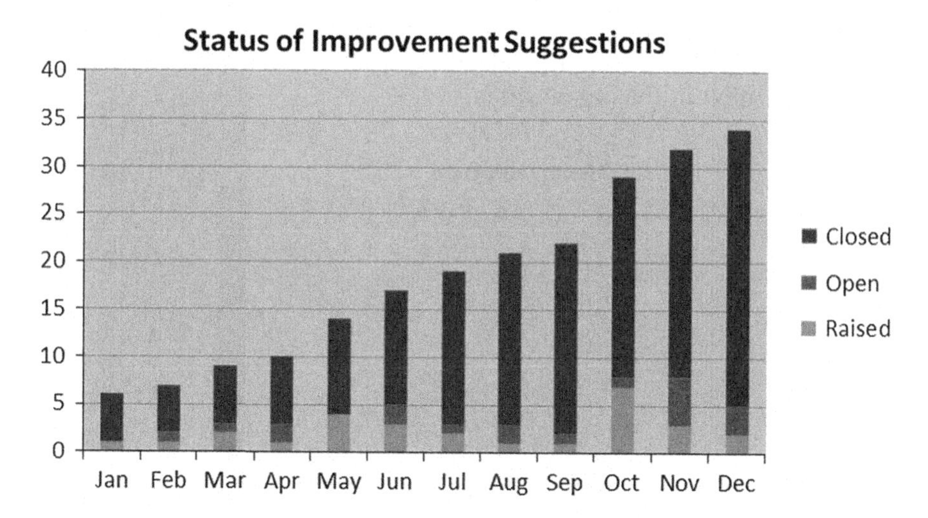

FIGURE 13.12
Status of Process Improvement Suggestions.

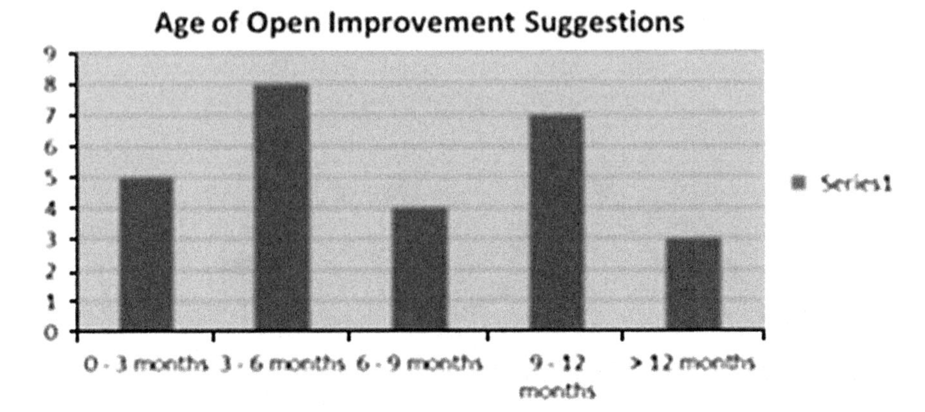

FIGURE 13.13
Age of Open Process Improvement Suggestions.

13.3.5 Human Resources and Training Metrics

These metrics give visibility into the human resources and training areas of an organization. They provide visibility into the current headcount (Fig. 13.14) of the organization per calendar month and the turnover of staff in the organization (Fig. 13.15). The human resources department will typically maintain measurements of the number of job openings to be filled per month, the arrival rate of resumes per month, the average number of interviews to fill one position, the percentage of employees that have received their annual appraisal, etc.

The goals of the HR department include "attract and retain the best employees," and this breaks down into the two obvious sub-goals of attracting the best employees and retaining them. The next chart gives visibility into the turnover of staff during the calendar year. It indicates the effectiveness of staff retention in the organization.

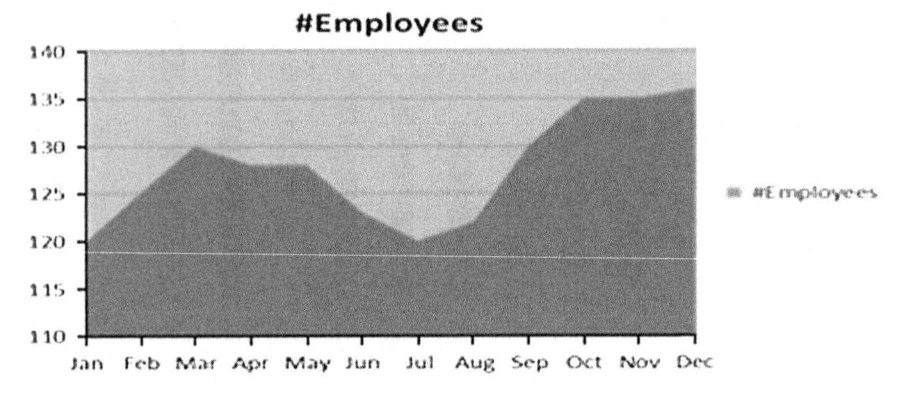

FIGURE 13.14
Employee Headcount in Current Year.

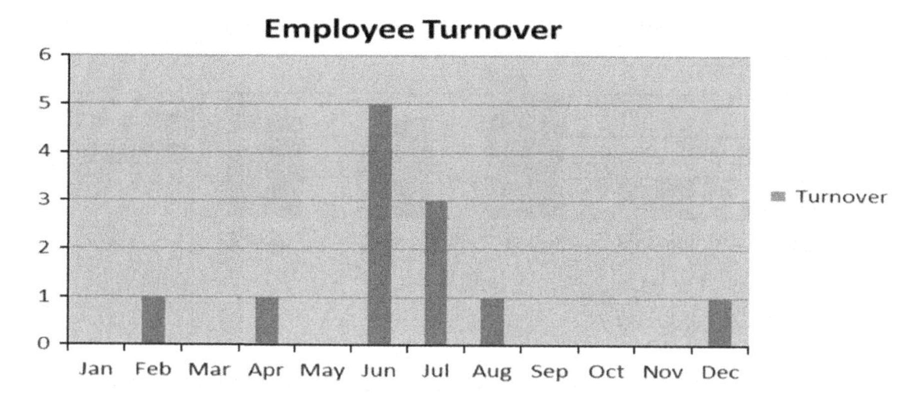

FIGURE 13.15
Employee Turnover in Current Year.

13.3.6 Project Management Metrics

The goal of project management is to deliver a high-quality product on time and on budget to the customer. The project management metrics provide visibility into the effectiveness of the project manager in delivering the project on time, on budget, and with the defined functionality.

The timeliness metric provides visibility into the extent to which the project has been delivered on time (Fig. 13.16), and the number of months over or under schedule per project in the organization is shown. The schedule timeliness metric is a lagging measure, as it indicates that the project has been delivered within the schedule or not after the event.

The on-time delivery of a project requires that the various milestones in the project be carefully tracked and corrective actions taken to address slippage in milestones during the project.

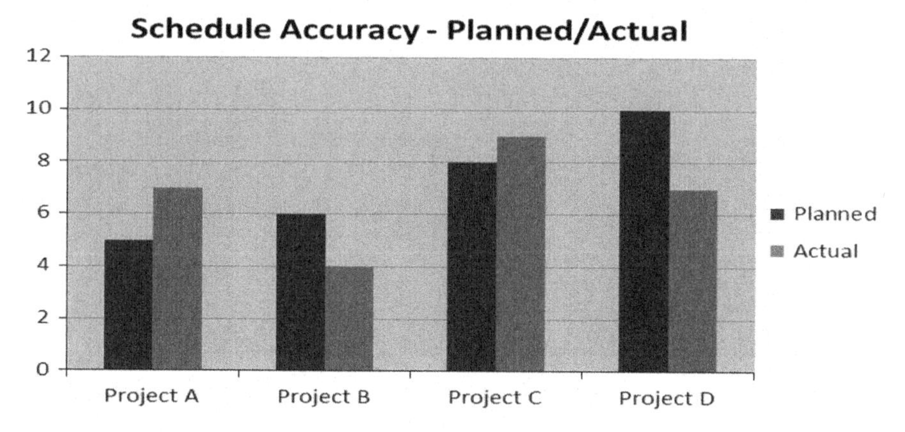

FIGURE 13.16
Schedule Timeliness Metric.

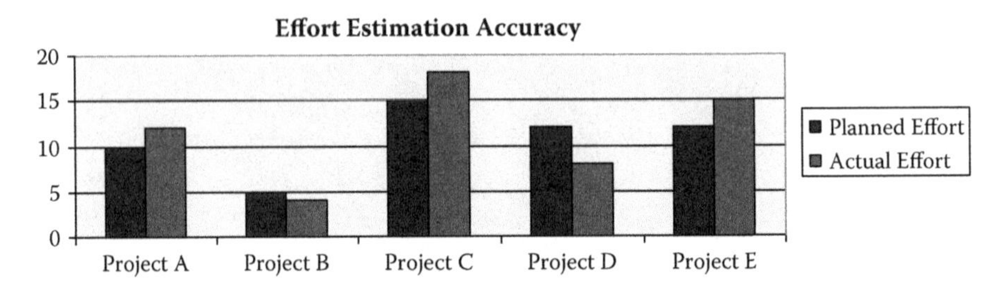

FIGURE 13.17
Effort Timeliness Metric.

The second metric provides visibility into the effort estimation accuracy of a project (Fig. 13.17). Effort estimation is a key component in calculating the cost of a project, and in preparing the schedule, and its accuracy is essential. The accurate estimate of effort and schedule is difficult, and often projects are delivered over budget and schedule.

The effort estimation chart is similar to the schedule estimation chart, except that the schedule metric is referring to elapsed calendar months, whereas the effort estimation chart refers to the planned number of person months required to carry out the work, and the actual number of person months that it actually took. Projects need an effective estimation methodology, and the project manager will use metrics to determine how accurate the estimation has actually been.

The next metric is related to the commitments that are made to the customer with respect to the content of a particular release, and it indicates the effectiveness of the projects in delivering the agreed requirements to the customer (Fig. 13.18). This chart could be adapted to include enhancements or fixes promised to a customer for a particular release of a product.

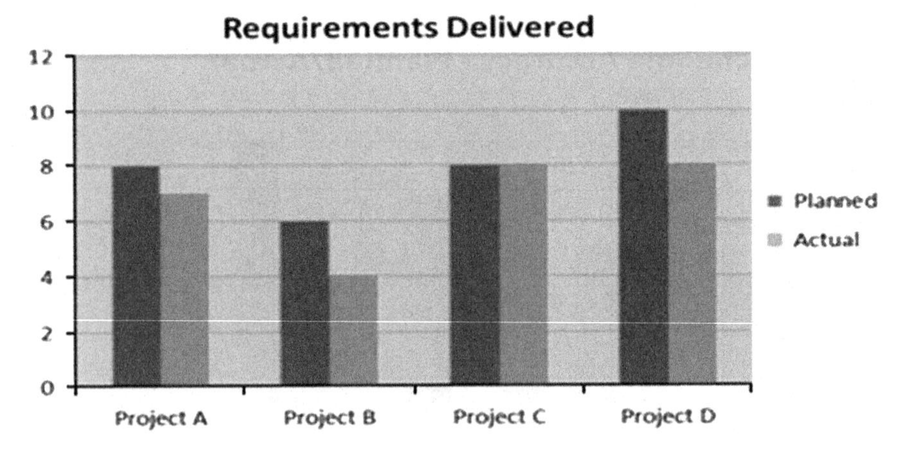

FIGURE 13.18
Requirements Delivered.

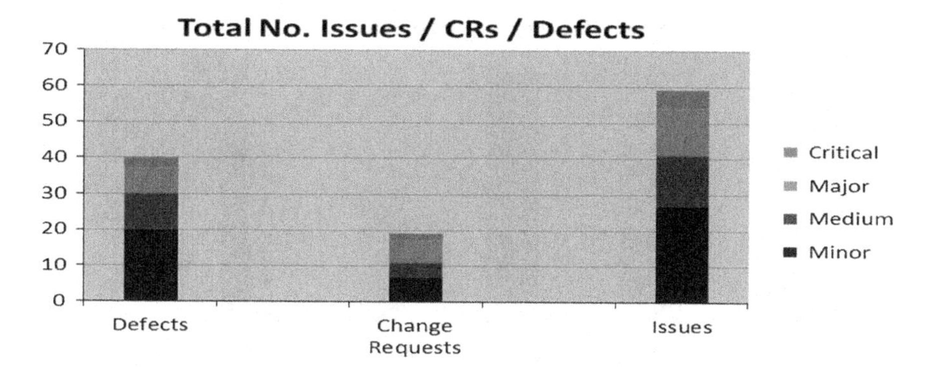

FIGURE 13.19
Total Number of Issues in Project.

13.3.7 Development Quality Metrics

These metrics give visibility into the development and testing of the product. Fig. 13.19 gives an indication of the quality of the product, and the quality of the definition of the initial requirements. It shows the total number of defects and the total number of change requests raised during the project, as well as details of their severities. The presence of a large number of change requests suggests that the initial definition of the requirement was incomplete, and that there is room for improvement in the requirements process.

Fig. 13.20 gives the status of open issues with the project, which gives an indication of the current quality of the project, and the effort required to achieve the desired quality in the software. This chart is not used in isolation, as the project manager will need to know the arrival rate of problems to determine the stability of the product.

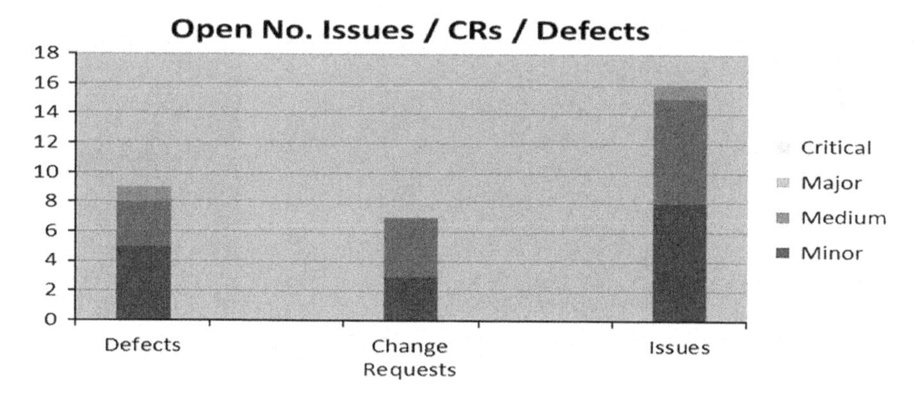

FIGURE 13.20
Open Issues in Project.

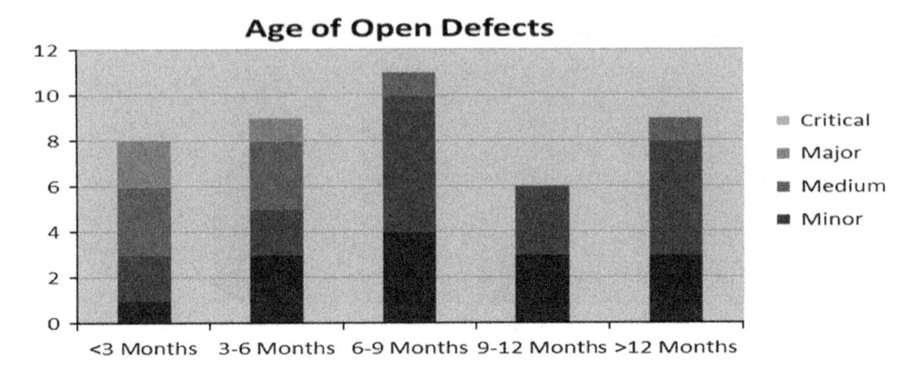

FIGURE 13.21
Age of Open Defects in Project.

The organization may decide to release a product with open problems provided that the associated risks with the known problems can be managed. It is important to perform a risk assessment to ensure that these may be managed, and the known problems (and work-arounds) should be documented in the release notes for the product.

The project manager will need to know the age of the open problems to determine the effectiveness of the project team in resolving problems in a timely manner. Fig. 13.21 presents the age of the open defects, and it highlights the fact that there is one major problem that has been open for over one year. The project manager needs to prevent this situation from arising, as critical and major problems need to be swiftly resolved.

The problem arrival rate enables the project manager to judge the stability of the product, and this (combined with other metrics) helps in judging whether the product is fit for purpose and ready for release to potential customers. Fig. 13.22 presents a sample problem arrival chart, and the chart

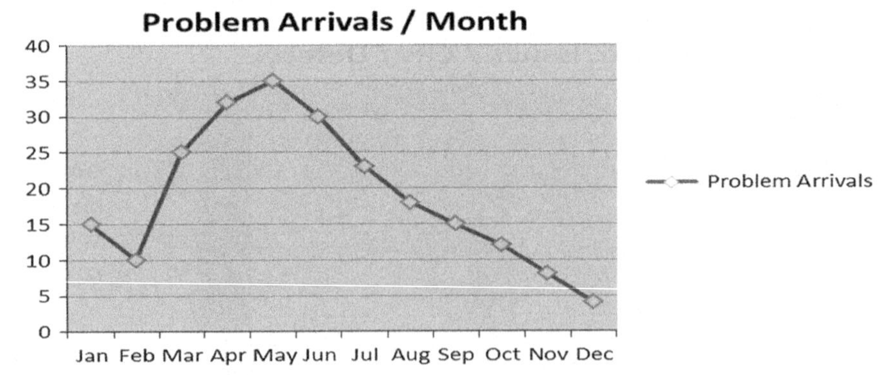

FIGURE 13.22
Problem Arrivals per Month.

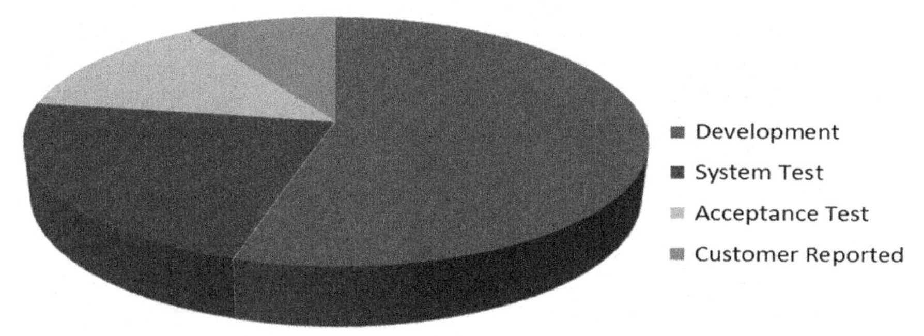

FIGURE 13.23
Phase Containment Effectiveness.

indicates positive trends with the arrival rate of problems falling to very low levels.

The project manager will need to do analysis to determine if there are other factors that could contribute to the fall in the arrival rate; for example, it may be the case that testing was completed in September, which would mean, in effect, that no testing has been performed since then, with an inevitable fall in the number of problems reported. The important point is not to jump to a conclusion based on a particular chart, as the circumstances behind the chart must be fully known and taken into account in order to draw valid conclusions.

Fig. 13.23 measures the effectiveness of the project in identifying defects in the development phase, and the effectiveness of the test groups in detecting defects that are present in the software product.

Various types of testing (e.g., unit, system, performance, usability, acceptance) are discussed in [ORg:19]. Fig. 13.23 indicates that the project had a phase containment effectiveness (PCE) of approximately 54%. That is, the developers identified 54% of the defects, the system-testing phase identified approximately 23% of the defects, acceptance testing identified approximately 14% of the defects and the customer identified approximately 9% of the defects. The objective is that the number of defects reported at acceptance test and after the product is officially released to customer should be minimal.

13.3.8 Quality Audit Metrics

These metrics provide visibility into the audit program, and include metrics for the number of audits planned and performed (Fig. 13.24), and the status of the audit actions (Fig. 13.25). Fig. 13.24 presents visibility into the number of audits carried out in the organization, and the number of audits that remain to be done.

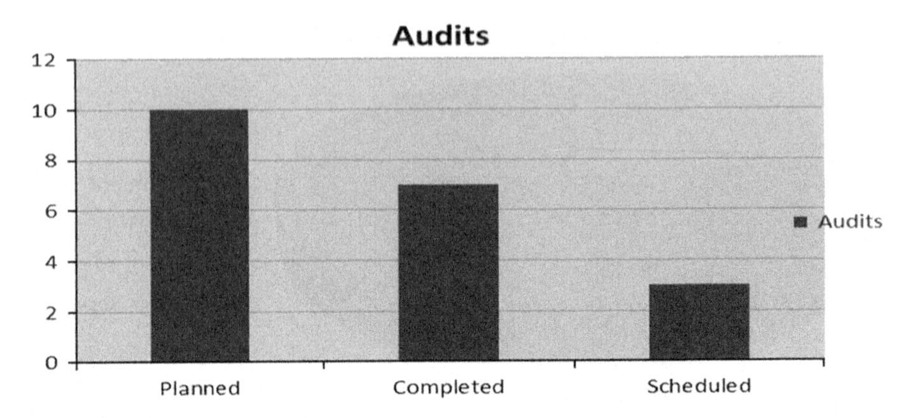

FIGURE 13.24
Annual Audit Schedule.

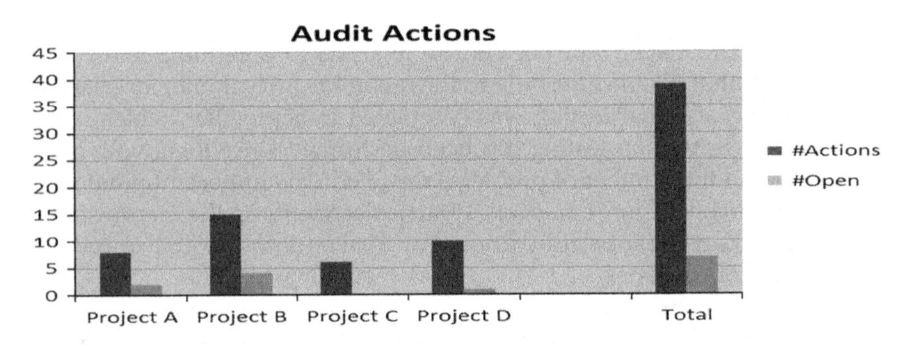

FIGURE 13.25
Status of Audit Actions.

It shows that the organization has an audit program, and gives information on the number of audits performed during a particular time period. The chart does not give a breakdown into the type of audits performed, e.g., supplier audits, project audits and audits of particular departments in the organization, but it could be adapted to provide this information.

Fig. 13.25 gives an indication of the status of the audit actions from the various audits. An auditor performs an audit and the results are documented in an audit report, and the audit actions then need to be completed by the affected individuals and groups.

Fig. 13.26 gives visibility into the type of actions raised during the audit of a particular area. They could potentially include entry and exit criteria, planning issues, configuration management issues, issues with compliance to the lifecycle or templates, traceability to the requirements, issues with the review of various deliverables, issues with testing or process improvement suggestions.

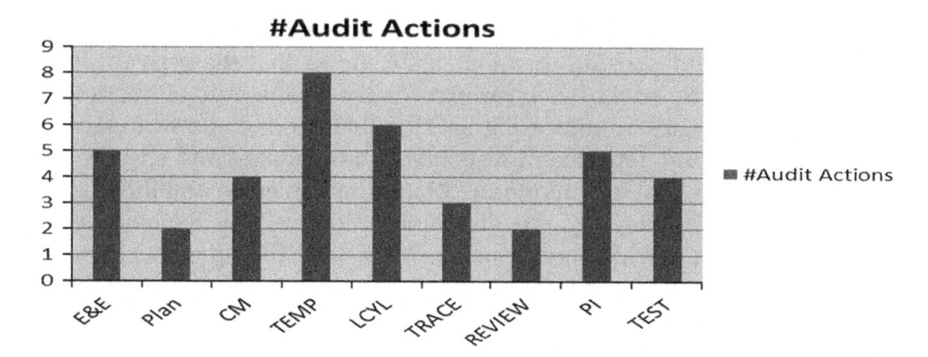

FIGURE 13.26
Audit Action Types.

13.3.9 Customer Care Metrics

The goals of the customer care group in an organization are to respond efficiently and effectively to customer problems, to ensure that their customers receive the highest standards of service from the company, and to ensure that its products function reliably at the customer's site. The organization will need to know its efficiency in resolving customer queries, the number of customer queries, the availability of its software systems at the customer site and the age of open queries. A customer query may result in a defect report where there is a defect with the software.

Fig. 13.27 presents the arrival and closure rate of customer queries (it could be developed further to include a severity attribute for the query). Quantitative goals are generally set for the resolution of queries (especially in the case of service-level agreements). A chart for the age of open queries (similar to Fig. 13.21) may be maintained. The organization will need to

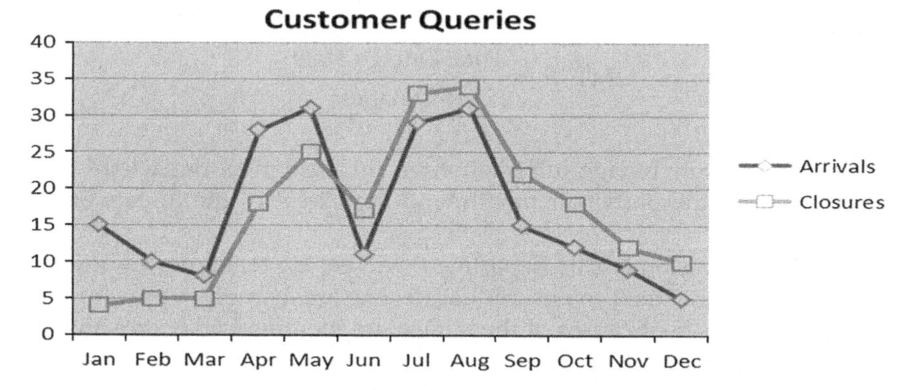

FIGURE 13.27
Customer Queries (Arrivals/Closures).

know the status of the backlog of open queries per month, and a simple trend graph would provide this. Fig. 13.27 shows that the arrival rate of queries: in the early part of the year exceeds the closure rate of queries per month. This indicates an increasing backlog that needs to be addressed.

The customer care department responds to any outages and ensures that the outage time is kept to a minimum. Many companies set ambitious goals for network availability, e.g., the "five nines initiative" has the objective of developing systems which are available 99.999% of the time, i.e., approximately 5 minutes of down time per year. The calculation of availability is from the formula:

$$\text{Availability} = \frac{\text{MTBF}}{\text{MTBF} + \text{MTTR}}$$

where the mean time between failure (MTBF) is the average length of time between outages.

$$\text{MTBF} = \frac{\text{Sample Interval Time}}{\#\,\text{Outages}}$$

The formula for MTBF above is for a single system only, and the formula is adjusted when there are multiple systems.

$$\text{MTBF} = \frac{\text{Sample Interval Time}}{\#\,\text{Outages}} \quad * \;\#\,\text{Systems}$$

The mean time to repair (MTTR) is the average length of time that it takes to correct the outage, i.e., the average duration of the outages that have occurred, and it is calculated from the following formula:

$$\text{MTTR} = \frac{\text{Total Outage Time}}{\#\,\text{Outages}}$$

Fig. 13.28 presents outage information on the customers impacted by the outage during the particular month, and the extent of the impact on the customer.

An effective customer care department will ensure that a post-mortem of an outage is performed to ensure that lessons are learned to prevent a reoccurrence. This root causes of the outage are identified and corrective actions are implemented to prevent a reoccurrence. Metrics to record system availability and outage time per month will often be maintained by the customer care group in the form of a trend graph.

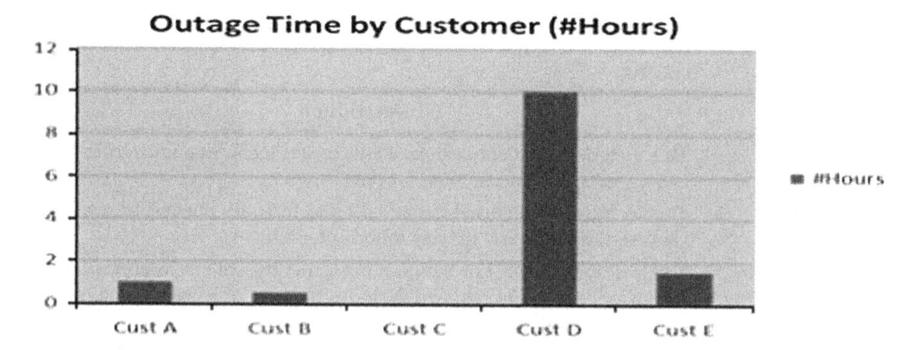

FIGURE 13.28
Outage Time per Customer.

Fig. 13.29 provides visibility on the availability of the system at the customer sites, and many systems are designed to be available 99.999% of the time. System availability and software reliability are discussed in more detail in [ORg:19].

13.3.10 Cost of Quality Metrics

Crosby argued that the most meaningful measurement of quality is the cost of poor quality [Crs:79], and that the emphasis of the improvement activities should be to reduce the *cost of poor quality* (COPQ). The cost of quality includes the cost of external and internal failure, the cost of providing an infrastructure to prevent the occurrence of problems, and the cost of the infrastructure to verify the correctness of the product.

The cost of quality was divided into four subcategories (Table 13.2) by Feigenbaum in the 1950s, and developed further by James Harrington of IBM.

The cost of quality graph (Fig. 13.30) will initially show high external and internal costs and low prevention costs, and the total quality costs will be high. However, as an effective quality system is put in place and becomes fully

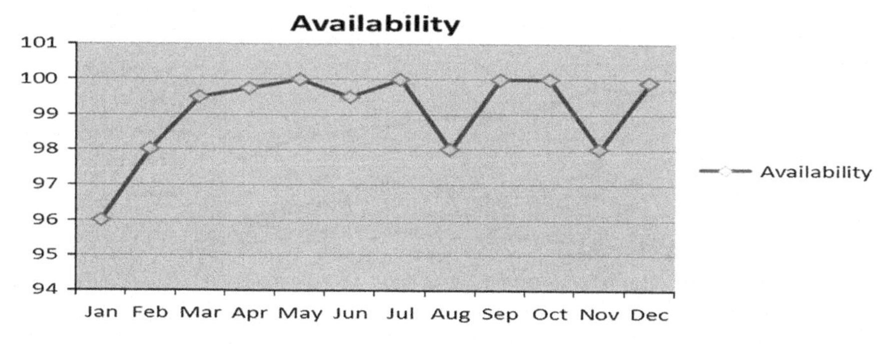

FIGURE 13.29
Availability of System per Month.

TABLE 13.2
Cost of Quality Categories

Type of Cost	Description
Cost external	This includes the cost of external failure and includes engineering repair, warranties, and a customer support function
Cost internal	This includes the internal failure cost and includes the cost of reworking and re-testing of any defects found internally.
Cost prevention	This includes the cost of maintaining a quality system to prevent the occurrence of problems, and includes the cost of software quality assurance, the cost of training, etc.
Cost appraisal	This includes the cost of verifying the conformance of a product to the requirements and includes the cost of provision of software inspections and testing processes.

operational, there will be a noticeable decrease in the external and internal cost of quality, and a gradual increase in the cost of prevention and appraisal.

The total cost of quality will substantially decrease, as the cost of provision of the quality system is substantially below the cost of internal and external failure. The COPQ curve will indicate where the organization is in relation to the cost of poor quality, and the organization will need to implement improvements to put an effective quality management system in place to minimize the cost of poor quality.

13.4 Implementing a Metrics Program

The metrics presented in the previous section may be adapted and tailored to meet the needs of organizations. The metrics are only as good as the

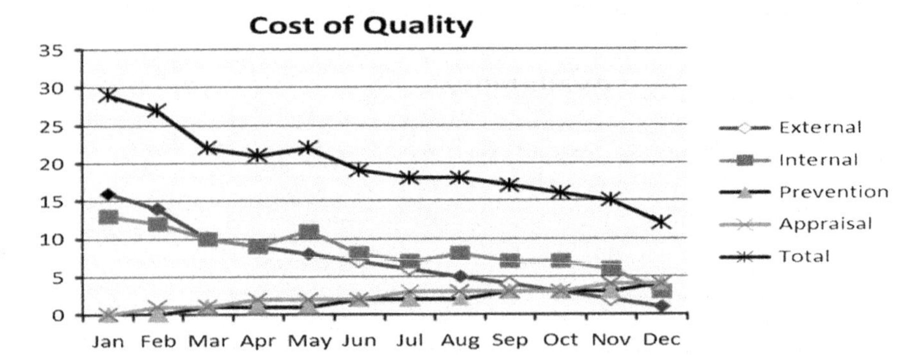

FIGURE 13.30
Cost of Poor Quality (COPQ).

TABLE 13.3

Implementing Metrics

Implementing Metrics in Organization
Define the business goals
Determine the key questions
Define the metrics
Identify tools to (semi-)automate metrics
Determine data that needs to be gathered
Identify and provide needed resources
Gather data and prepare metrics
Communicate the metrics and review monthly
Provide training

underlying data, and good data gathering is essential. The following are typical steps in the implementation of a metrics program (Table 13.3).

The business goals are the starting point in the implementation of a metrics program, as there is no sense in measurement for the sake of measurement, and so the metrics must be closely related to the business goals. The next step is to identify the relevant questions to determine the extent to which the business goal is being satisfied, and to define metrics that provide an objective answer to the questions.

The organization defines its business goals, and each department develops specific goals to meet the organization's goals. Measurement data will indicate the extent to which the goals are being achieved, and the data to be gathered needs to be determined as well as the methods by which the data will be recorded. The data gathered will provide the information needed to answer the questions related to the goals. A small organization may decide to record the data manually, but often automated or semi-automated tools will be employed. It is essential that the data collection and extraction are efficient, as otherwise the metrics program may fail.

The roles and responsibilities of staff with respect to the implementation and day-to-day operation of the metrics program need to be defined. Training is needed to enable staff to perform their roles effectively. Finally, a regular management review is needed, where the metrics and trends are presented, and actions identified and carried out to ensure that the business goals are achieved.

13.4.1 Data Gathering for Metrics

Metrics are only as good as the underlying data, and so data gathering is a key activity in a metrics program. The data is used to give an objective answer to the questions, and the business goals are usually expressed

TABLE 13.4

Goals and Questions

Goal	Reduce escaped defects from each lifecycle phase by 10%.
Questions	How many defects are identified within each lifecycle phase? How many defects are identified after each lifecycle phase is exited? What percentage of defects escaped from each lifecycle phase?

quantitatively for extra precision. Table 13.4 gives an example of questions related to a particular goal.

Table 13.5 is designed to determine the effectiveness of the software development process to enable the above questions to be answered. It includes a column for inspection data that records the number of *defects* recorded at the various inspections. The *defects* include the phase where the defect originated; for example, a defect identified in the coding phase may have originated in the requirements or design phase. This data is typically maintained in a spreadsheet, e.g., Excel (or a dedicated tool), and it needs to be kept up to date. It enables the PCE to be calculated for the various phases.

We will distinguish between a defect that is detected *in-phase* versus a defect that is detected *out-of-phase*. An in-phase defect is a problem that is detected in the phase in which it is created (e.g., usually by a software inspection). An out-of-phase defect is detected in a later phase (e.g., a problem with the requirements may be discovered in the design phase, which is a later phase from the phrase in which it was created).

The effectiveness of the requirements phase in Table 13.5 is judged by its success in identifying defects as early as possible, as the cost of correction of a requirements defect increases later in the cycle that it is identified. The

TABLE 13.5

Phase Containment Effectiveness

		Phase of Origin						
Phase	**Inspect Defects**	**Reqs**	**Design**	**Code**	**Accept Test**	**In-Phase Defects**	**Other Defects**	**% PCE**
Reqs	4		1	1		4	6	40%
Design	3					3	4	42%
Code	20					20	15	57%
Unit Test		2	2	10				
System Test		2	2	5				
Accept Test								

requirement of PCE is calculated to be 40%, i.e., the total number of defects identified in phase divided by the total number of defects identified. There were four defects identified at the inspection of the requirements, and six defects were identified outside of the requirements phase: one in the design phase, one in the coding phase, two in the unit testing phase and two at the system testing phase, i.e., 4/10 = 40%. Similarly, the code PCE is calculated to be 57%.

The overall PCE for the project is calculated to be the total number of defects detected in phase in the project divided by the total number of defects, i.e., 27/52 = 52%. Table 13.4 is a summary of the collected data and its construction consists of:

- Maintain inspection data of requirements, design and code inspections.
- Identify defects in each phase and determine their phase of origin.
- Record the number of defects in each phase per phase of origin.

The staff who perform inspections need to record the problems identified, whether it is a defect, and its phase of origin. Staff will need to be appropriately trained to do this consistently.

The above is just one example of data gathering, and in practice the organization will need to collect various data to enable it to give an objective answer to the extent that the particular goal is being satisfied.

13.5 Problem-Solving Techniques

Problem solving is a key part of quality improvement, and a *quality circle* (or problem-solving team) is a group of employees who do similar work and volunteer to come together on company time to identify and analyse work-related problems. Quality circles were originally described by Deming in the 1950s, and developed further by Ishikawa in Japan in the 1960s [Isi:85].

Various tools that assist problem solving include *process mapping, trend charts, bar charts, scatter diagrams, fishbone diagrams, histograms, control charts* and *Pareto charts* [BR:94]. These provide visibility into the problem and help to quantify the extent of the problem. The problem-solving team involves:

- Group of employees who do similar work
- Voluntarily meet regularly on company time
- Supervisor as leader
- Identify and analyse work-related problems

- Recommend solutions to management
- Implement solution where possible

The facilitator of the quality circle coordinates the activities, ensures that the team leaders and teams members receive sufficient training, and obtains specialist help where required. The quality circle facilitator has the following responsibilities:

- Focal point of quality circle activities
- Train circle leaders/members
- Coordinate activities of all the circle groups
- Assist in inter-circle investigations
- Obtain specialist help when required.

The circle leaders receive training in problem-solving techniques, and are responsible for training the team members. The leader needs to keep the meeting focused and requires skills in team building. The steps in problem solving are as follows:

- Select the problem
- State and restate the problem
- Collect the facts
- Brainstorm
- Choose course of action
- Present to management
- Measurement of success.

The benefits of a successful problem-solving culture in the organization include the following:

- Savings of time and money
- Increased productivity
- Reduced defects
- Fire prevention culture.

Various problem-solving tools are discussed in the following sections.

13.5.1 Fishbone Diagram

This well-known problem-solving tool consists of a cause-and-effect diagram that is in the shape of the backbone of a fish. The objective is to identify the various causes of some particular problem, and then these causes are broken down into a number of sub-causes. The various causes and sub-causes are analysed to determine the root cause of the particular problem, and actions to address the root cause are identified to prevent a re-occurrence of the problem. There are various categories of causes, and these may include people, methods and tools, and training.

The fishbone diagram offers a crisp mechanism to summarize the collective knowledge that a team has about a particular problem, and it facilitates the detailed exploration of the causes and identifying appropriate solutions.

The construction of a fishbone diagram involves a clear statement of the particular effect, and the effect is placed at the right-hand side of the diagram. The major categories of cause are drawn on the backbone of the fishbone diagram; brainstorming is used to identify causes; and these are then placed in the appropriate category. For each cause identified the various sub-causes may be identified by asking the question "Why does this happen?" This leads to a more detailed understanding of the causes and sub-causes of a particular problem.

Example 13.1

An organization wishes to determine the causes of a high number of customer-reported defects. There are various categories that may be employed such as people, training, methods, tools and environment. In practice, the fishbone diagram in Fig. 13.31 would be more detailed than that presented, as sub-causes would also be identified by a detailed examination

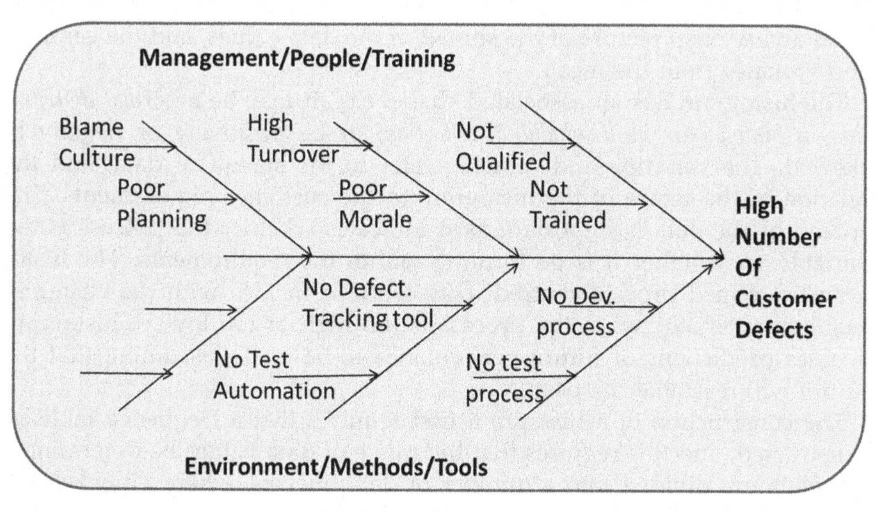

FIGURE 13.31
Fishbone Cause-and-Effect Diagram.

of the identified causes. The root cause(s) are determined from detailed analysis.

This example suggests that the organization has major work to do in several areas, and that an improvement program is required. The improvements needed include the implementation of a software development process and a software test process; the provision of training to enable staff to do their jobs more effectively; and the implementation of better management practices to motivate staff and to provide a supportive environment for software development.

The causes identified may be symptoms rather than actual root causes: for example, high staff turnover may be the result of poor morale and a "blame culture," rather than a cause in itself of poor-quality software. The fishbone diagram gives a better understanding of the possible causes of the high number of customer defects. A small subset of these causes is then identified as the root cause(s) of the problem following further discussion and analysis.

The root causes are then addressed by appropriate corrective actions (e.g., an appropriate software development process and test process are defined and providing training to all development staff on the new processes). The management attitude and organization culture will need to be corrected to enable a supportive software development environment to be put in place.

13.5.2 Histograms

A histogram is a way of representing data in bar chart format, and it shows the relative frequency of various data values or ranges of data values. It is typically employed when there are a large number of data values, and it gives a very crisp picture of the spread of the data values, and the centring and variance from the mean.

The histogram has an associated shape, e.g., it may be a *normal distribution*, a *bimodal* or *multi-modal distribution*, or be positively or negatively skewed. The variation and centring refer to the spread of data, and the relation of the centre of the histogram to the customer requirements. The spread of the data is important as it indicates whether the process is too variable, or whether it is performing within the requirements. The histogram is termed process centred if its centre coincides with the customer requirements; otherwise, the process is too high or too low. A histogram enables predictions of future performance to be made, assuming that the future will resemble the past.

The construction of a histogram first requires that a frequency table be constructed, and this requires that the range of data values be determined. The data are divided into a number of data buckets, where a bucket is a particular range of data values, and the relative frequency of each bucket is displayed in bar format. The number of class intervals or buckets is determined, and the class intervals are defined. The class intervals are mutually disjoint and span the range of the data values. Each data value belongs to exactly one class interval and the frequency of each class interval is determined.

The histogram is a well-known statistical tool and its construction is made more concrete with the following example.

Example 13.2

An organization wishes to characterize the behaviour of the process for the resolution of customer queries in order to achieve its customer satisfaction goal.

Goal

Resolve all customer queries within 24 hours.

Question

How effective is the current customer query resolution process?

What action is required (if any) to achieve this goal?

The data class size chosen for the histogram is 6 hours, and the data class sizes are of the same in standard histograms. The sample mean is 19 hours for this example. The histogram in Fig. 13.32 is based on query resolution data from 36 samples. The organization goal of customer resolution of all queries within 24 hours is not met, and the goal is satisfied in (25/36 = 70% for this particular sample).

Further analysis is needed to determine the reasons why 30% of the goals are outside the target 24-hour time period. It may prove to be impossible to meet the goal for all queries, and the organization may need to refine the

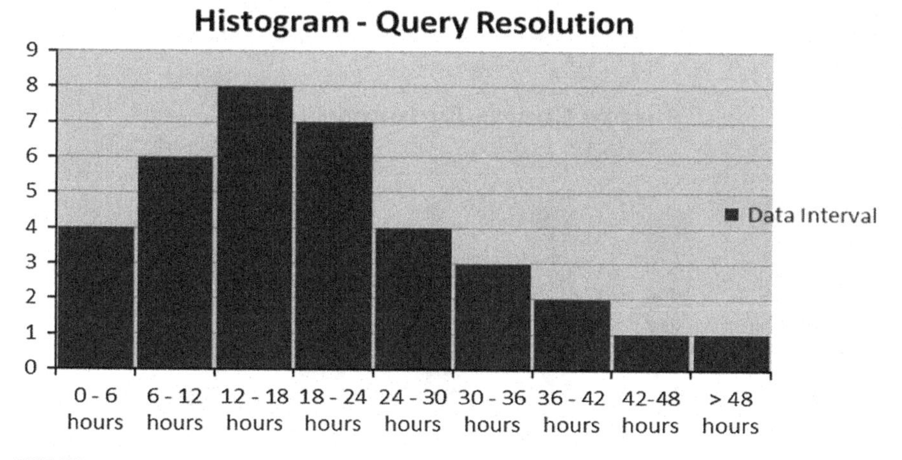

FIGURE 13.32
Histogram.

goal to state that instead all critical and major queries will be resolved within 24 hours.

13.5.3 Pareto Chart

The objective of a Pareto chart is to identify and focus on the resolution of problems that have the greatest impact (as *often 20% of the causes are responsible for 80% of the problems*). The problems are classified into various categories, and the frequency of each category of problem is determined. The Pareto chart is displayed in a descending sequence of frequency, with the most significant cause presented first, and the least significant cause presented last.

The Pareto chart is a key problem-solving tool, and a properly constructed chart will enable the organization to focus on and resolve the key causes of problems, and to verify their resolution. The effectiveness of the improvements may be judged at a later stage from the analysis of new problems and the creation of a new Pareto chart. The results should show improvements, with less problems arising in the category that was the major source of problems.

The construction of a Pareto chart requires the organization to decide on the problem to be investigated; to identify the causes of the problem via brainstorming; to analyse the historical or real time data; to compute the frequency of each cause; and finally to display the frequency in descending order for each cause category.

Example 13.3

An organization wishes to understand the various causes of outages, and to minimize their occurrence.

The Pareto chart in Fig. 13.33 includes data from an analysis of outages, where each outage is classified into a particular cause. The six causal categories identified are as follows: hardware, software, operator error, power

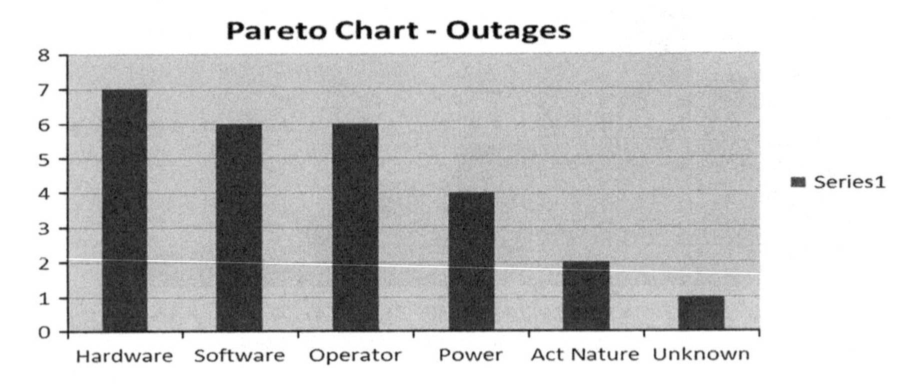

FIGURE 13.33
Pareto Chart Outages.

failure, an act of nature and unknown. The three main causes of outages are hardware, software and operator error, and analysis is needed to identify appropriate actions to address these. The hardware category may indicate that there are problems with the reliability of the system hardware, and that existing hardware systems may need improvement or replacement. There may be a need to address availability and reliability concerns with more robust hardware solutions.

The software category may be due to the release of poor-quality software, or to usability issues in the software, and this requires further investigation. Finally, operator issues may be due to lack of knowledge or inadequate training of the operators. An improvement plan needs to be prepared and implemented, and its effectiveness will be judged by a reduction in outages, and reductions of problems in the targeted category.

13.5.4 Trend Graphs

A trend graph monitors the performance of a variable over time, and it allows trends in performance to be identified, as well as allowing predictions of future trends to be made (assuming that the future resembles the past). Its construction involves deciding on the variable to measure, and gathering and plotting the data points.

Example 13.4

An organization plans to deploy an enhanced estimation process, and wishes to determine if estimation is actually improving with the new process.

The estimation accuracy determines the extent to which the actual effort differs from the estimated effort. A reading of 25% indicates that the project effort was 25% more than estimated, whereas a reading of –10% indicates that the actual effort was 10% less than estimated.

The trend chart in Fig. 13.34 indicates that initially that estimation accuracy is very poor, but then there is a gradual improvement coinciding with the implementation of the new estimation process.

It is important to analyse the performance trends in the chart. For example, the estimation accuracy for August (17% in the chart) needs to be investigated to determine the reasons why it occurred. It could potentially indicate that a project is using the old estimation process, or that a new project manager received no training on the new process). A trend graph is useful for noting positive or negative trends in performance, with negative trends analysed and actions identified to correct performance.

13.5.5 Scatter Graphs

The scatter diagram is used to determine whether there is a relationship or correlation between two variables, and where there is to measure the

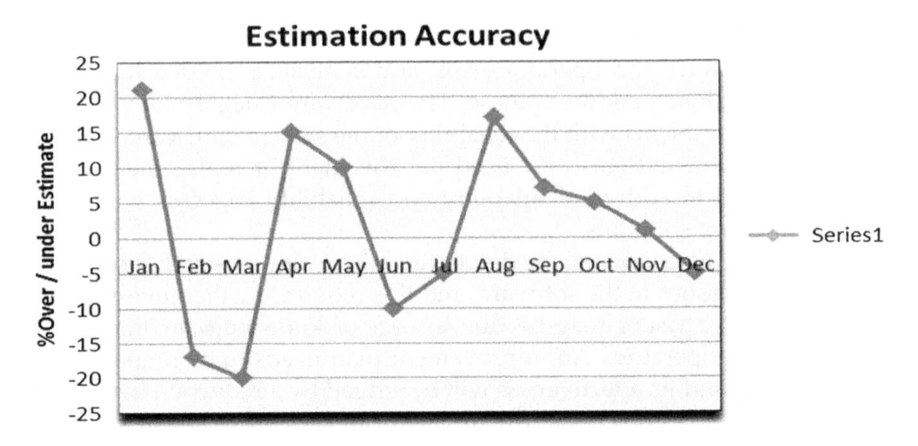

FIGURE 13.34
Trend Chart Estimation Accuracy.

relationship between them (Chapter 9). The results may be a positive correlation, negative correlation or no correlation. Correlation has a precise statistical definition and it provides a precise mathematical understanding of the extent to which the two variables are related or unrelated.

The scatter graph provides a graphical way to determine the extent that two variables are related, and it is often used to determine whether there is a connection between an identified cause and the effect. The construction of a scatter diagram requires the collection of paired samples of data, and the drawing of one variable as the x-axis, and the other as the y-axis. The data are then plotted and interpreted.

Example 13.5

An organization wishes to determine if there is a relationship between the inspection rate and the error density of defects identified.

The scatter graph in Fig. 13.35 provides evidence for the hypothesis that there is a relationship between the lines of code inspected, and the error density recorded (per KLOC). The graph suggests that the error density of defects identified during inspections is low if the speed of inspection is too fast, and the error density is high if the speed of inspection is below 300 lines of code per hour. A line can be drawn through the data that indicates a linear relationship.

13.5.6 Metrics and Statistical Process Control

The principles of statistical process control (SPC) are important in the monitoring and control of a process. It involves developing a control chart, which is a tool that may be used to control the process, with upper and

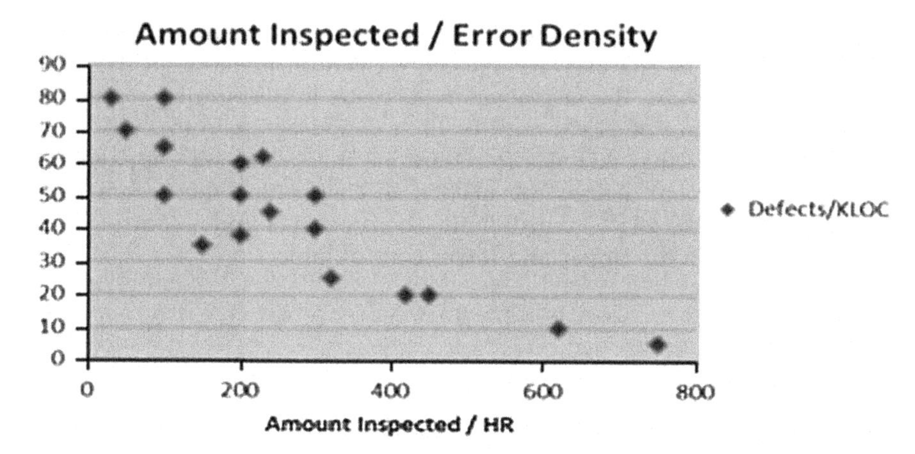

FIGURE 13.35
Scatter Graph Amount Inspected Rate/Error Density.

lower limits for process performance specified. The process is under control if it is performing within the lower and upper control limits.

Fig. 13.36 presents an example on breakthrough in performance of an estimation process, and is adapted from [Kee:00]. The initial upper and lower control limits for estimation accuracy are set at ±40%, and the performance of the process is within the defined upper and control limits.

However, the organization will wish to improve its estimation accuracy and this leads to the organization's revising the upper and lower control limits to ±25%. The organization will need to analyse the slippage data to

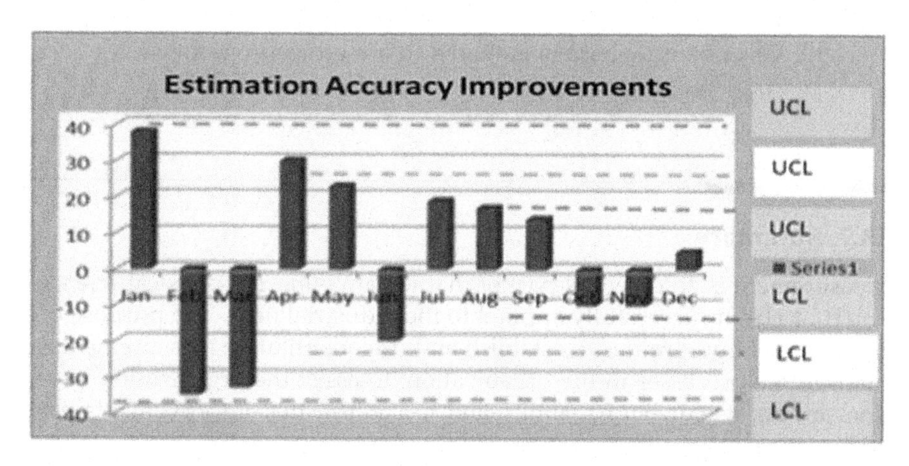

FIGURE 13.36
Estimation Accuracy and Control Charts.

determine the reasons for the wide variance in the estimation, and part of the solution will be the use of enhanced estimation methods in the organization. In this chart, the organization succeeds in performing within the revised control limit of ±25%, and the limit is revised again to ±15%.

This requires further analysis to determine the causes for slippage and further improvement actions are needed to ensure that the organization performs within the ±15% control limit. The process is under control if it is performing within the upper and lower control limits.

13.6 Review Questions

1. Describe the Goal, Question, Metric model.
2. Explain how the balanced scorecard may be used in the implementation of organization strategy.
3. Describe various problem-solving techniques.
4. What is a fishbone diagram?
5. What is a histogram and describe its applications?
6. What is a scatter graph?
7. What is a Pareto chart? Describe its applications.
8. Discuss how a metrics programme may be implemented.
9. What is statistical process control?
10. Describe how data is gathered in a metrics program.
11. How are metrics used?

13.7 Summary

Measurement is an essential part of mathematics and the physical sciences, and it has been successfully applied to the industrial field. The purpose of a measurement program is to establish and use quantitative measurements to manage the processes in the organization, to assist the organization in understanding its current capability and to confirm that improvements have been successful.

This chapter included a collection of sample metrics to give visibility into the performance of various areas in the organization, including financial, sales and marketing, customer satisfaction, HR, process improvement, project management, development and quality metrics, and customer care metrics.

The Goal, Question, Metric paradigm is a rigorous, goal-oriented approach to measurement in which goals, questions and measurements are closely integrated. The business goals are first defined, and then questions that relate to the achievement of the goal are identified, and for each question a metric that gives an objective answer to the particular question is defined.

Metrics plays a key role in problem solving, and various problem-solving techniques were discussed, which include histograms, Pareto charts, trend charts and scatter graphs. The measurement data are used to assist the analysis and determine the root cause of a particular problem, and to verify that the actions taken to correct the problem have been effective. Trends may be seen over time, and the analysis of the trends allows action plans to be prepared for continuous improvement.

Metrics may be employed to track the quality, timeliness, cost, schedule and effort of software projects. They provide an internal view of the quality of the software product, and care is needed before deducing the behaviour that a product will exhibit externally.

14

Matrix Theory

A *matrix* is a rectangular array of numbers that consists of horizontal rows and vertical columns. A matrix with m rows and n columns is termed an $m \times n$ matrix, where m and n are its dimensions. A matrix with an equal number of rows and columns (e.g., n rows and n columns) is termed a *square matrix*. The example matrix below is a square matrix with four rows and four columns (Fig. 14.1).

The entry in the ith row and the jth column of a matrix A is denoted by A[i, j], $A_{i,j}$, or a_{ij}, and the matrix A may be denoted by the formula for its (i, j)th entry, i.e., (a_{ij}) where i ranges from 1 to m and j ranges from 1 to n.

An $m \times 1$ matrix is termed a column vector, and a $1 \times n$ matrix is termed a row vector. Any row or column of an $m \times n$ matrix determines a row or column vector which is obtained by removing the other rows (respectively columns) from the matrix. For example, the row vector (11, −5, 5, 3) is obtained from the matrix example by removing rows 1, 2 and 4 of the matrix.

Two matrices A and B are equal if they are both of the same dimensions, and if $a_{ij} = b_{ij}$ for each $i = 1, 2, \ldots, m$ and each $j = 1, 2, \ldots, n$.

Matrices can be added and multiplied (provided certain conditions are satisfied). An additive identity matrix is a matrix such that the addition of it to any matrix A yields A, i.e., A + 0 = 0 + A = A for any matrix A; and similarly, a multiplicative identity matrix is such that AI = IA = A for any matrix A. Square matrices have multiplicative inverses (provided that their determinant is non-zero), and every matrix satisfies its characteristic polynomial (discussed in section 14.4).

It is possible to consider matrices with infinite rows and columns, and although it is not possible to write down such matrices explicitly it is still possible to add, subtract and multiply by a scalar provided there is a well-defined entry in each (i,j)th element of the matrix.

Matrices are an example of an abstract algebraic structure known as an *algebra*. We discussed classical algebra in Chapter 2, but we did not introduce abstract algebraic structures such as groups, rings, fields and vector spaces. The reader may wish to consult [ORg:20] for more information on these structures and their applications. The matrix algebra for $m \times n$ matrices A, B, C and scalers λ and μ satisfies the following properties:

1. A + B = B + A (Commutativity)
2. A + (B + C) = (A + B) + C (Associativity)

DOI: 10.1201/9781003308140-14

$$\begin{pmatrix} 6 & 0 & -2 & 3 \\ 4 & 2 & 3 & 7 \\ 11 & -5 & 5 & 3 \\ 3 & -5 & -8 & 1 \end{pmatrix}$$

FIGURE 14.1
Example of a 4 × 4 Square Matrix.

3. $A + 0 = 0 + A = A$ (Additive Identity)
4. $A + (-A) = (-A) + A = 0$ (Additive Inverse)
5. $\lambda(A + B) = \lambda A + \lambda B$
6. $(\lambda + \mu)A = \lambda A + \mu B$
7. $\lambda(\mu A) = (\lambda\mu)A$
8. $1A = A$

Matrices have many applications including their use in graph theory to keep track of the distance between pairs of vertices in the graph; a rotation matrix may be employed to represent the rotation of a vector in three-dimensional space. The product of two matrices represents the composition of two linear transformations, and matrices may be employed to determine the solution to a set of linear equations. They also arise in computer graphics and may be employed to project a three-dimensional image onto a two-dimensional screen. It is essential to employ efficient algorithms for matrix computation, and this is an area of research in the field of numerical analysis.

14.1 Two × Two Matrices

Matrices arose in practice as a means of solving a set of linear equations. One of the earliest examples of their use is in a Chinese text dating from between 300 BC and 200 AD. The Chinese text showed how matrices could be employed to solve simultaneous equations. Consider the set of equations:

$$ax + by = r$$
$$cx + dy = s$$

Then the coefficients of the linear equations in x and y above may be represented by the matrix A, where A is given by:

$$A = \begin{bmatrix} a & b \\ c & d \end{bmatrix}$$

The linear equations may be represented as the multiplication of the matrix A and a vector $\underline{x}$ resulting in a vector $\underline{v}$:

$$A\underline{x} = \underline{v}$$

The matrix representation of the linear equations and its solution are as follows:

$$\begin{bmatrix} a & b \\ c & d \end{bmatrix} \begin{bmatrix} x \\ y \end{bmatrix} = \begin{bmatrix} r \\ s \end{bmatrix}$$

The vector $\underline{x}$ may be calculated by determining the inverse of the matrix A (provided that its inverse exists). The vector $\underline{x}$ is then given by:

$$\underline{x} = A^{-1}\underline{v}$$

The solution to the set of linear equations is then given by:

$$\begin{bmatrix} x \\ y \end{bmatrix} = \begin{bmatrix} a & b \\ c & d \end{bmatrix}^{-1} \begin{bmatrix} r \\ s \end{bmatrix}$$

The inverse of a matrix A exists if and only if its *determinant* is non-zero, and if this is the case the vector $\underline{x}$ is given by:

$$\begin{bmatrix} x \\ y \end{bmatrix} = \frac{1}{\det A} \begin{bmatrix} d & -b \\ -c & a \end{bmatrix} \begin{bmatrix} r \\ s \end{bmatrix}$$

The determinant of a 2×2 matrix A is given by:

$$\det A = ad - cb$$

The determinant of a 2×2 matrix is denoted by:

$$\begin{vmatrix} a & b \\ c & d \end{vmatrix}$$

A key property of determinants is that

$$\det(AB) = \det(A).\det(B)$$

The transpose of a 2 × 2 matrix A (denoted by A^T) involves exchanging rows and columns and is given by:

$$A^T = \begin{bmatrix} a & c \\ b & d \end{bmatrix}$$

The inverse of the matrix A (denoted by A^{-1}) is given by:

$$A^{-1} = \frac{1}{\det A}\begin{bmatrix} d & -b \\ -c & a \end{bmatrix}$$

Further, $A.A^{-1} = A^{-1}.A = I$, where I is the identity matrix of the algebra of 2 × 2 matrices under multiplication. That is:

$$AA^{-1} = A^{-1}A = \begin{bmatrix} 1 & 0 \\ 0 & 1 \end{bmatrix}$$

The addition of two 2 × 2 matrices A and B is given by a matrix whose entries are the addition of the individual components of A and B. The addition of two matrices is commutative and we have:

$$A + B = B + A = \begin{bmatrix} a+p & b+q \\ c+r & d+s \end{bmatrix}$$

where A and B are given by:

$$A = \begin{bmatrix} a & b \\ c & d \end{bmatrix} B = \begin{bmatrix} p & q \\ r & s \end{bmatrix}$$

The identity matrix under addition is given by the matrix whose entries are all 0, and it has the property that $A + 0 = 0 + A = A$.

$$\begin{bmatrix} 0 & 0 \\ 0 & 0 \end{bmatrix}$$

The multiplication of two 2 × 2 matrices is given by:

$$AB = \begin{bmatrix} ap + br & aq + bs \\ cp + dr & cq + ds \end{bmatrix}$$

The multiplication of matrices is not commutative, i.e., AB ≠ BA. The multiplicative identity matrix I has the property that A.I = I.A = A and it is given by:

$$I = \begin{bmatrix} 1 & 0 \\ 0 & 1 \end{bmatrix}$$

A matrix A may be multiplied by a scaler λ, and this yields the matrix λA where each entry in A is multiplied by the scaler λ. That is, the entries in the matric λA are λa_{ij}.

14.2 Matrix Operations

More general sets of linear equations may be solved with $m \times n$ matrices (i.e., a matrix with m rows and n columns) or square $n \times n$ matrices. In this section, we extend the discussion on 2 × 2 matrices to consider operations on more general matrices.

The addition and subtraction of two matrices A and B are meaningful if and only if A and B have the same dimensions, i.e., they are both $m \times n$ matrices. In this case, A + B is defined by adding the corresponding entries:

$$(A + B)_{ij} = A_{ij} + B_{ij}$$

The additive identity matrix for the square $n \times n$ matrices is an $n \times n$ matrix whose entries are zero, i.e., $r_{ij} = 0$ for all i, j where $1 \leq i \leq n$ and $1 \leq j \leq n$.

The scalar multiplication of a matrix A by a scalar k is meaningful and the resulting matrix kA is given by:

$$(kA)_{ij} = ka_{ij}$$

The multiplication of two matrices A and B is meaningful if and only if the number of columns of A is equal to the number of rows of B, i.e., A is an $m \times n$ matrix and B is an $n \times p$ matrix and the resulting matrix AB is an $m \times p$ matrix (Fig. 14.2).

$$\begin{pmatrix} a_{11} & a_{12} & a_{13} & \cdots & a_{1n} \\ a_{21} & a_{22} & a_{23} & \cdots & a_{2n} \\ a_{31} & a_{32} & a_{33} & \cdots & a_{3n} \\ \cdots & \cdots & \cdots & \cdots & \cdots \\ \cdots & \cdots & \cdots & \cdots & \cdots \\ a_{m1} & a_{m2} & a_{m3} & \cdots & a_{mn} \end{pmatrix} \begin{pmatrix} b_{11} & b_{12} & \cdots & b_{1p} \\ b_{21} & b_{22} & \cdots & b_{2p} \\ b_{31} & b_{32} & \cdots & b_{3p} \\ \cdots & \cdots & \cdots & \cdots \\ \cdots & \cdots & \cdots & \cdots \\ \cdots & \cdots & \cdots & \cdots \\ b_{n1} & b_{n2} & \cdots & b_{np} \end{pmatrix} = \begin{pmatrix} c_{11} & c_{12} & \cdots & c_{1p} \\ c_{21} & c_{22} & \cdots & c_{2p} \\ c_{31} & c_{32} & \cdots & c_{3p} \\ \cdots & \cdots & \cdots & \cdots \\ \cdots & \cdots & \cdots & \cdots \\ c_{m1} & c_{m2} & \cdots & c_{mp} \end{pmatrix}$$

m rows, n columns *n rows, p columns* *m rows, p columns*

FIGURE 14.2
Multiplication of Two Matrices.

Let A = (a_{ij}) where i ranges from 1 to m and j ranges from 1 to n, and let B = (b_{jl}) where j ranges from 1 to n and l ranges from 1 to p. Then AB is given by (c_{il}) where i ranges from 1 to m and l ranges from 1 to p with c_{il} given by:

$$c_{il} = \sum_{k=1}^{n} a_{ik} b_{kl}.$$

That is, the entry (c_{il}) is given by multiplying the ith row in A by the lth column in B followed by a summation. Matrix multiplication is not commutative, i.e., AB ≠ BA.

The multiplicative identity matrix I is an $n \times n$ matrix and the entries are given by r_{ij} where $r_{ii} = 1$ and $r_{ij} = 0$ where $i \neq j$ (Fig. 14.3). A matrix that has non-zero entries only on the diagonal is termed a *diagonal matrix.* A triangular matrix is a square matrix in which all the entries above or below the main diagonal are zero. A matrix is an *upper triangular* matrix if all entries below the main diagonal are zero, and *lower triangular* if all of the entries above the main diagonal are zero. Upper triangular and lower triangular matrices form a subalgebra of the algebra of square matrices.

A key property of the identity matrix is that for all $n \times n$ matrices A we have:

$$AI = IA = A$$

$$\begin{pmatrix} 1 & 0 & 0 & \cdots & 0 \\ 0 & 1 & 0 & \cdots & 0 \\ 0 & 0 & 1 & \cdots & 0 \\ \cdots & \cdots & \cdots & \cdots & \cdots \\ \cdots & \cdots & \cdots & \cdots & \cdots \\ \cdots & \cdots & \cdots & \cdots & \cdots \\ 0 & 0 & 0 & \cdots & 1 \end{pmatrix}$$

FIGURE 14.3
Identity Matrix I_n.

The inverse of an $n \times n$ matrix A is a matrix A^{-1} such that:

$$A A^{-1} = A^{-1} A = I$$

The inverse A^{-1} exists if and only if the determinant of A is non-zero.

The *transpose* of a matrix $A = (a_{ij})$ involves changing the rows to columns and vice versa to form the transpose matrix A^T. The result of the operation is that the $m \times n$ matrix A is converted to the $n \times m$ matrix A^T (Fig. 14.4). It is defined by:

$$(A^T)_{ij} = (a_{ji}) \quad 1 \leq j \leq n \text{ and } 1 \leq i \leq m$$

A matrix is *symmetric* if it is equal to its transpose, i.e., $A = A^T$.

14.3 Determinants

The determinant is a function defined on square matrices and its value is a scalar. A key property of determinants is that a matrix is invertible if and only if its determinant is non-zero. The determinant of a 2×2 matrix is given by:

$$\begin{vmatrix} a & b \\ c & d \end{vmatrix} = ad - bc$$

$$\begin{pmatrix} a_{11} & a_{12} & a_{13} & \cdots & a_{1n} \\ a_{21} & a_{22} & a_{23} & \cdots & a_{2n} \\ a_{31} & a_{32} & a_{33} & \cdots & a_{3n} \\ \cdots & \cdots & \cdots & \cdots & \cdots \\ \cdots & \cdots & \cdots & \cdots & \cdots \\ a_{m1} & a_{m2} & a_{m3} & \cdots & a_{mn} \end{pmatrix}^T = \begin{pmatrix} a_{11} & a_{21} & a_{31} & \cdots & a_{m1} \\ a_{12} & a_{22} & a_{32} & \cdots & a_{m2} \\ a_{13} & a_{23} & a_{33} & \cdots & a_{m3} \\ \cdots & \cdots & \cdots & \cdots & \cdots \\ \cdots & \cdots & \cdots & \cdots & \cdots \\ \cdots & \cdots & \cdots & \cdots & \cdots \\ a_{1n} & a_{2n} & a_{3n} & \cdots & a_{mn} \end{pmatrix}$$

m rows, n columns *n rows, m columns*

FIGURE 14.4
Transpose of a Matrix.

$$\begin{pmatrix} a_{11} & a_{12} & \cdots & a_{1j} & \cdots & a_{1n} \\ a_{21} & a_{22} & \cdots & a_{2j} & \cdots & a_{2n} \\ a_{31} & a_{32} & \cdots & a_{3j} & \cdots & a_{3n} \\ \cdots & \cdots & \cdots & \cdots & \cdots & \cdots \\ a_{i1} & a_{i2} & \cdots & a_{ij} & \cdots & a_{3n} \\ \cdots & \cdots & \cdots & \cdots & \cdots & \cdots \\ a_{n1} & a_{n2} & \cdots & a_{mj} & \cdots & a_{nn} \end{pmatrix} =$$

i,j minor of A

FIGURE 14.5
Determining the (i, j) Minor of A.

The determinant of a 3×3 matrix is given by:

$$\begin{vmatrix} a & b & c \\ d & e & f \\ g & h & i \end{vmatrix} = aei + bfg + cdh - afh - bdi - ceg$$

Cofactors

Let A be an $n \times n$ matrix. For $1 \leq i, j \leq n$, the (i, j) *minor* of A is defined to be the $(n - 1) \times (n - 1)$ matrix obtained by deleting the ith row and jth column of A (Fig. 14.5).

The shaded row is the ith row and the shaded column is the jth column. These are both deleted from A to form the (i, j) minor of A, and this is $(n - 1) \times (n - 1)$ matrix.

The (i, j) *cofactor* of A is defined to be $(-1)^{i+j}$ times the determinant of the (i, j) minor. The (i, j) cofactor of A is denoted by $K_{ij}(A)$.

The cofactor matrix *Cof* A is formed in this way where the (i, j)th element in the cofactor matrix is the (i, j) cofactor of A.

Definition of Determinant

The determinant of a matrix is defined as:

$$\det A = \sum_{i=1}^{n} A_{ij} K_{ij}$$

Another words, the determinant of A is determined by taking any row of A and multiplying each element by the corresponding cofactor and adding results. The determinant of the product of two matrices is the product of their determinants.

$$\det(AB) = \det A \times \det B$$

Definition

The *adjugate* of A is the $n \times n$ matrix $Adj(A)$ whose (i, j) entry is the (j, i) cofactor $K_{ji}(A)$ of A. That is, the adjugate of A is the transpose of the cofactor matrix of A.

Inverse of A

The inverse of A is determined from the determinant of A and the adjugate of A.

INVERSE OF MATRIX

$$A^{-1} = \frac{1}{\det A}\text{Adj } A = \frac{1}{\det A}(\text{Cof } A)^T$$

A matric is invertible if and only if its determinant is non-zero, i.e., A is invertible if and only if $\det(A) \neq 0$.

Cramer's Rule

Cramer's rule is a theorem that expresses the solution to a system of linear equations with several unknowns using the determinant of a matrix. There is a unique solution if the determinant of the matrix is non-zero.

For a system of linear equations of the $Ax = v$ where x and v are n-dimensional column vectors, then if $\det A \neq 0$ then the unique solution for each x_i is:

CRAMER'S RULE

$$x_i = \frac{\det U_i}{\det A}$$

where U_i is the matrix obtained from A by replacing the ith column of A by the v-column.

Characteristic Equation

For every $n \times n$ matrix A there is a polynomial equation of degree n satisfied by A. The *characteristic polynomial* of A is a polynomial in x of degree n. It is given by:

CHARACTERISTIC POLYNOMIAL

$$cA(x) = \det(xI - A).$$

Cayley-Hamilton Theorem

Every matrix A satisfies its characteristic polynomial, i.e., $p(A) = 0$ where $p(x)$ is the characteristic polynomial of A.

14.4 Eigen Vectors and Values

A number λ is an eigen value of an $n \times n$ matrix A if there is a non-zero vector v such that the following equation holds:

EIGEN VALUES

$$Av = \lambda v$$

The vector v is termed an eigen vector and the equation is equivalent to:

$$(A - \lambda I)v = 0$$

This means that $(A - \lambda I)$ is a zero divisor and hence it is not an invertible matrix. Therefore:

$$\det(A - \lambda I) = 0$$

The polynomial function $p(\lambda) = \det(A - \lambda I)$ is called the characteristic polynomial of A, and it is of degree n. The characteristic equation is $p(\lambda) = 0$ and as the polynomial is of degree n there are at most n roots of the characteristic equation, and so there at most n eigen values.

The *Cayley-Hamilton theorem* states that every matrix satisfies its characteristic equation, i.e., the application of the characteristic polynomial to the matrix A yields the zero matrix.

$$p(A) = 0$$

14.5 Gaussian Elimination

Gaussian elimination with backward substitution is an important method used in solving a set of linear equations. A matrix is used to represent the set of linear equations, and Gaussian elimination reduces the matrix to a *triangular* or *reduced form*, which may then be solved by backward substitution.

The set of n linear equations (E_1 to E_n) below is solved by manipulating the equations to reduce the matrix to a triangular form. This reduced form is easier to solve and it provides exactly the same solution as the original set of equations. The set of equations is defined as:

$$\begin{array}{ll} E_1: & a_{11}x_1 + a_{12}x_2 + \ldots + a_{1n}x_n = b_1 \\ E_2: & a_{21}x_1 + a_{22}x_2 + \ldots + a_{2n}x_n = b_2 \\ \vdots & \quad\vdots \qquad\quad \vdots \qquad\qquad\qquad \vdots \\ E_n: & a_{n1}x_1 + a_{n2}x_2 + \ldots + a_{nn}x_n = b_n \end{array}$$

There are three operations permitted on the equations to transform the linear system into a reduced form, which are as follows:

a. Any equation may be multiplied by a non-zero constant.
b. An equation E_i may be multiplied by a constant and added to another equation E_j, with the resulting equation replacing E_j.
c. Equations E_i and E_j may be transposed with E_j replacing E_i and vice versa.

We illustrate the method of Gaussian elimination for solving a set of linear equations by considering an example taken from [BuF:89]. The solution to a set of linear equations with four unknowns using Gaussian elimination involves:

$$\begin{array}{ll} E_1: & x_1 + x_2 \qquad\qquad + 3x_4 = 4 \\ E_2: & 2x_1 + x_2 - x_3 + x_4 = 1 \\ E_3: & 3x_1 - x_2 - x_3 + 2x_4 = -3 \\ E_4: & -x_1 + 2x_2 + 3x_3 - x_4 = 4 \end{array}$$

First, the unknown x_1 is eliminated from E_2, E_3 and E_4 and this is done by replacing E_2 with $E_2 - 2E_1$; replacing E_3 with $E_3 - 3E_1$; and replacing E_4 with $E_4 + E_1$. The resulting system is:

$$
\begin{aligned}
E_1&: x_1 + x_2 \qquad\quad + 3x_4 = 4 \\
E_2&: \quad - x_2 - x_3 - 5x_4 = -7 \\
E_3&: \quad - 4x_2 - x_3 - 7x_4 = -15 \\
E_4&: \quad 3x_2 + 3x_3 + 2x_4 = 8
\end{aligned}
$$

The next step is then to eliminate x_2 from E_3 and E_4. This is done by replacing E_3 with $E_3 - 4E_2$ and replacing E_4 with $E_4 + 3E_2$. The resulting system is now in triangular form, and the unknown variables may be solved easily by backward substitution.

$$
\begin{aligned}
E_1&: x_1 + x_2 \qquad\quad + 3x_4 = 4 \\
E_2&: \quad - x_2 - x_3 - 5x_4 = -7 \\
E_3&: \quad 3x_3 + 13x_4 = 13 \\
E_4&: \quad - 13x_4 = -13
\end{aligned}
$$

We then determine x_1, x_2, x_3 and x_4 by backward substitution, and we see that $x_4 = 1$, $x_3 = 0$, $x_2 = 2$ and $x_1 = -1$ (the answer can be checked by replacing the values in the original equations).

The usual approach to Gaussian elimination is to do it with an augmented matrix. That is, the set of equations is an $n \times n$ matrix and it is augmented by the column vector to form the augmented $n \times n + 1$ matrix. Gaussian elimination is then applied to the matrix to put it into a triangular form, and it is then easy to solve the unknowns.

The other approach to solving a set of linear equation is to employ Cramer's rule, which was discussed in section 14.4.

14.6 Applications of Matrices

We discuss operations research in Chapter 15, and note the applications of matrices to the field of linear programming. Various situations in business and economics may be represented using matrices, where the solution is found by solving systems of linear equations.

The organization structure of many organizations today is a matrix rather than the traditional functional structure with single managerial responsibility for a functional area in the organization. A *matrix organization* is one in which there is dual or multiple managerial accountability and responsibility, and there are generally two chains of command with the first being along traditional functional lines and the second being along project or client (Fig. 14.6).

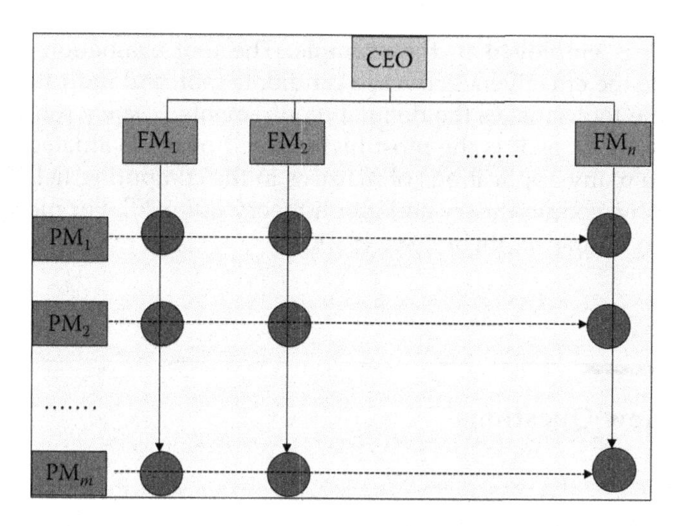

FIGURE 14.6
Matrix Organization.

That is, each person in a matrix organization is essentially reporting to two managers: their line manager for the functional area that they work in, and the project manager for the project that they are assigned to. The project is a temporary activity and so this reporting line ceases on project closure, whereas the functional line is more permanent (subject to the regular changes following company reorganizations as part of continuous improvement).

Another application of matrices in business is that of a decision matrix that allows an organization to make decisions objectively based on criteria. For example, the tool evaluation matrix in Table 14.1 lists all of the requirements vertically that the tool is to satisfy, and the candidate tools that are to be evaluated and rated against each requirement are listed horizontally. Various rating schemes may be employed, and a simple numeric

TABLE 14.1
Tool Evaluation Matrix

	Tool 1	**Tool 2**	...	**Tool** ***k***
Requirement 1	8	7		9
Requirement 2	4	6		8
....				
....				
Requirement *n*	3	6		8
Total	35	38	...	45

mechanism is employed in the example. The tool evaluation criteria are used to rate the effectiveness of each candidate tool, and indicate the extent to which the tool satisfies the defined requirements. The chosen tool in this example is Tool k as it is the most highly rated of the evaluated tools.

There are many applications of matrices in the computing field including cryptography, coding theory and graph theory [ORg:20]. For more detailed information on matrix theory, see [SS:13].

14.7 Review Questions

1. Show how 2 × 2 matrices may be added and multiplied.
2. What is the additive identity for 2 × 2 matrices? The multiplicative identity?
3. What is the determinant of a 2 × 2 matrix?
4. Show that a 2 × 2 matrix is invertible if its determinant is non-zero.
5. Solve the simultaneous equations using a 2 × 2 matrix.

$$\begin{aligned} x + 2y &= -1 \\ 4x - 3y &= 18 \end{aligned}$$

6. Describe matrix algebra for general matrices.
7. What is Cramer's rule?
8. Find the eigen values and corresponding eigen vectors of the matrix

$$\begin{bmatrix} 2 & 1 \\ 1 & 2 \end{bmatrix}$$

9. Find the characteristic equation of the matrix in Q9 and verify that the matrix satisfies the characteristic equation.
10. Show how Gaussian elimination may be used to solve a set of linear equations.

14.8 Summary

A matrix is a rectangular array of numbers that consists of horizontal rows and vertical columns. A matrix with m rows and n columns is termed an $m \times n$ matrix, and m and n are its dimensions. A matrix with an equal number of rows and columns (e.g., n rows and n columns) is termed a square matrix.

Matrices arose in practice as a means of solving a set of linear equations, and one of the earliest examples of their use is from a Chinese text dating between 300 BC and 200 AD.

Matrices of the same dimensions may be added, subtracted and multiplied by a scalar. Two matrices A and B may be multiplied provided that the number of columns of A equals the number of rows in B).

A square matrix has an inverse provided that its determinant is non-zero. The inverse of a matrix involves determining its determinant, constructing the cofactor matrix and transposing the cofactor matrix.

The solution to a set of linear equations may be determined by Gaussian elimination to convert the matrix to upper triangular form and then employing backward substitution.

Eigen values and eigen vectors lead to the characteristic polynomial and every matrix satisfies its characteristic polynomial.

15

Operations Research

Operations research is a multi-disciplinary field that is concerned with the application of mathematical and analytic techniques to assist in decision-making. It employs techniques such as mathematical modelling, statistical analysis and mathematical optimization as part of its goal to achieve optimal (or near optimal) solutions to complex decision-making problems. The modern field of operations research includes other disciplines such as computer science, industrial engineering, business practices in manufacturing and service companies, supply chain management and operations management.

Pascal did early work on operations research in the 17th century. He attempted to apply early work on probability theory to solve complex decision-making problems. Babbage's work on the transportation and sorting of mail contributed to the introduction of the uniform "Penny Post" in England in the 19th century. The origins of the operations research field are from the work of military planners during the First World War, and the field took off during the Second World War as it was seen as a scientific approach to decision-making using quantitative techniques. It was applied to strategic and tactical problems in military operations, where the goal was to find the most effective utilization of limited military resources through the use of quantitative techniques. It played an important role in solving practical military problems such as determining the appropriate convoy size in the submarine war in the Atlantic.

Numerous peacetime applications of operations research emerged after the Second World War, where operations research and management science was applied to many industries and occupations. It was applied to procurement, training, logistics and infrastructure in addition to its use in operations. The remarkable progress in the computing field means that operations research can now solve problems with hundreds of thousands of variables and constraints.

Operations research (OR) is the study of mathematical models for complex organizational systems, where a *model* is a mathematical description of a system that accounts for its known and inferred properties, and it may be used for the further study of its properties, and a *system* is a functionally related collection of elements such as a network of computer hardware and software. *Optimization* is a branch of operations research that uses mathematical techniques to derive values from system variables that will optimize system performance.

DOI: 10.1201/9781003308140-15

Operations research has been applied to a wide variety of problems including network optimization problems, designing the layouts of the components on a computer chip, supply chain management, critical path analysis during project planning to identify key project activities that effect the project timeline, scheduling project tasks and personnel and so on. Several of the models used in operations research are described in Table 15.1.

Mathematical programming involves defining a mathematical model for the problem, and using the model to find the optimal solution. A mathematical model consists of variables, constraints the objective function to be maximized or minimized, and the relevant parameters and data. The general form is:

$$\text{Min or Max} f(x_1, x_2, \ldots.. x_n) \quad \text{(Objective function)}$$
$$g(x_1, x_2, \ldots.. x_n) \leq (\text{or} >, \geq, = <)\, b_i \quad \text{(Constraints)}$$
$$x \in X$$
$$f, g \text{ are linear and } X \text{ is continuous (for linear programming } LP)$$

A feasible solution is an assignment of values to the variables such that the constraints are satisfied. An optimal solution is one whose objective function exceeds all other feasible solutions (for maximization optimization). We now discuss linear programming in more detail.

15.1 Linear Programming

Linear programming (LP) is a mathematical model for determining the best possible outcome (e.g., maximizing profit or minimizing cost) of a particular problem. The problem is subject to various constraints such as resources or costs, and the constraints are expressed as a set of linear equations and linear inequalities. The best possible outcome is expressed as a linear equation. For example, the goal may be to determine the number of products that should be made to maximize profit subject to the constraint of limited available resources.

The constraints for the problem are linear, and they specify regions that are bounded by straight lines. The solution will lie somewhere within the regions specified, and a feasible region is a region where all of the linear inequalities are satisfied. Once the feasible region is found the challenge is then to find where the best possible outcome may be maximized in the feasible region, and this will generally be in one of the corners of the region. The steps involved in developing a linear programming model include:

TABLE 15.1

Models Used in Operations Research

Model	Description
Linear programming	These problems aim to find the best possible outcome (where this is expressed as a linear function) such as to maximize profit or minimize cost subject to various linear constraints. The function and constraints are linear functions of the decision variables, and modern software can solve problems containing millions of variables and thousands of constraints.
Network flow programming	This is a special case of the general linear programming problem, and includes problems such as the transportation problem, the shortest path problem, the maximum flow problem and the minimum cost problem. There are very efficient algorithms available for these (faster and more efficient than standard linear programming).
Integer programming	This is a special case of the general linear programming problem, where the variables are required to take on discrete values.
Non-linear programming	The function and constraints are non-linear, and these are much more difficult to solve than linear programming. Many real-world applications require a non-linear model, and often the solution is approximated with a linear model.
Dynamic programming	A dynamic programming (DP) model describes a process in terms of states, decisions, transitions and a return. The process begins in some initial state, a decision is made leading to a transition to a new state, the process continues through a sequence of states until final state is reached. The problem is to find a sequence that maximizes the total return.
Stochastic processes	A stochastic process models practical situations where the attributes of a system randomly change over time (e.g., number of customers at an ATM machine, the share price, etc.), and the state is a snapshot of the system at a point in time that describes its attributes. Events occur that change the state of the system.
Markov chains	A stochastic process that can be observed at regular intervals (such as every day or every week) can be described by a matrix, which gives the probabilities of moving to each state from every other state in one time interval. The process is called a Markov chain when this matrix is unchanging over time.
Markov processes	A Markov process is a continuous-time stochastic process in which the duration of all state changing activities is exponentially distributed.
Game theory	Game theory is the study of mathematical models of strategic interaction among rational decision-makers. It is concerned with logical decision-making by humans, animals and computers.

(*Continued*)

TABLE 15.1 (Continued)

Models Used in Operations Research

Model	Description
Simulation	Simulation is a general technique for estimating statistical measures of complex systems.
Time series and forecasting	A time series is a sequence of observations of a periodic random variable, and is generally used as input to an OR decision model.
Inventory theory	Aims to optimize inventory management, e.g., determining when and how much inventory should be ordered.
Reliability theory	Aims to model the reliability of a system from probability theory.

STEPS IN DEVELOPING A LINEAR PROGRAMMING PROBLEM

Formulation of the problem
Solution of the problem
Interpretation of the solution

Linear programming models seek to select the most appropriate solution from the alternatives that are available subject to the specified constraints. Often, graphical techniques are employed to sketch the problem and the regions corresponding to the constraints.

The graphical techniques identify the feasible region where the solution lies, and then the maximization or minimization function is employed within the region to search for the optimal value. The optimal solution will generally lie at one or more of the corner points of the feasible region.

15.1.1 Linear Programming Example

We consider an example in an industrial setting where a company is trying to decide how many of each product it should make to maximize profits subject to the constraint of limited resources.

Square Deal Furniture produces two products namely chairs and tables, and it needs to decide on how many of each to make each month in order to maximize profits. The amount of time to make tables and chairs and the maximum hours available to make each product as well as the profit contribution of each product is summarized in Table 15.2. There are additional constraints that need to be specified:

TABLE 15.2

Square Deal Furniture

	Table	Chair	Hours Available
Carpentry	3 hours	4 hours	2400
Painting	2 hours	1 hour	1000
Profit contribution	7 Euros	5 Euros	

- At least 100 tables must be made.
- The maximum number of chairs to be made is 450.

We use variables to represent tables and chairs and formulate an objective function to maximize profits subject to the constraints.

$$T = \text{Number of Tables to make}$$
$$C = \text{Number of Chairs to make}$$

The objective function (to maximize profits) is then specified as

$$\text{Maximise the value of } 5C + 7T$$

The constraints on the hours available for carpentry and painting may be specified as:

$$3T + 4C \leq 2400 \quad \text{(carpentry time available)}$$
$$2T + C \leq 1000 \quad \text{(painting time available)}$$

The constraints that at least 100 tables must be made and the maximum number of chairs to be made is 450 may be specified as:

$$T \geq 100 \quad \text{(number of tables)}$$
$$C \leq 450 \quad \text{(number of chairs)}$$

Finally, it is not possible to produce a negative number of chairs or tables and this is specified as:

$$T \geq 0 \quad \text{(non-negative)}$$
$$C \geq 0 \quad \text{(non-negative)}$$

The model is summarized as:

Max 5C + 7T	(Maximization problem)
$3T + 4C \le 2400$	(carpentry time available)
$2T + C \le 1000$	(painting time available)
$T \ge 100$	(number of tables)
$C \le 450$	(number of chairs)
$T \ge 0$	(non-negative)
$C \ge 0$	(non-negative)

We graph the LP model and then use the graph to find a feasible region for where the solution lies, and we then identify the optimal solution. The feasible region is an area where all of the constraints for the problem are satisfied, and the optimal solution lies at one or more of the corner points of the feasible region.

First, for the constraints on the hours available for painting and carpentry $3T + 4C \le 2400$ and $2T + C \le 1000$, respectively, we draw the two lines $3T + 4C = 2400$ and $2T + C = 1000$. We choose two points on each line and then join both points to form the line, and we choose the intercepts of both lines as the points.

For the line $3T + 4C = 2400$ when T is 0 then C is 600 and when C is 0 then T is 800. Therefore, the points (0, 600) and (800, 0) are on the line $3T + 4C = 2400$. For the line $2T + C = 1000$ when $T = 0$ then $C = 1000$ and when $C = 0$ then $T = 500$. Therefore, the points (0, 1000) and (500, 0) are on the line $2T + C = 1000$.

Fig. 15.1 is the first step in developing a graphical solution and we note that for the first two constraints $3T + 4C \le 2400$ and $2T + C \le 1000$ that the solution lies somewhere in the area bounded by the lines $3T + 4C = 2400$, $2T + C = 1000$, the T axis and the C axis.

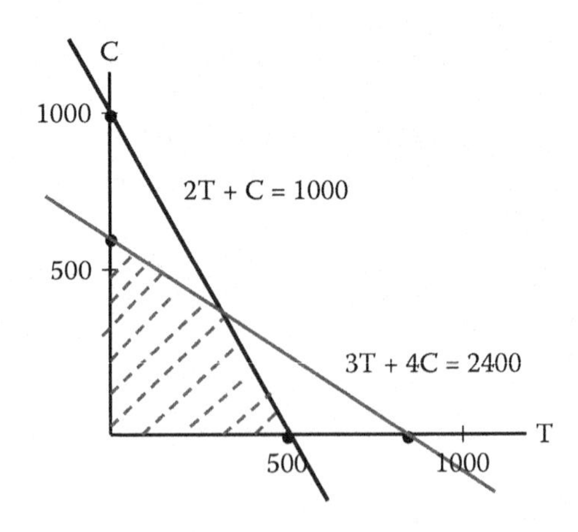

FIGURE 15.1
Linear Programming – Developing a Graphical Solution.

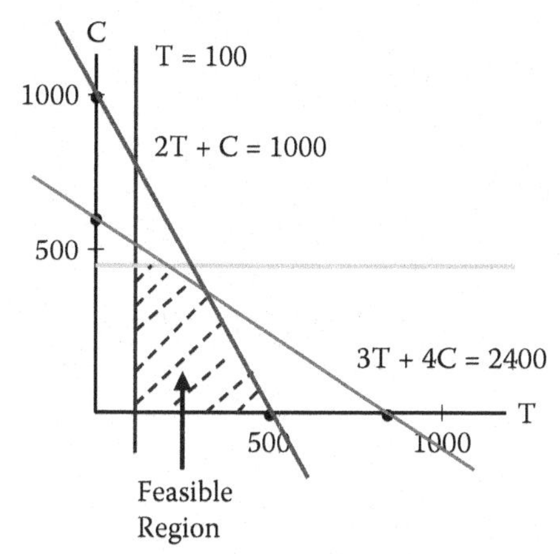

FIGURE 15.2
Feasible Region of Solution.

Next, we add the remaining constraints (T ≥ 100, C ≤ 450, T ≥ 0, C ≥ 0) to the graph, and this has the effect of reducing the size of the feasible region in Fig. 15.1 (which placed no restrictions on T and C). The feasible region can be clearly seen in Fig. 15.2, and the final step is to find the optimal solution in the feasible region that maximizes the profit function 5C +7T.

Fig. 15.3 shows how we find the optimal solution by drawing the line 7T + 5C = k within the feasible region, and this forms a family of parallel lines where the slope of the line is $-^5/_7$ and k represents the profit. Each point in the feasible region is on one of the lines in the family, and to determine the equation of that line we just input the point into the equation 7T + 5C = k. For example, the point (200, 0) is in the feasible region and it satisfies the equation 7T + 5C = 7 ∗ 200 + 5 ∗ 0 = 1400.

We seek to maximize k and it is clear that the largest value of k that is maximal is at one of the corner points of the feasible region. This is the point

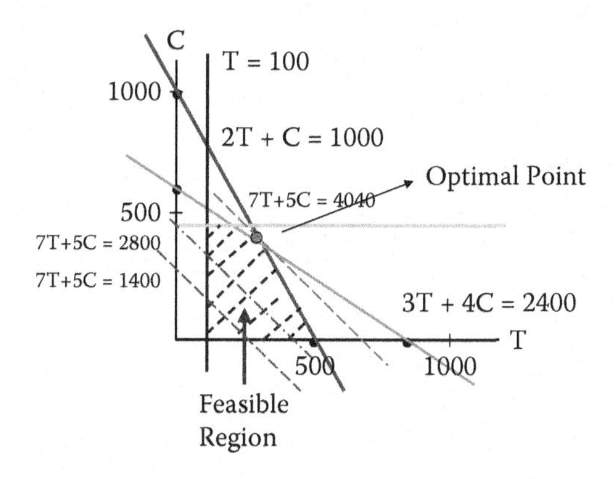

FIGURE 15.3
Optimal Solution.

of intersection of the lines 2T + C = 1000 and 3T + 4C = 2400, and we solve for T and C to get T = 320 and C = 360. This means that the equation of the line containing the optimal point is 7T + 5C = 2240 + 1800 = 4040. That is, its equation is 7T + 5C = 4040 and so the maximum profit is €4040.

15.1.2 General Formulation of LP Problem

The more general formulation of the linear programming problem can be stated as follows. Find variables $x_1, x_2, \ldots, x_n$ to optimize (i.e., maximize or minimize) the linear function:

$$Z = c_1x_1 + c_2x_2 + \ldots + c_nx_n$$

where the problem is subject to the linear constraints:

$$\begin{array}{llllll} a_{11}x_1 & + a_{12}x_{12} & + \ldots., & a_{1j}x_j & + \ldots. + & a_{1n}x_n \quad (\leq \; = \; \geq) b_1 \\ a_{21}x_1 & + a_{22}x_{12} & + \ldots., & a_{2j}x_j & + \ldots. + & a_{2n}x_n \quad (\leq \; = \; \geq) b_2 \\ \vdots & \vdots & & \vdots & & \vdots \qquad\qquad \vdots \\ a_{m1}x_1 & + a_{m2}x_{12} & + \ldots., & a_{mj}x_j & + \ldots. + & a_{mn}x_n \quad (\leq \; = \; \geq) b_m \end{array}$$

and to non-negative constraints on the variables such as:

$$x_1, x_2, \ldots, x_n \geq 0$$

where a_{ij}, b_j, and c_i are constants and x_i are variables.

The variables $x_1, x_2, \ldots., x_n$ whose values are to be determined are called decision variables.

The coefficients $c_i, c_2, \ldots, c_n$ are called cost (profit) coefficients.

The constraints $b_1, b_2, \ldots, b_m$ are called the requirements.

A set of real values $(x_1, x_2, \ldots, x_n)$ which satisfies the constraints (including the non-negative constraints) is said to be a *feasible* solution.

A set of real values $(x_1, x_2, \ldots, x_n)$ which satisfies the constraints (including the non-negative constraints) and optimizes the objective function is said to be an *optimal* solution.

There may be no solution, a unique solution or multiple solutions.

The constraints may also be formulated in terms of matrices as follows:

$$\begin{pmatrix} a_{11} & a_{12} & a_{13} & \cdots & a_{1n} \\ a_{21} & a_{22} & a_{33} & \cdots & a_{2n} \\ a_{31} & a_{32} & a_{33} & \cdots & a_{3n} \\ \cdots & \cdots & \cdots & \cdots & \cdots \\ \cdots & \cdots & \cdots & \cdots & \cdots \\ a_{m1} & a_{m2} & a_{m3} & \cdots & a_{mn} \end{pmatrix} \begin{pmatrix} x_1 \\ x_2 \\ x_3 \\ \cdots \\ \cdots \\ x_n \end{pmatrix} (\leq \; = \; \geq) \begin{pmatrix} b_1 \\ b_2 \\ b_3 \\ \cdots \\ \cdots \\ b_m \end{pmatrix}$$

This may also be written as AX (≤ = ≥) B and the optimization function may be written as Z = CX where C = $(c_1, c_2, \ldots, c_n)$ and X = $(x_1, x_2, \ldots, x_n)^T$ and X ≥ 0.

Linear programming problems may be solved by graphical techniques (when there are a small number of variables), or analytic techniques using matrices. There are techniques that may be employed to find the solution of the LP problem that are similar to finding the solution to a set of simultaneous equations using Gaussian elimination. Matrices were discussed in Chapter 14.

15.2 Cost Volume Profit Analysis

A key concern in business is profitability, and management needs to decide on the volume of products to produce, as well as the costs and total revenue. Cost volume profit analysis (CVPA) is a useful tool in the analysis of the relationship between the costs, volume, revenue and profitability of the products produced. The relationship between revenue and costs at different levels of output can be displayed graphically, with revenue behaviour and cost behaviour shown graphically.

The *breakeven point* (BP) is where the total revenue is equal to the total costs, and breakeven analysis is concerned with identifying the volume of products that need to be produced to break even.

Example (CVPA)

Pilar is planning to set up a business that makes pottery cups, and she has been offered a workshop to rent for €800 per month. She estimates that she needs to spend €10 on the materials to make each pottery cup, and that she can sell each cup for €25. She estimates that if she is very productive she can make 500 pottery cups in a month.

Prepare a table that shows the profit or loss that Pilar makes based on the sales of 0, 100, 200, 300, 400 and 500 pottery cups.

Solution (CVPA)

Each entry in the table consists of the revenue for the volume sold, the material costs per volume of the pottery cups, the fixed cost of renting the workshop per month, the total cost per month and the net income per month (Table 15.3).

The total sale (revenue) is determined from the volume of sales multiplied by the unit sales price of a pottery cup (€25). There are two types of cost that may be incurred namely *fixed costs* and *variable costs*.

Fixed costs are incurred irrespective of the volume of items produced, and so the cost of renting of the workshop is a fixed cost. Variable costs are

TABLE.15.3

Projected Profit or Loss per Volume

#Cups	0	100	200	300	400	500
Revenue (Sales)	0	2500	5000	7500	10,000	12,500
Materials (Var Cost)	0	1000	2000	3000	4000	5000
Workshop (Fix Cost)	800	800	800	800	800	800
Total cost	800	1800	2800	3800	4800	5800
Net income	−€800	€700	€2200	€3700	€5200	€6700

constant per unit of output, and include the direct material and labour costs, and so the total variable cost increases as the volume increases. That is, the total variable cost is directly related to the volume of items produced, and Table 15.4 summarizes the revenue and costs.

We may represent the relationships between volume, cost and revenue graphically, and use it to see the relationship between revenue and costs at various levels of output (Fig. 15.4). We may then use the graph to determine the breakeven point for when the total revenue is equal to the total cost.

We may also determine the breakeven point algebraically by letting X represent the volume of cups produced for breakeven. Then breakeven is when the total revenue is equal to the total cost. That is,

$$\begin{aligned} SP * X &= FC + VC * X \\ \Rightarrow 25X &= 800 + 10X \\ \Rightarrow 15X &= 800 \\ \Rightarrow X = 800/15 &= 53.3 \end{aligned}$$

The breakeven amount in revenue is 25 ∗ 53.3 = €1333.32.

Next, we present an alternate way of calculating the breakeven point in terms of *contribution margin* and sales. The contribution margin is the monetary value that each extra unit of sales makes towards profitability, and it is given by the selling price per unit minus the variable cost per unit.

TABLE.15.4

Revenue and Costs

Item	Amount
Total revenue (TR)	SP ∗ X
Total variable cost (TVC)	VC ∗ X
Fixed cost (FC)	FC
Total cost (TC)	FC + TVC = FC + (VC ∗ X)
Net income (profit)	TR – TC (SP ∗ X) – FC – (VC ∗ X)

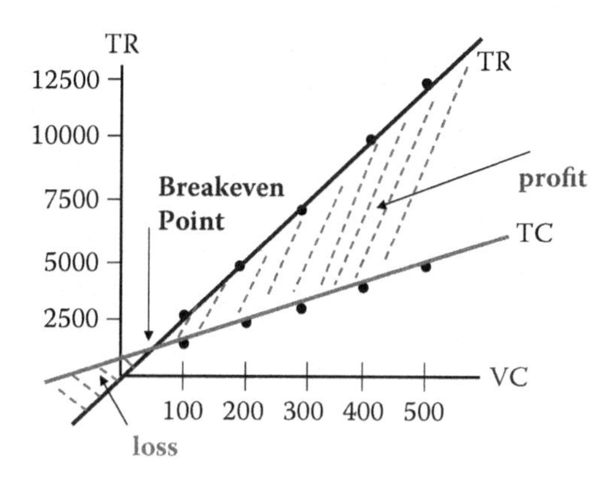

FIGURE 15.4
Breakeven Point.

Each additional pottery cup sold increases the revenue by €25 whereas the increase in costs is just €10 (the materials required). This, the contribution margin per unit is the selling price minus the variable cost per unit (i.e., SP – VC = €15), so the total contribution margin is the total volume of units sold multiplied by the contribution margin per unit (i.e., X ∗ (SP – VC) = 15X).

The breakeven volume is reached when the total contribution margin covers the fixed cost (i.e., the cost of renting the workshop which is €800 is covered by the total contribution). That is, the breakeven volume is reached when:

$$X * (SP - VC) = FC$$

$$X = \frac{FC}{SP - VC}$$

$$X = \frac{800}{(25 - 10)} = \frac{800}{15} = 53.3$$

Example

Suppose that the rent of the workshop is increased to €1200 per month, and that it also costs €12 (more than expected) to make each cup, and that she can sell each cup for just €20. What is her new breakeven volume and revenue?

Solution

The breakeven volume is reached when:

$$
\begin{aligned}
X * (SP - VC) &= FC \\
X * (20 - 12) &= 1200 \\
8X &= 1200 \\
X &= 150
\end{aligned}
$$

Further, the breakeven revenue is:

$$
\begin{aligned}
&X * SP \\
&= 150 * 20 \\
&= €3000.
\end{aligned}
$$

15.3 Game Theory

Game theory is the study of mathematical models of strategic interaction among rational decision-makers, and it was originally applied to zero-sum games where the gains or losses of each participant are exactly balanced by those of the other participants.

Modern game theory emerged as a field following John von Neumann's 1928 paper on the theory of games of strategy [VN:28]. The Rand corporation investigated possible applications of game theory to global nuclear strategy in the 1950s. Game theory has been applied to many areas including economics, biology and the social sciences. It is an important tool in situations where a participant's best outcome depends on what other participants do, and their best outcome depends on what he/she does. We illustrate the idea of game theory through the following example.

Example (Game Theory)

We consider an example of two television networks that are competing for an audience of 100 million on the 8 to 9 pm nighttime television slot. The networks announce their schedule ahead of time, but do not know of the other network's decision until the program begins. A certain number of viewers will watch Network 1, with the remainder watching Network 2. Market research has been carried out to show the expected number of viewers for each network based on what will be shown by the networks (Table 15.5).

Problem to Solve (Viewing Figures)

Determine the best strategy that both networks should employ to maximize their viewing figures.

Table 15.5 shows the number of viewers of Network 1 for each type of film that also depends on the type of film that is being shown by Network 2. For

TABLE 15.5
Network Viewing Figures

	Network 2		
Network 1	**Western**	**Soap Opera**	**Comedy**
Western	35	15	60
Soap opera	45	58	50
Comedy	38	14	70

example, if Network 1 is showing a western while Network 2 is showing a comedy then Network 1 will have 60 million viewers, and Network 2 will have 100 – 60 = 40 million viewers. However, if Network 2 was showing a soap opera then the viewing figures for Network 1 are 15 million, and 100 – 15 = 85 million will be tuned into Network 2.

Solution (Game Theory)

Network 1 is a *row player* whereas Network 2 is a *column player*, and the table above is termed a *payoff matrix*. This is a *constant-sum game* (as the outcome for both players always adds up to a constant 100 million).

The approach to finding the appropriate strategy for Network 1 is to examine each option. If Network 1 decides to show a western then it can get as many as 60 million viewers if Network 2 shows a comedy, or as few as 15 million viewers if Network 2 shows a soap opera. That is, this choice cannot guarantee more than 15 million viewers. Similarly, if Network 1 shows a soap opera it may get as many as 58 million viewers if Network 2 shows a soap opera as well, or as few as 45 million viewers should Network 2 show a western. That is, this choice cannot guarantee more than 45 million viewers. Finally, if Network 1 shows a comedy it would get 70 million viewers if Network 2 is showing a comedy as well, or as few as 14 million viewers if Network 2 is showing a soap opera. That is, this option cannot guarantee more than 14 million viewers. Clearly, the best option for Network 1 would be to show a soap opera, as at least 45 million viewers would tune into Network 1 irrespective of what Network 2 does.

Another words, the strategy for Network 1 (being a row player) is to determine the row minimum of each row, and then to choose the row with the largest row minimum.

Similarly, the best strategy for Network 2 (being a column player) is to determine the column maximum of each column, and then to choose the column with the smallest column maximum. For Network 2 the best option is to show a western and so 45 million viewers will tune into Network 1 to watch the soap opera, and 55 million will tune into Network 2 to watch a western.

It is clear from the table that the two outcomes satisfy the following inequality:

$$\text{Max}_{(\text{rows})} \text{ (row minimum)} \leq \text{Min}_{(\text{cols})} \text{ (col maximum)}$$

This choice is simultaneously best for Network 1 and Network 2 (as max(row minimum) = min (col maximum)), and this is called a *saddle point*, and the common value of both sides of the equation is called the *value* of the game. An equilibrium point of the game is where there is a choice of strategies for both players where neither player can improve their outcome by changing their strategy, and a saddle point of a game is an example of an equilibrium point.

Example (Prisoner Dilemma)

The police arrest two people who they know have committed an armed robbery together. However, they lack sufficient evidence for a conviction for armed robbery, but they have sufficient evidence for a conviction of two years for the theft of the getaway car. The police make the following offer to each prisoner:

> *If you confess to your part in the robbery and implicate your partner and he does not confess, then you will go free and he will get ten years. If you both confess you will both get five years. If neither of you confess you will get each get two years for the theft of the car.*

Model the prisoners' situation as a game and determine the rational (best possible) outcome for each prisoner.

Solution (Prisoner's Dilemma)

There are four possible outcomes for each prisoner:

- Go Free (He confesses. Other does not)
- 2-year sentence (Neither confess)
- 5-year sentence (Both confess)
- 10 years (He does not confess. Other does)

Each prisoner has a choice of confessing or not, and Table 15.6 summarizes the various outcomes depending on the choices that the prisoners make. The first entry in each cell of the table is the outcome for prisoner 1 and the second entry is the outcome for prisoner 2. For example, the cell with the entries 10, 0 states that prisoner 1 is sentenced for 10 years and prisoner 2 goes free.

TABLE 15.6

Outcomes in Prisoners' Dilemma

		Prisoner 2	
		Confess	**Refuse Confess**
Prisoner 1	Confess	5, 5	0, 10
	Refuse Confess	10, 0	2, 2

It is clear from Table 12.6 that if both prisoners confess they both will receive a 5-year sentence; if neither confess then they will both receive a 2-year sentence; if prisoner 1 confesses and prisoner 2 does not then prisoner 1 goes free whereas prisoner 2 gets a 10-year sentence; and, finally, if prisoner 2 confesses and prisoner 1 does not then prisoner 2 goes free and prisoner 1 gets a 10-year sentence.

Each prisoner evaluates his two possible actions by looking at the outcomes in both columns, as this will show which action is better for each possible action of their partner. If prisoner 2 confesses then prisoner 1 gets a 5-year sentence if he confesses or a 10-year sentence if he does not confess. If prisoner 2 does not confess then prisoner 1 goes free if he confesses or 2 years if he does not confess. Therefore, prisoner 1 is better off confessing irrespective of the choice of prisoner 2. Similarly, prisoner 2 comes to exactly the same conclusion as prisoner 1, and so the best outcome for both prisoners is to confess to the crime, and both will go to prison for 5 years.

The paradox in the prisoner's dilemma is that two individuals acting in their own self-interest do not produce the optimal outcome. Both parties choose to protect themselves at the expense of the other, and as a result both find themselves in a worse state than if they had cooperated with each other in the decision-making process and received 2 years.

15.4 Queueing Theory

The term "queue" refers to waiting in line for a service, such as waiting in line at a bakery or a bank, and *queueing theory* is the mathematical study of waiting lines or queues. The origins of queueing theory are in work done by Erlang at the Copenhagen Telephone Exchange in the early 20th century where he modelled the number of telephone calls arriving as a Poisson process.

Queueing theory has been applied to many fields including telecommunications and traffic management. This section aims to give a flavour and a very short introduction to queueing theory, and it has been adapted from [Dei:90]. The interested reader may consult the many other texts available for more detailed information [Gro:08].

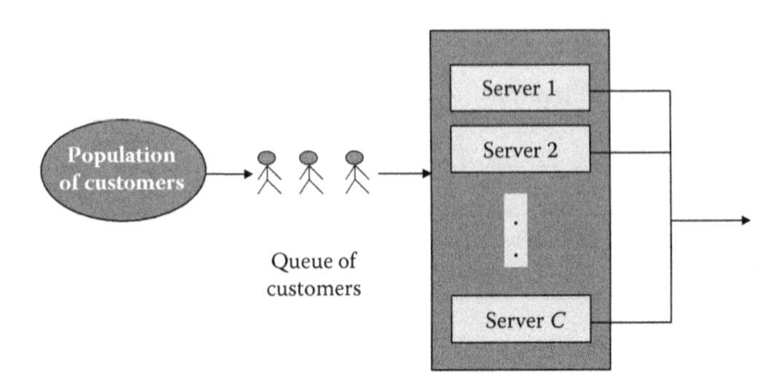

FIGURE 15.5
Basic Queueing System.

A supermarket may be used to illustrate the ideas of queueing theory, as it has a large population of customers some of whom may enter the supermarket and queueing system (the checkout queues). Customers will generally wait for a period of time in a queue before receiving service at the checkout, and they wait for a further period of time for the actual service to be carried out. Each service facility (the checkouts) contains identical servers, and each server is capable of providing the desired service to the customer (Fig. 15.5).

Clearly, if there are no waiting lines then immediate service is obtained. However, in general, there are significant costs associated with the provision of an immediate service, and so there is a need to balance cost with a certain amount of waiting.

Some queues are *bounded* (i.e., they can hold only a fixed number of customers), whereas others are *unbounded* and can grow as large as is required to hold all waiting customers. The customer source may be finite or infinite, and where the customer source is finite but very large it is often considered to be infinite.

Random variables (described by probability distribution functions) arise in queueing problems, and these include the random variable q, which represents the time that a customer spends in the queue waiting for service; the random variable s, which represents the amount of time that a customer spends in service; and the random variable w, which represents the total time that a customer spends in the queueing system (Fig. 15.6). Clearly,

$$w = q + s$$

It is assumed that the customers arrive at a queueing system one at a time at random times ($t_0 < t_1 < \ldots < t_n$) with the random variable $\tau_k = t_k - t_{k-1}$ representing the *interarrival times* (i.e., it measures the times between successive arrivals). It is assumed that these random variables are independent and identically distributed, and it is usually assumed that arrival form a Poisson arrival process.

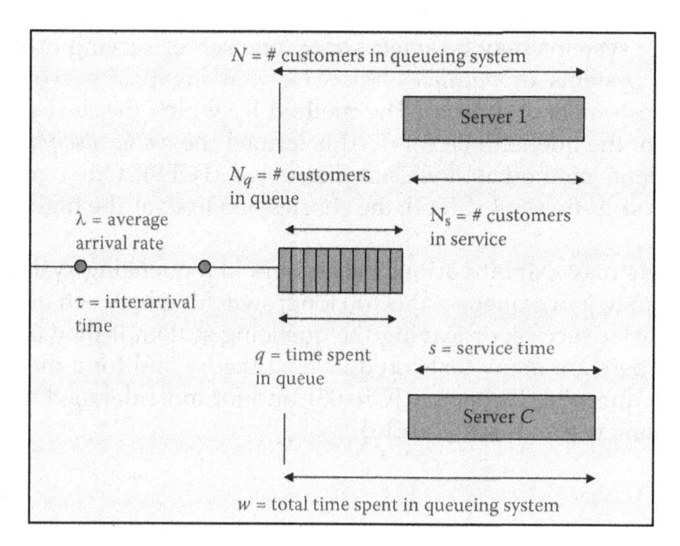

FIGURE 15.6
Sample Random Variables in Queueing Theory.

A Poisson arrival process is characterized by the fact that the interarrival times are distributed exponentially. That is:

$$P(\tau \leq t) = 1 - e^{-\lambda t}$$

Further, the probability that exactly n customers will arrive in any time interval of length t is given by:

$$\frac{e^{-\lambda t}(\lambda t)^n}{n!} \quad (\text{where } n = 0, 1, 2, \ldots)$$

where λ is a constant average arrival rate of customers per unit time, and the number of arrivals per unit time is Poisson distributed with mean λ.

Similarly, it is usual to assume in queueing theory that the service times are random with μ denoting the average service rate, and we let s_k denote the service time that the kth customer requires from the system. The distribution of service times is given by:

$$W_s(t) = P(s \leq t) = 1 - e^{-\mu t}$$

The capacity of the queues may be *infinite* (where every arriving customer is allowed to enter the queueing system no matter how many waiting customers are present), or finite (where arriving customers may wait only if there is still room in the queue).

Queueing systems may be *single server* (one server serving one customer at a time) systems or *multiple servers* (several identical servers that can service c customers at a time). The method by which the next customer is chosen from the queue to be serviced is termed the *queue discipline*, and the most common method is *first-come-first-served* (FCFS). Other methods include the last-in-first-out (LIFO); the shortest job first; or the highest priority job next.

Customers may exhibit various behaviours in a queueing system such as deciding not to join a queue if it is too long; switching between queues to try to obtain faster service; or leaving the queueing system if they have waited too long. There are many texts on queueing theory and for a more detailed account on queueing theory see [Gro:08], and for more detailed information on operations research see [Tah:16].

15.5 Review Questions

1. What is operations research?
2. Describe the models used in operations research.
3. What is linear programming and describe the steps in developing a linear programming model?
4. What is cost volume profit analysis?
5. Suppose the fixed costs are rent of £1,500 per month and that the cost of making each item is £20 and it may then be sold for £25. How many items must be sold to breakeven and what is the breakeven revenue?
6. What is game theory?
7. What is a zero-sum game?
8. What is queueing theory?

15.6 Summary

Operations research is a multi-disciplinary field concerned with the application of mathematical and analytic techniques to assist in decision-making. It employs mathematical modelling, statistical analysis and mathematical optimization to achieve optimal (or near optimal) solutions to complex decision-making problems. The modern field of operations research includes other disciplines such as computer science, industrial engineering, business practices in manufacturing and service companies, supply chain management and operations management.

Linear programming is a mathematical model for determining the best possible outcome such as maximizing profit or minimizing cost of a particular problem. The problem is subject to various constraints such as resources or costs, and the constraints are expressed as a set of linear equations and linear inequalities. The best possible outcome is expressed as a linear equation. For example, the goal may be to determine the number of products that should be made to maximize profit subject to the constraint of limited available resources.

CVPA is used in the analysis of the relationship between the costs, volume, revenue and profitability of the products produced. The relationship between revenue and costs at different levels of output can be displayed graphically, with revenue behaviour and cost behaviour shown graphically.

Game theory is the study of mathematical models of strategic interaction among rational decision-makers. Von Neumann was the founder of modern game theory with his 1928 paper on the theory of games of strategy. The Rand Corporation applied game theory to global nuclear strategy in the 1950s, and the original applications of game theory was to zero-sum games where the gains or losses of each participant are exactly balanced by those of the other participants.

16

Basic Financial Statements

The *balance sheet* is a snapshot of the net worth of a company, and it lists the financial assets and liabilities of a company on a particular date. It is a summary of everything that is owned by the company less everything that it owes. The shareholder funds show where the money has come from, and how it is invested in the company, and it represents the net worth of the company.

The balance sheet gives a snapshot of the assets and liabilities (i.e., the net asset value) of the company on a particular date, but it gives no information on the earnings of the company during a period. The *profit and loss account* is the earnings statement of the company for the period, and it shows how much the company has earned during the period. The sales and expenditure are matched in the accounting period.

The balance sheet is prepared on a certain date, so it is a snapshot of the health of a company on that date in time. Long-term assets are referred to as *fixed assets* on the company balance sheet, whereas short-term assets are referred to as *current assets*. Short-term liabilities are referred to as *current liabilities* on the company balance sheet.

The cash accounts record all the cash in the company including the cashier's department and bank accounts, and the cash flow forecast indicates how much cash will be required to meet the company's requirements in the year ahead.

16.1 Balance Sheet

We illustrate the idea of a balance sheet by going through the example of a person trying to determine what they are worth at a point in time. This involves going through everything that they own and everything that they owe. This is essentially what a balance sheet is, i.e., a summary listing of everything that is owned less everything that is owed.

Let's consider a homeowner who purchased a house several years ago for €200,000 (it may now be worth more with inflation and so we need to decide on whether to value it at the price paid or its current value). Next, we consider the contents of the home and this involves itemizing the contents of the house and determining their value (either the original price paid or

DOI: 10.1201/9781003308140-16

the replacement value). The homeowner may own one or more cars, and have some cash in the bank as well as salary that is due as well as more long-term investments such as shares and a life insurance policy.

Next, we make a list of everything that is owed by the homeowner including long-term loans such as the mortgage on the home and the car loan, as well as bills that need to be paid and an overdraft from the bank. This leads to the summary statement (Fig. 16.1).

It is clear from the summary of the net worth that several items fall naturally into groups. These include items that have been bought and will be used for long time periods (e.g., house, contents and vehicles) and these are known as long-term assets. The second group of assets are those that we expect to change on a short-term basis (less than 12 months and often daily) and are referred to as short-term assets (e.g., salary and cash). The third group of assets contain investments of one kind or another and these have a monetary value (usually long term but they can be cashed in if required).

Similarly, we distinguish between debts that are due to be repaid in the short term (less than 12 months) which are termed short-term liabilities

Net Worth – Dec 2020	
House	€200,000
Contents	€34,000
Vehicles	€8,000
Salary Owed	€3,500
Cash	€1,500
Shares	€2,200
Life Insurance	€1,800
Total Owned	€251,000
Bills to Pay	€1,200
Bank Overdraft	€2,300
Car Loan outstanding	€3,500
Mortgage	€40,000
Total Owed	€47,000
Net Worth	€204,000

FIGURE 16.1
Net Worth of Homeowner.

(e.g., bills to pay and bank overdraft), and debts that are due to be repaid in a longer period of time which are termed long-term liabilities (e.g., mortgage and car loan) (Fig. 16.2).

A major difference between a company balance sheet and a personal balance sheet is that the shareholder funds are also shown in the company balance sheet. The shareholders are the owners of the company and they are the people who originally provided the funds to buy the assets and build the company. That is, they provided €204,000 in finance over the years (Fig. 16.3).

The balance sheet is prepared on a certain date and so is a snapshot of the company on that date in time. Long-term assets are referred to as fixed assets on the company balance sheets, whereas short-term assets are

Net Worth – Dec 2020	
LONG-TERM ASSETS	
House	€200,000
Contents	€34,000
Vehicles	€8,000
INVESTMENTS	
Shares	€2,200
Life Insurance	€1,800
SHORT-TERM ASSETS	
Salary Owed	€3,500
Cash	€1,500
Total Assets	€251,000
SHORT-TERM LIABILITIES	
Bills to Pay	€1,200
Bank Overdraft	€2,300
LONG-TERM LIABILITIES	
Car Loan outstanding	€3,500
Mortgage	€40,000
Total Liabilities	€47,000
Net Assets	€204,000

FIGURE 16.2
Personal Balance Sheet.

Balance Sheet of Company X– Dec 2020		
LONG-TERM ASSETS		
House	€200,000	
Plant/Equipment	€34,000	
Vehicles	€8,000	
	----------	€242,000
INVESTMENTS		
Shares	€2,200	
Life Insurance	€1,800	
	---------	€4,000
SHORT-TERM ASSETS		
Debtors	€3,500	
Cash	€1,500	
	----------	€5,000

Total Assets		€251,000
SHORT-TERM LIABILITIES		
Creditors	€1,200	
Bank Overdraft	€2,300	
	---------	€3,500
Total Assets less Short-Term Liabilities (Capital Employed)		€247,500
LONG-TERM LIABILITIES		
Loans	€3,500	
Mortgage	€40,000	€43,500

Net Assets (Assets less liabilities)		€204,000
SHAREHOLDER FUNDS		
Ordinary Shares	€60,000	
Capital Reserve	€75,000	
Revenue Reserve	€69,000	

Total Shareholder Funds		€204,000

FIGURE 16.3
Balance Sheet of Company X.

referred to as current assets. Short-term liabilities are referred to as current liabilities on the company balance sheet.

The valuation of assets in a business context is often based on the price originally paid less an amount to reflect wear and tear over time (i.e., depreciation) rather than the current market value of the asset. *Depreciation* refers to the process where the value of an asset is written off over its lifetime.

Intangible assets are assets that are not physical in form and refer to assets such as goodwill, brand values or intellectual property. They are difficult to place a value on, and often their value is seen on the sale of a company where the price paid is above the book value of the company. Goodwill is, in a sense, a measure of the reputation and capability of a company (e.g., in terms of the personal worth of an individual, it could represent the skills and expertise of the individual). Brand values refer to the value that brand names have, and these are difficult to place a value on. Clearly, a popular well-known brand has value as customers have familiarity with the company and its products. Intellectual property refers to the value of patents, trademarks and copyrights.

Current assets are short-term assets and include debtors, cash and inventory. Current assets earn nothing until they are converted into saleable products and sold to the customers, so current assets are kept as low as possible. It can be difficult to calculate the value of the inventory, especially when dealing with work-in-progress inventory items but also with finished inventory items. Debtors are those individuals or organizations that owe money to the company.

Current liabilities refer to debts that must be paid in the short term (i.e., within the next 12 months). They include creditors, the bank overdraft, taxation and unpaid dividends. *Creditors* are those individuals or organizations that the company owes money to (e.g., suppliers). The total assets less the current liabilities is often referred to as the *capital employed*, and this represents the amount of money tied up in the business and how the company is investing the money. The difference between the current assets and the current liabilities is referred to as *working capital* (also known as net working capital), and is a measure of the company's liquidity and its short-term financial health. For example, if the company's current assets do not exceed its current liabilities then it may have trouble growing or paying back its creditors.

The long-term liabilities include medium- and long-term borrowings such as loans, mortgages and bonds. A bond is a fixed interest investment where a bondholder lends money to the business at a fixed rate of interest for a defined period of time. The business pays interest regularly during the period and the capital is repaid on the maturity date. Some bonds have options to be converted to equity (i.e., shares in the company) at some point depending on several conditions such as share price.

The *shareholder funds* consist of ordinary shares, preference shares, capital reserves and revenue reserves. The holders of the ordinary shares are the shareholders or owners of the company. Shareholders in a publicly quoted company are free to sell their shares at any time, and the nominal price of the share may differ from the share price due to demand or otherwise for shares in the company. Shareholders generally receive a dividend (usually an interim and final dividend) from the company in line with company performance for the period, but dividends are not guaranteed. Preference shares provide greater security than ordinary shares in that the dividend is usually guaranteed. The revenue reserves refer to the retained profits for the period that are not given to the shareholders, and are instead retained in the business.

16.2 Profit and Loss Account

The balance sheet gives an account of the assets and liabilities (i.e., the net asset value) of the company on a particular date, but it gives no information on the earnings of the company during a period. The profit and loss account is the earnings statement of the company during the period.

Suppose that a consultant earns €70,000 per year but that he incurred legitimate business expenses as a result of carrying the work. He incurred direct costs of €4750 in travel as a result of travelling to his clients, and as he maintains an office in his home he incurred €800 in telephone expenses and €350 in office expenses during the period. This is summarized in Fig. 16.4.

Gross profit is the profit after the deduction of all of the direct costs from the sales income, whereas *trading profit* is given by the deduction of all direct costs and other expenses from the sales income. Next, we consider a more detailed example where the company makes and sells goods (Fig. 16.5).

The *cost of sales* is the cost of actually making the finished product, which includes the materials and direct labour and expenses. The support costs that are required to keep the business going are listed under overheads, and include the cost of telephone, office costs and administration.

The sales for the period include the cash sales and the invoiced sales, and as customers are generally given 30-day credit to pay an invoice the actual

Statement of Earnings - Dec 2020		
Sale of Services		€70,000
Less Direct Costs		€4,750
GROSS PROFIT		€65,250
Less Expenses		
Phone Calls	€800	
Office Costs	€350	
		€1150
TRADING PROFIT (Operating profit)		€64,100
Less Tax		€16,500
PROFIT after Tax		€47,600

FIGURE 16.4
Simple Profit and Loss Account.

Profit and Loss Account – Company X - Dec 2020		
Sales		€200,000
Less Cost of Sales		
Materials	€60,000	
Direct Labour	€48,000	
Direct Expenses	€5,500	
Factory Expenses	€6,500	€120,000
GROSS PROFIT		€80,000
Less Overheads		
Phone Costs	€1550	
Office Costs	€6500	
Administration	€5450	
Sales / Advertisement	€4500	
Delivery	€3500	
Professional Fees	€6500	
Office Equipment/Depreciation	€2500	€30,500
TRADING PROFIT (Operating profit)		€49,500
Interest Paid		€6500
PRE-TAX PROFITS		€43,000
Taxation		€15000
AFTER-TAX PROFITS		€28,000
Interim Dividend	€4500	
Final Dividend	€8500	€13000
RETAINED PROFITS for year		€15000
Retained profit previous years		€50,000
Retained profit carried forward		€65,000

FIGURE 16.5
Profit and Loss Account for Company X.

sales are the invoiced sales and not the cash received in the period. The *accruals principle* is fundamental to the way that accounts are presented, and are employed to match income and expenditure in the accounting period.

It may be necessary to make adjustments to the figures for sales and expenditure to ensure that their values correspond to the accounting period. For example, the entry of €60,000 for materials would need to be adjusted to €50,000 if only €50,000 of the material was needed for making the products for sale, and as there is now €10,000 available for the next accounting period this would be reflected in next year's accounts.

The trading profit (*operating profit*) is an indication of how well management has managed the capital employed in the company (i.e., the funds locked into the company for the management to use). The ratio of the operating profit to the amount of capital employed is important, and the trading profit should at least cover interest and tax.

A company rarely pays out all of its after tax profits and it usually retains some of the profits to expand the business, and the rest of the profits are paid out as dividend payments to the shareholders. The dividends usually consist of an interim and final dividend payment to the shareholders, who also approve the company accounts and the profits to be retained by the company.

16.3 Managing Cash

The personal management of cash is essential in the management of home finances such as payments of utility bills, mortgages, personal loans and rent. There are negative consequences for failure to pay back personal debt, and these include potential legal consequences as well as the risk of repossession of the individual's home or property. Similarly, the effective management of cash is essential in business, as a business requires sufficient cash at hand to be able to pay its employees, suppliers or the tax due to the relevant authorities. Poor cash management may lead to negative consequences such as litigation or bankruptcy.

The cash accounts record all the cash in the company including the cashier's department and bank accounts, and the cash flow forecast indicates how much cash will be required to meet the company's needs in the year ahead. Cash is distinct from profits in that profit reflects the difference in net worth of a company or individual from the beginning of the year to the end of the year, whereas cash management is the process of managing the company's financial needs to ensure that the cash resources required at a point in time are always available.

Fig. 16.6 presents a very simple personal cash account based on the profit and loss account in Fig. 16.4. The figures in both accounts are the same but in real life, this would rarely be the case, as all phone bills may not have been received for the period (usually an accrual would be made for this in the P&L accounts, and so while the profit figure would be the same there may be more cash on hand). Further, expenditure may have been made on other assets that do not appear on the P&L account (they belong to the balance sheet), so adjustments would need to be made for depreciation in year 1 of the asset in the P&L account (depreciation occurs under expenses in the P&L account, and as it does not involve the movement of cash it does not appear in the cash account). Another words, the cash on hand at the end of year could differ significantly from the position in the P&L account, and all entries that involve the movement of cash will appear in the cash account.

We now adjust the simple cash account to give a more realistic picture of the cash account for personal finances (Fig. 16.7).

Simple Cash Account (Personal)- Dec 2020			
CASH IN		**CASH OUT**	
Opening Balance	0	Travel	€4,750
Salary/Sales	€70,000	Phone Calls	€800
		Office Cost	€350
		Tax	€16,500
TOTALS	€70,000		€22,400

FIGURE 16.6
Personal Cash Account.

Detailed Cash Account (Personal)- Dec 2020			
CASH IN		**CASH OUT**	
Opening Balance	0	Travel	€4,750
Salary/Sales	€70,000	Phone Calls	€800
Cash from Car loan	€10,000	Office Cost	€350
		Tax	€16,500
		New Car	€25,500
		New furniture	€4,600
		Personal Exps	
		Mortgage	€15,500
		Food	€8,500
		Clothing	€2,500
		Holiday	€6,500
		Interest Loan	€2,500
		Interest Overdraft	€1,500
TOTALS	€80,000		€89,500

FIGURE 16.7
Detailed Personal Cash Account.

Fig. 16.4 indicates that a profit of €47,600 was made during the period, but it is clear from the cash account that much more than that was spent during the year, resulting in being overdrawn at the bank by €9500. This may or may not be a problem, provided that the overdraft has been agreed in advance with the bank, and provided that the interest and loan repayments can be met in future periods. Further, the opening balance for the cash out position for the future accounting period is (€9500). The phenomenon of a company or person making a profit and yet going bankrupt is all too common, and the current year includes "once off" expenditure on a new car and new furniture. However, there is always the potential danger of a downturn or reduced revenue in future periods that could reduce the ability to service the repayments on the outstanding loans.

It is therefore important to organize and plan the cash requirements, and it may be prudent to spread out the purchase of long-term assets such as a car and furniture over a number of years. That is, it is important to anticipate cash flow problems that may arise as they need to be managed, so the planning and timing of expenditure are very important. It is therefore important to maintain a monthly or quarterly forecast of the cash account, and this would include the predicted incoming or outgoing cash for the month/quarter. The predicted net cash position per month/quarter may indicate potential problems, and it is important to avoid these before a commitment is made.

Fig. 16.8 presents a quarterly cash flow forecast for the year, and an overdraft limit of €5000 has been agreed with the bank. Spending exceeds

Cash Flow Forecast (Personal)- 2020

	Q1	Q2	Q3	Q4	Tot Yr
Opening Balance	0	(€4225)	(€4750)	(3575)	
CASH IN					
Salary/Sales	€20,000	€20,000	€17,500	€12,500	€70,000
Cash from Car loan	€10,000				€10,000
Total Cash In	€30,000	€20,000	€17,500	€12,500	€80,000
CASH OUT					
Phone Calls	€200	€200	€200	€200	€800
Travel	€1500	€1250	€1000	€1000	€4,750
Office Cost	€100	€50	€100	€100	€350
Tax	-	€7,500	-	€9,000	€16,500
Capital Equipment					
New Car	€25,500	-	-	-	€25,500
New furniture	-	-	€4,600	-	€4,600
Personal Exps					
Mortgage	€3,500	€4,000	4,000	€4,000	€15,500
Food	€2,000	€2,000	€2,000	€2,500	€8,500
Clothing	€500	€1000	€500	€500	€2,500
Holiday	-	€3,500	€3,000	-	€6,500
Interest Loan	€625	€625	€625	€625	€2500
Sub-Total (cash out)	€33925	€20125	€16025	€17925	€88000
Net Cash	(€3925)	(€4350)	(€3275)	(€9000)	(€8000)
Interest Overdraft	€300	€400	€300	€500	€1,500
Closing Balance	(€4225)	(€4750)	(3575)	(€9500)	(€9500)

FIGURE 16.8
Personal Cash Flow Forecast.

cash received for the first two quarters, and while the net cash position in quarter 3 is positive it is not sufficient to clear the deficits from the previous two quarters. The closing balance at the end of the fourth quarter is €9500, which exceeds the overdraft limit agreed, and the revenue/sales in the fourth quarter is significantly less than the previous three quarters. There is, unless there are good reasons for the drop in sales revenue, a need to be more prudent in spending. It may be that a less expensive car may have been more appropriate, or the purchase of office furniture should have been deferred for a year. For a nice introduction to basic financial statements, see [Gas:91].

16.4 Review Questions

1. What is a balance sheet?
2. Explain the difference between an asset and a liability.
3. What is a profit and loss account?
4. Explain the difference between income and expenses.
5. What is the difference between profit and loss?
6. Explain why managing cash is so important for a business.
7. What is the difference between a fixed asset and a current asset?
8. What is the difference between a creditor and a debtor?
9. What is trading profit (operating profit)?
10. Explain the term "capital employed."
11. Explain the term "shareholder funds."

16.5 Summary

The balance sheet is a snapshot of the net worth of a company and lists the financial assets and liabilities at a company on a particular date, and it is a summary of everything that is owned by the company less everything that it owes. The shareholder funds show where the money has come from and how it is invested in the company and represents the net worth of the company.

The balance sheet gives a snapshot of the assets and liabilities (i.e., the net asset value) of the company on a particular date, but it gives no information on the earnings of the company during a period. The profit and loss account is the earnings statement of the company for the period, and it shows how much the company has earned during the period.

The cash accounts record all the cash in the company including the cashier's department and bank accounts, and the cash flow forecast indicates how much cash will be required to meet the company's needs in the year ahead.

17

Calculus I

Newton and Leibniz independently developed calculus in the late 17th century.[1] Calculus plays a key role in describing how rapidly things change, and it may be employed to determine the velocity and acceleration of moving bodies as well as to calculate the area of a region under a curve or between two curves. It may be used to determine the volumes of solids, computing the length of a curve and in finding the tangent to a curve. It is an important branch of mathematics concerned with limits, continuity, derivatives and integrals of functions.

The concept of a *limit* is fundamental in calculus. Let f be a function defined on the set of real numbers, then the limit of f at a is l (written as $\lim_{x \to a} f(x) = l$) if given any real number $\varepsilon > 0$ then there exists a real number $\delta > 0$ such that $|f(x) - l| < \varepsilon$ whenever $|x - a| < \delta$. The idea of a limit can be seen in Fig. 17.1.

The function f defined on the real numbers is *continuous* at a if $\lim_{x \to a} f(x) = f(a)$. The set of all continuous functions on the closed interval $[a, b]$ is denoted by $C[a, b]$.

If f is a function defined on an open interval containing x_0 then f is said to be *differentiable* at x_0 if the limit exists. Whenever this limit exists it is denoted by $f'(x_0)$ and is called the *derivative* of f at x_0. Differential calculus is concerned with the properties of the derivative of a function. The derivative of f at x_0 is the slope of the tangent line to the graph of f at $(x_0, f(x_0))$ (Fig. 17.2).

$$\lim_{x \to x_0} \frac{f(x) - f(x_0)}{x - x_0}$$

It is easy to see that if a function f is differentiable at x_0 then f is continuous at x_0.

Rolle's Theorem

Suppose $f \in C[a, b]$ and f is differentiable on (a, b). If $f(a) = f(b)$ then there exists c such that $a < c < b$ with $f'(c) = 0$.

Mean Value Theorem

Suppose $f \in C[a, b]$ and f is differentiable on (a, b). Then there exists c such that $a < c < b$ and

DOI: 10.1201/9781003308140-17

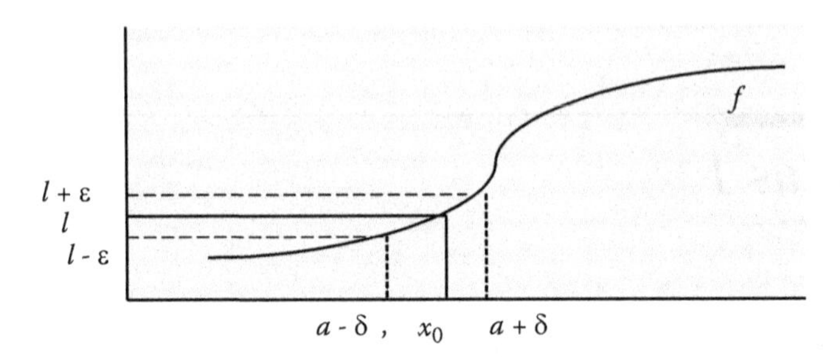

FIGURE 17.1
Limit of a Function.

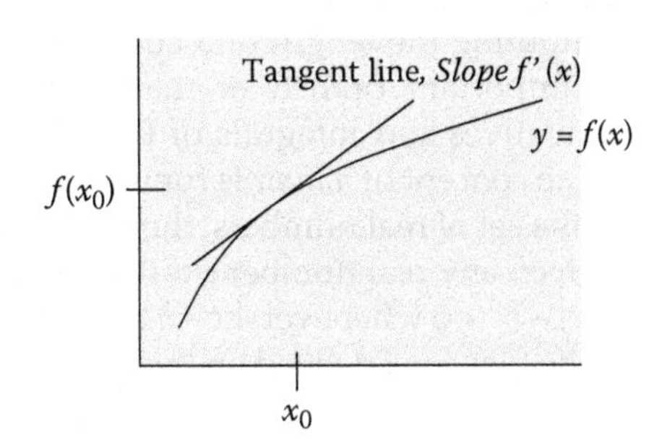

FIGURE 17.2
Derivative as a Tangent to Curve.

$$f'(c) = \frac{f(b) - f(a)}{b - a}$$

Proof

The mean value theorem is a special case of Rolle's theorem and the proof involves defining the function $g(x) = f(x) - rx$ where $r = (f(b) - f(a))/(b - a)$.

It is easy to verify that $g(a) = g(b)$. Clearly, g is differentiable on (a, b) and so by Rolle's theorem there is a c in (a, b) such that $g'(c) = 0$. Therefore, $f'(c) - r = 0$ and so $f'(c) = r = f(b) - f(a)/(b - a)$ as required.

Interpretation of the Mean Value Theorem

The mean value theorem essentially states that there is at least one point c on the curve $f(x)$ between a and b such that the slope of the chord is same as the tangent $f'(c)$ (Fig. 17.3).

Intermediate Value Theorem

Suppose $f \in C[a, b]$ and K is any real number between $f(a)$ and $f(b)$. Then there exists c in (a, b) for which $f(c) = K$.

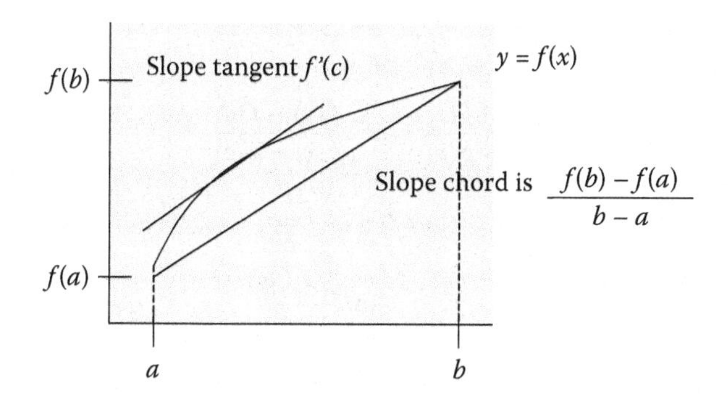

FIGURE 17.3
Interpretation of Mean Value Theorem.

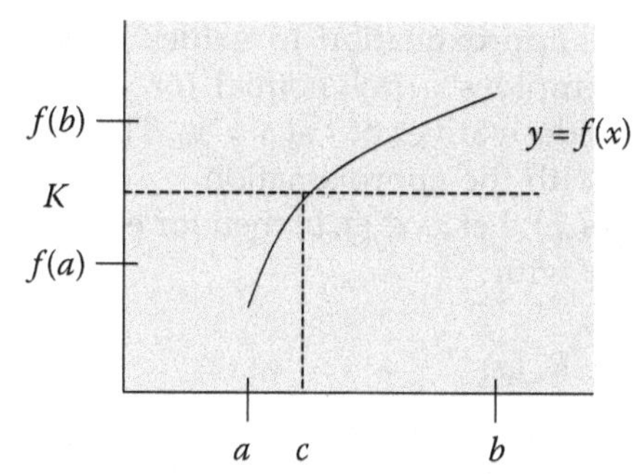

FIGURE 17.4
Interpretation of Intermediate Value Theorem.

Proof

The proof of this relies on the completeness property of the real numbers. It involves considering the set S in $[a, b]$ such that $f(x) \leq K$ and noting that this set is non-empty since $a \in S$ and bounded above by b. Therefore, the supremum[2] $\sup S = c$ exists and it is straightforward to show (using ε and δ arguments and the fact that f is continuous) that $f(c) = K$ (Fig. 17.4).

L'Hôpital's Rule

Suppose that $f(a) = g(a) = 0$ and that $f'(a)$ and $g'(a)$ exist and that $g'(a) \neq 0$. Then L'Hôpital's rule states that:

$$\lim_{x \to a} \frac{f(x)}{g(x)} = \frac{f'(a)}{g'(a)}$$

Proof

$$\lim_{x \to a} \frac{f(x)}{g(x)} = \lim_{x \to a} \frac{f(x) - f(a)}{g(x) - g(a)}$$

$$= \lim_{x \to a} \frac{\frac{f(x) - f(a)}{x - a}}{\frac{g(x) - g(a)}{x - a}}$$

$$= \frac{\lim_{x \to a} \frac{f(x) - f(a)}{x - a}}{\lim_{x \to a} \frac{g(x) - g(a)}{x - a}}$$

$$= \frac{f'(a)}{g'(a)}$$

Taylor's Theorem

The Taylor series is concerned with the approximation to values of the function f near x_0. The approximation employs a polynomial (or power series) in powers of $(x - x_0)$ as well as the derivatives of f at $x = x_0$. There is an error term (or remainder) associated with the approximation.

Suppose $f \in C^n[a, b]$ and f^{n+1} exists on (a, b). Let $x_0 \in (a, b)$ then for every $x \in (a, b)$ there exists $\xi(x)$ between x_0 and x with

$$f(x) = P_n(x) + R_n(x)$$

where $P_n(x)$ is the nth Taylor polynomial for f about x_0 and $R_n(x)$ is the called the remainder term associated with $P_n(x)$. The infinite series obtained by taking the limit of $P_n(x)$ as $n \to \infty$ is termed the Taylor series for f about x_0.

$$P_n(x) = f(x_0) + f'(x_0)(x - x_0) + \frac{f''(x_0)}{2!}(x - x_0)^2 + \ldots + \frac{f^n(x_0)}{n!}(x - x_0)^n$$

The remainder term is given by:

$$R_n(x) = \frac{f^{n+1}(\xi(x))}{(n + 1)!}(x - x_0)^{n+1}$$

17.1 Differentiation

Mathematicians of the 17th century were working on various problems concerned with motion. These included problems such as determining the velocity of objects on or near the earth, as well as the motion of the planets. They were also interested in changes of motion, i.e., in the acceleration of these moving bodies.

Velocity is the rate at which distance changes with respect to time, and the average speed during a journey is the distance travelled divided by the elapsed time. However, since the speed of an object may be variable over a period of time, there is a need to be able to determine its velocity at a specific time instance. That is, there is a need to determine the rate of change of distance with respect to time at any time instant.

The direction in which an object is moving at any instant of its flight was also studied. For example, the direction in which a projectile is fired determines the horizontal and vertical components of its velocity. The direction in which an object is moving can vary from one instant to another.

The problem of finding the maximum and minimum values of a function was also studied, e.g., the problem to determine the maximum height that a bullet reaches when it is fired. Other problems studied include problems to determine the lengths of paths, the areas of figures and the volume of objects.

Newton and Leibnitz (Figs. 17.5 and 17.6) showed that these problems could be solved by means of the concept of the derivative of a function, i.e., the rate of change of one variable with respect to another.

Rate of Change

The average rate of change and instantaneous rate of change are of practical interest. For example, if a motorist drives 200 miles in 4 hours then the average speed is 50 miles per hour, i.e., the distance travelled divided by the elapsed time. The actual speed during the journey may vary as if the driver stops for lunch then the actual speed is zero for the duration of lunch.

FIGURE 17.5
Issac Newton.

FIGURE 17.6
Wilhelm Gottfried Leibniz.

The actual speed is the instantaneous rate of change of distance with respect to time. This has practical implications as motorists are required to observe speed limits, and a speed camera may record the actual speed of a vehicle with the driver subject to a fine should the permitted speed limit be exceeded. The actual speed is relevant in a car crash as speed is a major factor in road fatalities.

In calculus, the term Δx means a change in x and Δy means the corresponding change in y. The derivative of f at x is the instantaneous rate of change of f, and f is said to be differentiable at x. It is defined as:

$$\frac{dy}{dx} = \lim_{\Delta x \to 0} \frac{\Delta y}{\Delta x} = \lim_{\Delta x \to 0} \frac{f(x + \Delta x) - f(x)}{\Delta x}$$

In the formula, Δy is the increment $f(x+\Delta x) - f(x)$.

The average velocity of a body moving along a line in the time interval t to $t + \Delta t$ where the body moves from position $s = f(t)$ to position $s+\Delta s$ is given by:

$$V_{av} = \frac{\text{displacement}}{\text{Time Travelled}} = \frac{\Delta s}{\Delta t} = \frac{f(t + \Delta t) - f(t)}{\Delta t}$$

The instantaneous velocity of a body moving along a line is the derivative of its position $s = f(t)$ with respect to t. It is given by:

$$v = \frac{ds}{dt} = \lim_{\Delta t \to 0} \frac{\Delta s}{\Delta t} = f'(t)$$

17.1.1 Rules of Differentiation

Table 17.1 presents several rules of differentiation.

TABLE 17.1

Rules of Differentiation

No.	Rule	Definition
1.	Constant	The derivative of a constant is 0. That is, for $y = f(x) = c$ (a constant value) we have $dy/dx = 0$.
2.	Sum	$d/dx\ (f + g) = df/dx + dg/dx$
3.	Power	The derivative of $y = f(x) = x^n$ is given by $dy/dx = nx^{n-1}$.
4.	Scalar	If c is a constant and u is a differentiable function of x then $dy/dx = c\ du/dx$ where $y = cu(x)$.
5.	Product	The product of two differentiable functions u and v is differentiable and $\frac{d}{dx}(uv) = v\frac{du}{dx} + u\frac{dv}{dx}$
6.	Quotient	The quotient of two differentiable functions u and v is differentiable (where $v \neq 0$) and $\frac{d}{dx}\left(\frac{u}{v}\right) = \frac{v\frac{du}{dx} - u\frac{dv}{dx}}{v^2}$
7.	Chain rule	Suppose $h = g \circ f$ is the composite of two differentiable functions $y = g(x)$ and $x = f(t)$. Then h is a differentiable function of t whose derivative at each value of t is: $h'(t) = (g \circ f)'(t) = g'(f(t))f'(t)$ This may also be written as: $\frac{dy}{dt} = \frac{dy}{dx}\frac{dx}{dt}$

Derivatives of Well-Known Functions

The following are the derivatives of some well-known functions including basic trigonometric functions, the exponential function and the natural logarithm function.

i. $^d/_{dx}$ Sin$x$ = Cos x
ii. $^d/_{dx}$ Cos x = –Sin x
iii. $^d/_{dx}$ Tan x = Sec^2x
iv. $^d/_{dx}\ e^x = e^x$
v. $^d/_{dx} \ln x = 1 / x$ (where $x > 0$)
vi. $^d/_{dx} a^x = \ln(a)\ a^x$
vii. $^d/_{dx} \log_a x = 1 / x \ln(a)$
viii. $^d/_{dx}$ arcsin $x = 1 / \sqrt{(1 - x^2)}$
ix. $^d/_{dx}$ arccos $x = -1 / \sqrt{(1 - x^2)}$
x. $^d/_{dx}$ arctan $x = 1 / (1 + x^2)$

Increasing and Decreasing Functions

Suppose that a function f has a derivative at every point x of an interval I. Then

i. f increases on I if $f'(x) > 0$ for all x in I.
ii. f decreases on I if $f'(x) < 0$ for all x in I.

The geometric interpretation of the first derivative test is that it states that differentiable functions increase on intervals where their graphs have positive slopes, and decrease on intervals where their graphs have negative slopes.

If f' changes from positive to negative values as x passes from left to right through point c then the value of f at c is a *local maximum* value of f. Similarly, if f' changes from negative to positive values as x passes from left to right through point c then the value of f at c is a *local minimum* value of f (Fig. 17.7).

The graph of a differentiable function $y = f(x)$ is concave down in an interval where f' decreases and concave up in an interval where f' increases. This may be defined by the second interval test for concavity. Another words, the graph of $y = f(x)$ is concave down in an interval where $f'' < 0$ and concave up in an interval where $f'' > 0$.

A point on the curve where the concavity changes from concave up to concave down or vice versa is termed a point of inflection. That is, at a *point of inflection c*, we have that f' is positive on one side and negative on the other side. At the point of inflection c, we have the value of the second derivative is zero, i.e., $f''(c) = 0$, or in other words f' goes through a local maximum or minimum.

17.2 Integration

The derivative is a functional operator that takes a function as an argument and returns a function as a result. The inverse operation involves

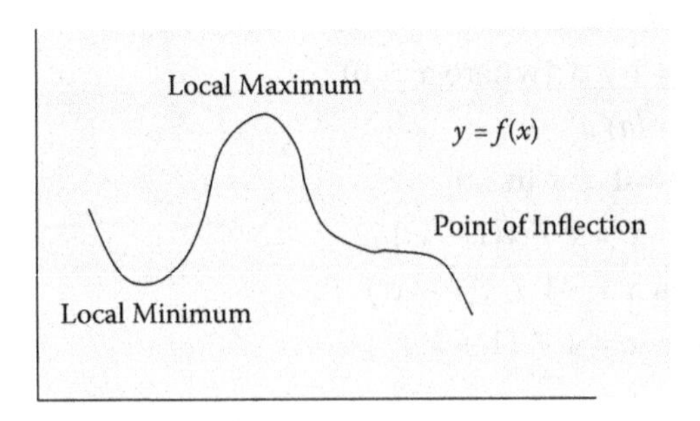

FIGURE 17.7
Local Minima and Maxima.

determining the original function from the known derivative, and integral calculus is the branch of the calculus concerned with this problem. The integral of a function consists of all those functions that have it as a derivative.

Integration is applicable to problems involving area and volume. It is the mathematical process that allows the area of a region with curved boundaries to be determined, and it also allows the volume of a solid to be determined.

The problem of finding functions whose derivatives are known involves finding a function $y = F(x)$ whose derivative is given by the differential equation:

$$\frac{dy}{dx} = f(x)$$

The solution to this differentiable equation over the interval I is F if F is differentiable at every point of I and for every x in I we have:

$$\frac{d}{dx}F(x) = f(x)$$

Clearly, if $F(x)$ is a particular solution to $d/dx\ F(x) = f(x)$ then the general solution is given by:

$$y = \int f(x)dx = F(x) + k$$

since

$$d/dx(F(x) + k) = f(x) + 0 = f(x).$$

Rules of Integration

The following are rules of integration including the integrals of some well-known functions such as basic trigonometric functions and the exponential function.

Table 17.2 presents several rules of integration.

It is easy to check that the integration has been carried out correctly. This is done by computing the derivative of the function obtained and checking that it is the same as the function to be integrated.

Often, the goal will be to determine a particular solution satisfying certain conditions rather than the general solution. The general solution is first determined, and then the constant k that satisfies the particular solution is calculated.

TABLE 17.2

Rules of Integration

No.	Rule	Definition
1.	Constant	$\int (u'(x)\, dx = u(x) + k$
2.	Sum	$\int (u(x) + v(x))\, dx = \int u(x)\, dx + \int v(x)dx$
3.	Scalar	$\int au(x)\, dx = a \int u(x)\, dx$ (where a is a constant)
4.	Power	$\int x^n dx = \frac{x^{n+1}}{n+1} + k$ (where $n \neq -1$)Z
5.	Cos	$\int \cos x\, dx = \sin x + k$
6.	Sin	$\int \sin x\, dx = -\cos x + k$
7.	$\sec^2 x$	$\int \sec^2 x\, dx = \tan x + k$
8.	e^x	$\int e^x dx = e^x + k$
9.	Log	$\int {}^1/_x\, dx = \ln x + k$

The *substitution method* is a useful method that is often employed in performing integration, and its effect is to potentially change an unfamiliar integral into one that is easier to evaluate. The procedure to evaluate $\int f(g(x))\, g'(x)dx$ where f' and g' are continuous functions is as follows:

1. Substitute $u = g(x)$ and $du = g'(x)dx$ to obtain $\int f(u)du$.
2. Integrate with respect to u.
3. Replace u by $g(x)$ in the result.

The method of *integration by parts* is a rule of integration that transforms the integral of a product of functions into simpler integrals. It is a consequence of the product rule for differentiation.

$$\int u\ dv = uv - \int v\ du$$

$$\int f(x)g'(x)dx = f(x)g(x) - \int f'(x)g(x)dx$$

17.2.1 Definite Integrals

A definite integral defines the area under the curve $y = f(x)$, and the area of the region between the graph of a non-negative continuous function $y = f(x)$ for the interval $a \leq x \leq b$ of the x-axis is given by the definite integral.

The sum of the area of the rectangles approximates the area under the curve and the more rectangles that are used the better the approximation (Fig. 17.8).

The definition of the area of the region beneath the graph of $y = f(x)$ from a to b is defined to be the limit of the sum of the area of the rectangles as the width of the rectangles become smaller and smaller, and the number of

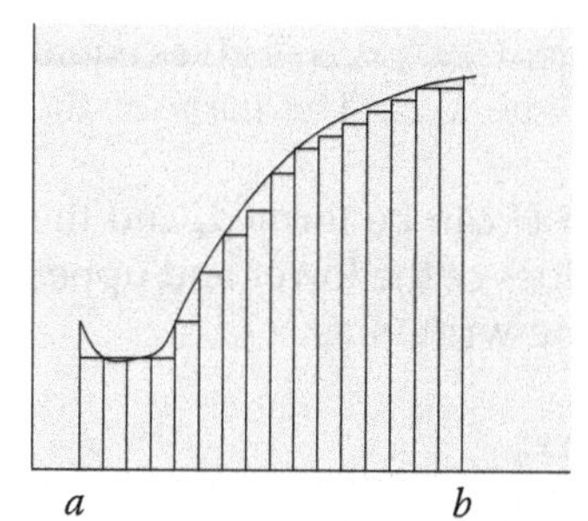

FIGURE 17.8
Area under the Curve.

rectangles used approaches infinity. The limit of the sum of the area of the rectangles exists for any continuous function.

The approximation of the area under the graph $y = f(x)$ between $x = a$ and $x = b$ is done by dividing the region into n strips with each strip of uniform width given by $\Delta x = (b - a)/n$ and drawing lines perpendicular to the x-axis (Fig. 17.9). Each strip is approximated with an inscribed rectangle where the base of the rectangle is on the x-axis to the lowest point on the curve above (lower Riemann sum). We let c_k be a point in which f takes on its minimum value in the interval from x_{k-1} to x_k and the height of the rectangle is $f(c_k)$. The sum of these areas is the approximation of the area under the curve and is given by:

$$S_n = f(c_1)\Delta x + f(c_2)\Delta x + \ldots + f(c_n)\Delta x$$

The area under the graph of a non-negative continuous function f over the interval $[a, b]$ is the limit of the sums of the area of inscribed rectangles of equal base length as n approaches infinity.

$$\begin{aligned} A &= \lim_{n\to\propto} S_n \\ &= \lim_{n\to\propto} f(c_1)\Delta x + f(c_2)\Delta x + \ldots + f(c_n)\Delta x \\ &= \lim_{n\to\infty} \textstyle\sum_{k=1}^{n} f(c_k)\Delta x \end{aligned}$$

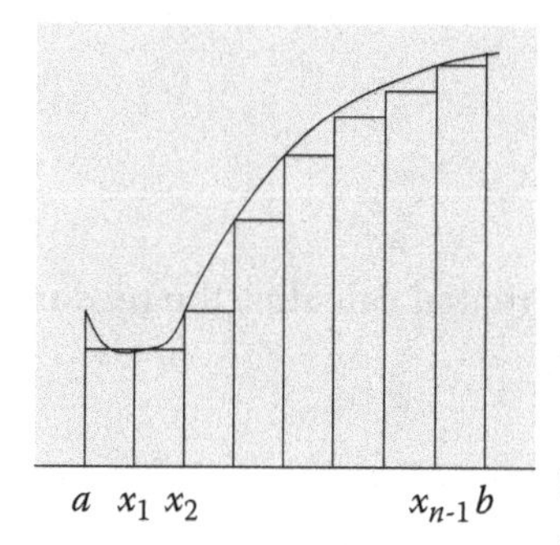

FIGURE 17.9
Area under the Curve – Lower Sum.

It is not essential that the division of $[a, b]$ into $a, x_1, x_2, \ldots, x_{n-1}, b$ gives equal subintervals $\Delta x_1 = x_1 - a$, $\Delta x_2 = x_2 - x_1, \ldots, \Delta x_n = b - x_{n-1}$. The *norm* of the subdivision is the largest interval length.

The lower Riemann sum L and the upper sum U can be formed, and the more finely divided that $[a, b]$ is the closer the values of the lower and upper sum U and L. The upper and lower sums may be written as:

$$L = \min_1 \Delta x_1 + \min_2 \Delta x_2 + \ldots + \min_n \Delta x_n$$
$$U = \max_1 \Delta x_1 + \max_2 \Delta x_2 + \ldots + \max_n \Delta x_n$$
$$\lim_{\text{norm}\to 0} U - L = 0 \text{ (i.e. , } \lim_{\text{norm}\to 0} U = \lim_{\text{norm}\to 0} L)$$

Further, if $S = \Sigma f(c_k) \Delta x_k$ (where c_k is any point in the subinterval and $\min_k \leq f(c_k) \leq \max_k$) we have:

$$L \leq S \leq U$$
$$\lim_{\text{norm}\to 0} L = \lim_{\text{norm}\to 0} S = \lim_{\text{norm}\to 0} U$$

Integral Existence Theorem (Riemann Integral)

If f is continuous on $[a, b]$ then

$$\int_a^b f(x)dx = \lim_{\text{norm}\to 0} \sum f(c_k)\Delta x_k$$

exists and is the same number for any choice of the numbers c_k.

Properties of Definite Integrals

Table 17.3 presents some algebraic properties of the definite integral.

17.2.2 Fundamental Theorems of Integral Calculus

Table 17.4 presents two fundamental theorems of integral calculus.

That is, the procedure to calculate the definite integral of f over $[a, b]$ involves just two steps:

i. Find an antiderivative F of f
ii. Calculate $F(b) - F(a)$

For a more detailed account of integral and differential calculus, the reader is referred to [Fin:88].

TABLE 17.3

Properties of Definite Integral

Properties of Definite Integral
(*i*) $\int_a^a f(x)dx = 0$
(*ii*) $\int_b^a f(x)dx = -\int_a^b f(x)dx$
(*iii*) $\int_a^b kf(x)dx = k\int_a^b f(x)dx$
(*iv*) $\int_a^b f(x)dx \geq 0$ if $f(x) \geq 0$ on $[a, b]$
(*v*) $\int_a^b f(x)dx \leq \int_a^b g(x)dx$ if $f(x) \leq g(x)$ on $[a, b]$
(*vi*) $\int_a^b f(x)dx + \int_b^c f(x)dx = \int_a^c f(x)dx$
(*vii*) $\int_a^b \{f(x) + g(x)\}dx = \int_a^b f(x)dx + \int_a^b g(x)dx$
(*viii*) $\int_a^b \{f(x) - g(x)\}dx = \int_a^b f(x)dx - \int_a^b g(x)dx$

TABLE 17.4

Fundamental Theorems of Integral Calculus

Theorem
First Fundamental Theorem: (Existence of Anti-Derivative) If f is continuous on $[a, b]$ then $F(x)$ is differentiable at every point x in $[a, b]$ where $F(x)$ is given by: $F(x) = \int_a^x f(t)dt$ If f is continuous on $[a, b]$ then there exists a function $F(x)$ whose derivative on $[a, b]$ is $f. \frac{dF}{dx} = \frac{d}{dx}\int_a^x f(t)dt = f(x)$ **Second Fundamental Theorem:** (Integral Evaluation Theorem) If f is continuous on $[a, b]$ and F is any anti-derivative of f on $[a, b]$ then: $\int_a^b f(x)dx = F(b) - F(a)$

17.3 Review Questions

1. Explain the concept of the limit of a function.
2. Explain the concept of continuity.
3. Explain the difference between average velocity and instantaneous velocity, and explain the concept of the derivative of a function.

4. Determine the following:
   a. $\lim_{x \to 0} \text{Sin } x$
   b. $\lim_{x \to 0} x \text{ Cos } x$
   c. $\lim_{x \to -\infty} |x|$
5. Determine the derivative of the following functions:
   a. $y = x^3 + 2x + 1$
   b. $y = x^2 + 1$ where $x = (t+1)^2$
   c. $y = \text{Cos } x^2$
6. Determine the integral of the following functions:
   a. $\int (x^2 - 6x)\, dx$
   b. $\int \sqrt{(x - 6)}\, dx$
   c. $\int (x^2 - 4)^2\, 3x^3 dx$
7. Explain the significance of the fundamental theorems of calculus.

17.4 Summary

This chapter provided a brief introduction to calculus including limits, continuity, differentiation and integration. Newton and Leibniz developed the calculus independently of each other in the late 17th century. Differentiation plays a key role in describing how rapidly things change and in finding tangents to curves. Integration may be employed to calculate areas of regions under curves and volumes of solids.

In calculus, the term Δx means a small change in x and Δy means the corresponding change in y. The derivative of f at x is the instantaneous rate of change of f, and is defined as

$$\frac{dy}{dx} = \lim_{\Delta x \to 0} \frac{\Delta y}{\Delta x} = \lim_{\Delta x \to 0} \frac{f(x + \Delta x) - f(x)}{\Delta x}$$

Integration is the inverse operation of differentiation, and it involves determining the original function from the known derivative. The integral of a function consists of all those functions that have the function as a derivative.

Integration is applicable to problems involving area and volume, and it allows the area of a region with curved boundaries to be determined.

Notes

1 The question of who invented the calculus led to a bitter controversy between Newton and Leibniz with the latter accused of plagiarising Newton's work. Newton, an English mathematician and physicist, was the giant of the late 17th century, and Leibnitz was a German mathematician and philosopher. Today, both Newton and Leibniz are credited with the independent development of calculus.
2 The supremum is the least upper bound and the infimum is the greatest lower bound.

18

Calculus II

This chapter considers several applications of calculus including the use of differentiation to deal with problems involving the rate of change, and we show how velocity, speed and acceleration may be determined, as well as solving maxima/minima problems. We show how integration may be used to solve problems involving area, volume and length of a curve.

The definite integral may be used to determine the area under a curve as well as to compute the area bounded by two curves. We show how the volume of a solid of a known cross-sectional area may be determined, and show how to compute the volume of a solid generated by rotating a region around the x-axis or y-axis.

We show how the length of a curve may be determined and we present the formulae for the Trapezoidal Rule and Simpson's Rule which are used to approximate definite integrals.

Finally, we introduce Fourier series, LaPlace transforms and differential equations. A Fourier series consists of the sum of a possibly infinite set of sine and cosine functions. The LaPlace transform is an integral transform that takes a function f and transforms it into another function F by means of an integral. An equation that contains one or more derivatives of an unknown function is termed a differential equation, and many important problems in engineering and physics involve determining a solution to these equations.

18.1 Applications of Calculus

There are rich applications of calculus in science and engineering, and we consider several applications of differentiation and integration. We discuss problems involving velocity and acceleration of moving objects; problems to determine the rate at which one variable changes from the rate at which another variable is known to change; and maxima and minima problems that are solved with differentiation. We then solve problems involving area and volume that are solved by integration.

Differentiation may be used to determine the speed, velocity and acceleration of bodies. Velocity is the rate of change of position with respect to time and acceleration is given by the rate of change of velocity with respect to time. The speed is the magnitude of the velocity. This is expressed

DOI: 10.1201/9781003308140-18

mathematically by letting $s(t)$ be a function giving the position of an object at time t, and the velocity, speed and acceleration are then given by

Velocity of object at time $t = v(t) = s'(t)$
Acceleration of object at time $t = a(t) = v'(t) = s''(t)$
Speed of object at time $t = |v(t)|$

Example

A ball is dropped from a height of 64 feet (Imperial system) and its height above the ground after t seconds is given by the equation $s(t) = -16t^2 + 64$. Determine

a. The velocity when it hits the ground.
b. The average velocity during its fall.

Solution

The ball hits the ground when $s(t) = 0$, i.e.,

$$\begin{aligned} 0 &= -16t^2 + 64 \\ 16t^2 &= 64 \\ t^2 &= 4 \\ t &= 2 \end{aligned}$$

The velocity is given by $s'(t) = -32t$ and so the velocity when the ball hits the ground is equal to $-32 * 2 = -64$ ft/sec.

The average velocity is given by distance travelled/time

$$= (s(2) - s(0))/2 - 0 = -64/2 = -32 \text{ ft/sec.}$$

Occasionally, problems arise where one variable is known to change at a certain rate and the problem is to determine the rate of change on another variable. For example, how fast does the height of the water level drop when a cylindrical tank is drained at a rate of 3 L/sec?

Example

How fast does the water level in a cylindrical tank drop when water is removed at the rate of 3 L/sec?

Solution

The radius of the tank is r (a constant), the height is h and the volume of water is V (which are both changing over time and so V and h are considered differentiable functions of time).

$$dV/dt = -3 (3 \text{ L/sec is being removed})$$

We wish to determine ${}^{dh}/_{dt}$ and we can determine this using the formula for the volume of a cylinder.

$$V = \pi r^2 h$$

Then $$dV/dt = \pi r 2dh/dt = -3 \qquad \text{(since } r \text{ is a constant)}$$

$$dh/dt = -3/\pi r^2$$

That is, the water level is dropping at the constant rate of $3/\pi r^2$ L/sec.

Maxima and minima problems refer to problems where the goal is to maximize or minimize a function on a particular interval, where the function is continuous and differentiable on the interval and the function does not attain its maximum or minimum at the endpoints of the interval. Then, we know that the maximum or minimum is at an interior point of the interval where the derivative is 0.

Example

Find two positive integers whose sum is 20 and whose product is as large as possible.

Solution

Let x be one of the numbers then the other number is $20 - x$ and so the product of both numbers is given by

$$f(x) = x(20 - x) = 20x - x^2$$

The objective is to determine the value of x that will maximize the product (i.e., the value of $f(x)$ in the interval $0 \leq x \leq 20$). The function $f(x)$ is continuous and differentiable, and attains a local maximum where its derivative is 0. The derivative is given by

$$f'(x) = 20 - 2x$$

The derivative is 0 when $20 - 2x = 0$ or when $x = 10$, and the maximum value of $f(x)$ is $200 - 100 = 100$.

Problems involving area and volume may be solved with integration. The definite integral may be applied to problems to determine the area below the curve as in the following example:

Example

Find the area under the curve $x^2 - 4$ and the x-axis from $x = -2$ to $x = 2$.

Solution

$$\int_{-2}^{2} (x^2 - 4)dx$$
$$= \frac{x^3}{3} - 4x\Big|_{-2}^{2}$$
$$= \left(\frac{8}{3} - 8\right) - \left(\frac{-8}{3} + 8\right)$$
$$= -\frac{32}{3}$$

The area between two curves $y = f(x)$ and $y = g(x)$ where $f(x) \geq g(x)$ on the interval $[a, b]$ is given by

$$A = \int_{a}^{b} (f(x) - g(x))dx$$

Example

Find the area of the region bounded by the curve $y = 2 - x^2$ and the line $y = -x$ on the interval $[-1, 2]$.

Solution

We take $f(x) = 2 - x^2$ and $g(x) = -x$ and it can be seen by drawing both curves that $f(x) \geq g(x)$ on the interval $[-1, 2]$. Therefore, the area between both curves is given by

$$A = \int_{-1}^{2} ((2 - x^2) - (-x))dx$$
$$= \int_{-1}^{2} ((2 + x - x^2)dx$$
$$= 2x + 1/2x^2 - 1/3x^3 \Big|_{-1}^{2}$$
$$= 10/3 + 7/6$$
$$= 27/6$$
$$= 4.5$$

The *volume* of a solid of known *cross-functional area* $A(x)$ from $x = a$ to $x = b$ is given by

$$V = \int_{a}^{b} A(x)dx$$

Example

Find the volume of the pyramid that has a square base that is 3 m on a side and 3 m high. The area of a cross-section of the pyramid is given by $A(x) = x^2$. Find the volume of the pyramid.

Solution

The volume is given by

$$V = \int_0^3 x^2 dx$$
$$= 1/3x^3 |_0^3$$
$$= 9\,\text{m}^2$$

The volume of a solid created by *revolving the region* bounded by $y = f(x)$ and $x = a$ to $x = b$ about the x-axis is given by

$$V = \int_a^b \pi (f(x))^2 dx$$

Example

Find the volume of the sphere generated by rotating the semi-circle $y = \sqrt{(a^2 - x^2)}$ about the x-axis (between $x = -a$ and $x = a$).

Solution

The volume is given by

$$V = \int_{-a}^a \pi (a^2 - x^2) dx$$
$$= \pi(a^2x - x^3/3) |_{-a}^a$$
$$= 4/3\pi a^3$$

The *length* of a *curve* $y = f(x)$ from $x = a$ to $x = b$ is given by

$$L = \int_a^b \sqrt{1 + (f(x))^2}\, dx$$

Example

Find the length of the curve $y = f(x) = x^{3/2}$ from (0, 0) to (4, 8).

Solution

The derivative $f'(x)$ is given by: $f'(x) = {}^3/_2 x^{1/2}$ and so $(f'(x))^2 = {}^9/_4 x$.

The length is then given by

$$L = \int_0^4 \sqrt{1 + 9/4x}\,dx$$
$$= 2/3\ 4/9(1 + 9/4x)^{3/2} |_0^4$$
$$= 8/27(10^{3/2} - 1)$$

There are various rules for approximating definite integrals including the Trapezoidal Rule and Simpson's Rule. The *Trapezoidal Rule* approximates short stretches of the curve with line segments and the sum of the areas of the trapezoids under the curve is calculated and used as the approximate to the definite integral. *Simpson's Rule* approximates short stretches of the curve with parabolas and the sum of the areas under the parabolic arcs is calculated and used as the approximate to the definite integral.

The approximation of the Trapezoidal Rule to the definite integral is given by the following formula:

$$\int_a^b f(x)dx \approx \frac{h}{2}(y_0 + 2y_1 + 2y_2 + \ldots + 2y_{n-1} + y_n)$$

where there are n sub-intervals and $h = (b - a)/n$.

Example

Determine an approximation to the definite integral $\int^2{}_1 x^2 dx$ using the Trapezoidal Rule with $n = 4$, and compare to the exact value of the integral.

Solution

The approximate value is given by

$$1/4/2(1 + 2 * 1.5625 + 2 * 2.25 + 2 * 3.0625 + 4)$$
$$= 1/8(18.75)$$
$$= 2.34$$

The exact value is given by: $\int_1^2 x^2 dx = x^3/3 |_1^2 = 2.33$

The approximation of Simpson's Rule to the definite integral is given by the following formula:

$$\int_a^b f(x)dx \approx \frac{h}{3}(y_0 + 4y_1 + 2y_2 + +4y_3 \ldots +2y_{n-2} + 4y_{n-1} + y_n)$$

where there are n sub-intervals (n is even) and $h = (b - a)/n$.

18.2 Fourier Series

Fourier series is named after Jean Baptiste Fourier, a 19th-century French mathematician, who was active in developing theories of heat and in developing expansions of functions as trigonometric series that are used to solve practical problems in physics. A Fourier series breaks down any periodic function into a simple series of sine and cosine waves, and it consists of the sum of a possibly infinite set of sine and cosine functions. The Fourier series of a periodic function f (with period $2l$) on the interval $-l \le x \le l$ defines a function f whose value at each point is the sum of the series for that value of x.

The sine and cosine functions are periodic functions.

$$f(x) = \frac{a_0}{2} + \sum_{m=1}^{\infty} \left[a_m \cos \frac{m\pi x}{l} + b_m \sin \frac{m\pi x}{l} \right]$$

Note 1 (Period of Function)

A function f is periodic with period $T > 0$ if $f(x + T) = f(x)$ for every value of x. The sine and cosine functions are periodic with period 2π, i.e., $\sin(x + 2\pi) = \sin(x)$ and $\cos(x + 2\pi) = \cos(x)$. The functions $\sin m\pi x/l$ and $\cos m\pi x/l$ have period $T = 2l/m$.

Note 2 (Orthogonality)

Two functions f and g are said to be orthogonal on $a \le x \le b$ if:

$$\int_a^b f(x)g(x)dx = 0$$

A set of functions is said to be mutually orthogonal if each distinct pair in the set is orthogonal. The functions $\sin m\pi x/l$ and $\cos m\pi x/l$ where $m = 1, 2, \ldots$ form a mutually orthogonal set of functions on the interval $-l \le x \le l$ and they satisfy the following orthogonal relations as specified in Table 18.1.

TABLE 18.1

Orthogonality Properties of Sine and Cosine

Orthogonality Properties of Sine and Cosine
$\int_{-l}^{l} \cos \frac{m\pi x}{l} \sin \frac{n\pi x}{l} dx = 0 \quad \text{all } m, n$
$\int_{-l}^{l} \cos \frac{m\pi x}{l} \cos \frac{n\pi x}{l} dx = \begin{cases} 0 & m \ne n \\ l & m = n \end{cases}$
$\int_{-l}^{l} \sin \frac{m\pi x}{l} \sin \frac{n\pi x}{l} dx = \begin{cases} 0 & m \ne n \\ l & m = n \end{cases}$

The orthogonality property of the set of sine and cosine functions allows the coefficients of the Fourier series to be determined. Thus, the coefficients a_n and b_n for the convergent Fourier series $f(x)$ are given by

$$a_n = \frac{1}{l}\int_{-l}^{l} f(x)\cos\frac{n\pi x}{l}dx \quad n = 0, 1, 2, \ldots.$$

$$b_n = \frac{1}{l}\int_{-l}^{l} f(x)\sin\frac{n\pi x}{l}dx \quad n = 1, 2, \ldots.$$

The values of the coefficients a_n and b_n are determined from the integrals and the ease of computation depends on the particular function f involved.

The values of a_n and b_n depend only on the value of $f(x)$ in the interval $-l \le x \le l$. The terms in the Fourier series are periodic with period $2l$ and the function converges for all x whenever it converges on $-l \le x \le l$. Further, its sum is a periodic function with period $2l$ and therefore $f(x)$ is determined for all x by its values in the interval $-l \le x \le l$.

$$f(x) = \frac{a_0}{2} + \sum_{m=1}^{\infty}\left(a_m \cos\frac{m\pi x}{l} + b_m \sin\frac{m\pi x}{l}\right)$$

Often, the period of the function is 2π and so the formula for the Fourier series on the interval $-\pi \le x \le \pi$ defines a function f whose value at each point is the sum of the series for that value of x.

$$f(x) = \frac{a_0}{2} + \sum_{m=1}^{\infty}(a_m \cos mx + b_m \sin mx)$$

The coefficients a_n and b_n are given by

$$a_n = \frac{1}{\pi}\int_{-\pi}^{\pi} f(x)\cos nx\ dx \quad n = 0, 1, 2, \ldots.$$

$$b_n = \frac{1}{\pi}\int_{-\pi}^{\pi} f(x)\sin nx\ dx \quad n = 1, 2, \ldots.$$

The ease of determining the coefficients a_n and b_n depends on the function $f(x)$. If $f(x)$ is *even* the b_n's are all zero, and so all that is required is to calculate the a_n's, and so an even function only has cosine terms in its Fourier expansion. Similarly, if $f(x)$ is *odd* the a_n's are all zero, and so all that is required is to calculate the b_n's, and so an odd function only has sine terms in its Fourier expansion.

18.3 The Laplace Transform

An integral transform takes a function f and transforms it into another function F by means of an integral. Often, the objective is to transform a problem for f into a simpler problem, and then to recover the desired function from its transform F. Integral transforms are useful in solving differential equations, and an integral transform is a relation of the form:

$$F(s) = \int_{\alpha}^{\beta} K(s, t) f(t) dt$$

The function F is said to be the transform of f and the function K is called the kernel of the transformation.

The Laplace transform is named after the well-known 18th-century French mathematician and astronomer, Pierre Laplace. The *Laplace transform* of f (denoted by $\mathcal{L}\{f(t)\}$ or $F(s)$) is given by

$$\mathcal{L}\{f(t)\} = F(s) = \int_{0}^{\infty} e^{-st} f(t) dt$$

The kernel $K(s, t)$ of the transformation is e^{-st} and the Laplace transform is defined over an integral from 0 to infinity. This is defined as a limit of integrals over finite intervals as follows:

$$\int_{a}^{\infty} f(t) dt = \lim_{A \to \infty} \int_{a}^{A} f(t) dt$$

Theorem (Sufficient Condition for Existence of Laplace Transform)
Suppose that f is a piecewise continuous function on the interval $0 \leq x \leq A$ for any positive A and $|f(t)| \leq Ke^{at}$ when $t \geq M$ where a, K, M are constants and $K, M > 0$, then the Laplace transform $\mathcal{L}\{f(t)\} = F(s)$ exists for $s > a$.

The following examples are Laplace transforms of some well-known elementary functions:

$$L\{1\} = \int_{0}^{\infty} e^{-st} dt = \frac{1}{s}, \quad s > 0$$

$$L\{e^{at}\} = \int_{0}^{\infty} e^{-st} e^{at} dt = \frac{1}{s - a} \quad s > 0$$

$$L\{\sin at\} = \int_{0}^{\infty} e^{-st} \sin at \; dt = \frac{a}{s^2 + a^2} \quad s > 0$$

Example

Find the Laplace transform of the function $f(t) = t$.

Solution

$$F(s) = \int_0^{\infty} e^{-st} t \; dt$$

We use integration by parts to evaluate the integral and get:

$$\int_0^{\infty} e^{-st} t \; dt = \frac{-te^{-st}}{s} \Big|_0^{\infty} + \frac{1}{s}\int_0^{\infty} e^{-st} dt$$

$$= -\left(\frac{t}{s} + \frac{1}{s^2}\right) e^{-st} \Big|_0^{\infty}$$

$$= \begin{cases} 1/s^2 & s > 0 \\ \infty & s \leq 0 \end{cases}$$

18.4 Differential Equations

Many important problems in engineering and physics involve determining a solution to an equation that contains one or more derivatives of the unknown function. Such an equation is termed a differential equation, and the study of these equations began with the development of the calculus by Newton and Leibnitz.

Differential equations are classified as ordinary or partial based on whether the unknown function depends on a single independent variable or on several independent variables. In the first case, only ordinary derivatives appear in the differential equation and it is said to be an *ordinary differential equation*. In the second case, the derivatives are partial and the equation is termed *a partial differential equation*.

For example, Newton's second law of motion ($F = ma$) expresses the relationship between the force exerted on an object of mass m and the acceleration of the object. The force vector is in the same direction as the acceleration vector. It is given by the ordinary differential equation.

The next example is that of a second-order partial differential equation. It is the wave equation and is

$$m\frac{d^2x(t)}{dt^2} = F(x(t))$$

which is used for the description of waves (e.g., sound, light and water waves) as they occur in physics. It is given by

$$a^2 \frac{\partial^2 u(x, t)}{\partial x^2} = \frac{\partial^2 u(x, t)}{\partial t^2}$$

There are several fundamental questions with respect to a given differential equation. First, there is the question as to the existence of a solution to the differential equation. Second, if it does have a solution then is this solution unique. The third question is how to determine a solution to a particular differential equation.

Differential equations are classified as to whether they are linear or non-linear. The ordinary differential equation $F(x, y, y', \ldots, y^{(n)}) = 0$ is said to be *linear* if F is a linear function of the variables $y, y', \ldots, y^{(n)}$. The general ordinary differential equation is of the form:

$$a_0(x)y^{(n)} + a_1(x)y^{(n-1)} + , \ldots, a_n(x)y = g(x)$$

A similar definition applies to partial differential equations and an equation is *non-linear* if it is not linear. For more detailed information on differential equations, see [BoD:92].

18.5 Review Questions

1. What is the difference between velocity and acceleration?
2. How fast does the radius of a spherical soap bubble change when air is blown into it at the rate of 10 cm^3/sec?
3. Find the area under the curve $y = x^3 - 4x$ and the x-axis between $x = -2$ and 0.
4. Find the area between the curves $y = x - 2x$ and $y = x^{1/2}$ between $x = 2$ and 4.
5. Determine the volume of the figure generated by revolving the line $x + y = 2$ about the x-axis bounded by $x = 0$ and $y = 0$.
6. Determine the length of the curve $y = {}^1/_3\,(x^2+2)^{3/2}$ from $x = 0$ to $x = 3$.
7. What is a periodic function and give examples.
8. Describe applications of Fourier series, Laplace transforms and differential equations.

18.6 Summary

This chapter provided a short account of applications of calculus to calculate the velocity and acceleration of moving bodies and problems involving rates of change and maxima/minima problems.

We showed that integration allows the area under a curve to be calculated and the area of the region between two curves to be computed. We discussed numerical analysis, Fourier series, Laplace transforms and differential equations.

Numerical analysis is concerned with devising methods for approximating solutions to mathematical problems. Often an exact formula is not available for solving a particular problem, and numerical analysis provides techniques to approximate the solution in an efficient manner. We discussed the Trapezoidal Rule and Simpson's Rule, which provide an approximation to the definite integral.

A Fourier series consists of the sum of a possibly infinite set of sine and cosine functions. A differential equation is an equation that contains one or more derivatives of an unknown function.

This chapter has sketched some important results in the calculus, and the reader is referred to [Fin:88] for more detailed information.

19

A Brief Introduction to Economics

Economics is the social science that studies the production, distribution and consumption of goods and services. The word "economy" is derived from two Greek words (*oikos* meaning house) and (*nemein* meaning to manage), so the original meaning of the term was household management. Households have limited resources and the management of them requires many decisions and a certain level of organization. Households and firms are the basic unit of the economy, and are concerned with the problem of satisfying unlimited desires with limited (or scarce) resources.

The meaning of the word "economics" has changed over time, and today it refers to the study of how societies make choices on what, how and to whom to produce given their limited resources. The key economic problem is how to reconcile the conflict between people's virtually unlimited desires with limited resources and means of production.

Economics includes many economic models, and these are typically mathematical models presented in visual charts, or in formal mathematical notation using algebra and basic calculus. The mathematical models are simplifications of the economic reality, and they allow understanding and predictions (forecasting) of economic behaviour.

Economics consists of two major fields: *macroeconomics*, which is concerned with how the overall economy works and so it is concerned with the behaviour of the entire economy; and *microeconomics*, which is concerned with the behaviour of individuals and businesses and their interactions, and especially how supply and demand interact in individual markets.

Macroeconomics is the branch of economics that studies how an overall economy behaves. It analyses the entire economy including production consumption and savings, as well as issues affecting the economy (e.g., it studies Gross Domestic Product [GDP], the rate of economic growth, inflation, deflation, employment and unemployment, overheating, bubbles, economic shocks, recessions and depressions). It looks at the output of the whole economy and how the economy allocates its scarce resources to maximize production.

Microeconomics analyses individual elements in an economy such as households and businesses, as well as buyers and sellers. Microeconomics looks at what happens when individuals make certain choices, where the individuals are buyers, sellers and business owners. They interact with each other according to the laws of supply and demand for scarce resources, and microeconomics studies the interactions of individuals and firms.

DOI: 10.1201/9781003308140-19

Adam Smith (Fig. 19.1) was an 18th-century economist and philosopher, and he is considered the father of economics. Smith argued for a minimal role for government in the regulation of markets, and he argued that each person by looking out for himself (or herself) inadvertently creates the best outcome for all.

That is, by selling products that people want to buy, a butcher, baker and brewer all hope to make money, and if they are effective in meeting their customers' needs they will earn financial rewards as well as create wealth for the entire nation. A wealthy nation is one in which people are working productively to better themselves to address their financial needs.

Smith published his influential book *The Wealth of Nations* in 1776 [Smi:76], and he argued that humans' natural tendency towards self-interest results in prosperity for all. Smith's thesis is that giving people the freedom to produce and exchange goods as they please, as well as opening the market to domestic and foreign competition, promotes greater prosperity

FIGURE 19.1
Adam Smith. Public Domain.

for society. *Another words, every person by looking out for themselves helps to create the best outcome for all.*

Smith's ideas on the importance of free markets, assembly line production methods and GDP formed the basis for theories of classical economics. Smith advocated for minimal government intervention in the market (*laissez fare*), but he recognized the need for institutions (e.g., justice) to protect free trade, and he stressed the importance of competition.

Smith argued that countries would mutually benefit from trade between each other, and his ideas on trade led to an increase in imports and exports between countries. He showed that assembly lines could be massively more productive than individual labour. He noted that one person executing the 18 steps required to produce a pin is able to produce only a handful of pins each week, whereas if an assembly line of 10 people carries out the 18 tasks then thousands of pins can be produced per week. Smith argued that division of labour and specialization leads to prosperity, and he noted that there is an "invisible hand" (or market force) turning the wheels of the economy that keeps the economy functioning.

He examined the evolution of society from hunter-gatherer societies without property rights, to nomadic societies with shifting residences, to feudal societies with laws to protect the property rights of the privileged class, to modern societies with free trade. Smith argued that GDP should be measured by the level of production and commerce rather than the existing system of mercantilism, which measured a country's wealth by its gold and silver deposits. Smith's ideas were to influence Marx and Ricardo in the 19th century, and Keynes/Friedman in the 20th century.

John Maynard Keynes published *The General Theory of Employment, Interest and Money* in the 1930s [Key:36], which provided an explanation for the fallout of the great depression in terms of unemployment and unsold goods. Keynesian economics argues that the business cycle can be managed by active government intervention in the economy. This is done through *fiscal policy* (spending more during recessions in order to stimulate demand) and *monetary policy* (stimulating demand with lower interest rates).

Milton Friedman founded the Monetarist school of economics, and it argues that the role of government is to control inflation by controlling the money supply. Monetarists reject the Keynesian belief that governments can manage demand, and they argue that attempts to do so result in inflation.

19.1 Macroeconomics

Macroeconomics is the branch of economics that studies how an overall economy behaves. It looks at the overall big picture and analyses the entire economy including production, consumption and savings, as well as issues affecting the economy such as inflation and unemployment. It looks at how the economy allocates its scarce resources in order to maximize production.

Macroeconomics examines questions such as: What causes unemployment or inflation? How can they be controlled? What is the role of government in managing the economy? What is needed to stimulate economic growth? How is the economy currently performing? How can the performance of the economy be improved?

Macroeconomists create formal mathematical models to study and predict various economic factors such as inflation, unemployment, economic growth and GDP. For example, economic growth is often modelled as a function of physical and human capital, the labour force and technology. The government then uses the macroeconomic models and their forecasts in managing their fiscal and monetary policies.

Government budgets and economic policy have a major impact on business and consumers, so macroeconomics and their associated models provide a useful insight into how economies function, as well as the impacts and consequences of particular economic policies and decisions. Economic models are simplifications of the complexities of the real world, and all models are subject to limitations. There is the well-known aphorism that *"All models are wrong, but some are useful."*

The emergence of modern economics and macroeconomics is due to the British economist, John Maynard Keynes, whose ideas fundamentally changed economic thinking in the 1930s (Fig. 19.2). Keynes argued that total spending in the economy (*aggregate demand*) determines the total level of economic activity, and that inadequate spending may lead to negative

FIGURE 19.2
John Maynard Keynes. Public Domain.

economic consequences such as periods of high unemployment. He argued for the use of *fiscal and monetary tools to mitigate and manage economic recessions* and depressions, and his approach is described in his influential book [Key:36]. In a nutshell, Keynes argues that governments should spend money they don't have in times of crisis, which is paid back by taxpayers in the future.

Keynesian remained the dominant economic theory embraced by western countries up to the late 1970s, when the ideas of Milton Friedman and other monetarists became dominant. However, Keynesian economics became popular again in dealing with the 2007–2008 financial crisis, and the COVID-19 pandemic in 2020 led to Keynesian economics being employed on a grand scale around the world. The latter led to national bonds being issued to borrow money on a vast scale to be paid back by future income of taxpayers. It led to massive government intervention in the economy with entire industries bailed out, and the emphasis moved away from balancing the budget and controlling inflation to saving the economies of the world.

19.2 Microeconomics

The word "microeconomics" is derived from the Greek word *mikro* meaning small and economics. Microeconomics is a branch of economics that analyses the behaviour of individual elements in an economy such as households and businesses as well as buyers and sellers. It studies how businesses make decisions in the allocation of scarce resources, as well as the interactions between businesses and consumers.

That is, microeconomics looks at what happens when individuals make certain choices, where the individuals are buyers, sellers and business owners. These actors interact with each other according to the laws of supply and demand for scarce resources, and microeconomics studies the interactions of these individuals and firms.

Microeconomics examines the effect of actions on supply and demand. Demand is an economic principle that refers to a consumer's desire to purchase goods and services, and willingness to pay a certain price for the good or service. In general, the demand for a good or service falls when the price increases, and increases when the price falls. There are many microeconomic models such as the supply demand model, which is a model of price determination in a perfectly competitive market. The unit price of a good is the price at which the quantity demanded by consumers is equal to the quantity supplied by producers (Fig. 19.3).

Microeconomics helps in answering questions such as: How do companies decide what to charge for new gadgets? Why are some people willing to pay more than others? How does consumer behaviour play into how

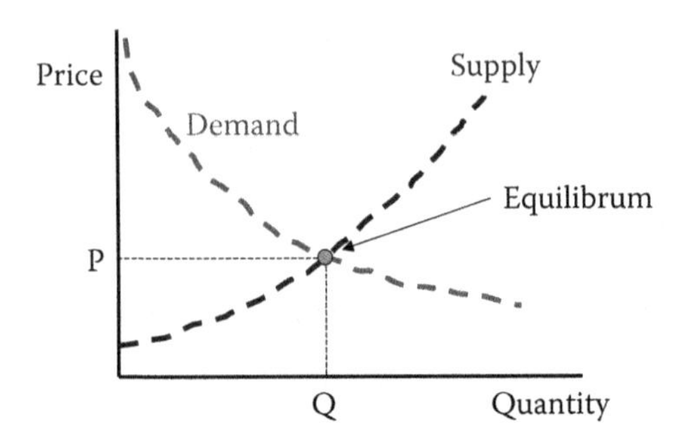

FIGURE 19.3
Law of Supply and Demand.

companies price their products? Microeconomics assumes the following principles:

- People are rational in decision-making and aim to maximize utility.
- Businesses make decisions based on the competition that they face.
- Both individuals and consumers take opportunity cost into account when making decisions.

The concept of *utility* is fundamental in microeconomics and it represents the benefits/happiness that an individual gains from the purchase of a particular good or service. The more benefit that the product provides the more that the consumer is generally willing to pay for it, and consumers assign different levels of utility to different goods. This leads to different levels of demand for different goods.

The quantity of a product that consumers are willing to buy at a given price is termed the *demand*, whereas the quantity of a product that suppliers are willing to sell at a given price is termed the *supply*. The market brings buyers and sellers together (i.e., demand and supply meet in the market), and the role of price is to coordinate the quantity of the product that the sellers wish to sell with the quantity that the buyers wish to purchase. That is, when demand and supply come together in the market place, this determines the price and quantity of goods sold.

The marginal revenue (MR) is informally defined as the increase in total revenue (TR) as a result of the sale of one extra unit of output, and the precise definition is that it is the rate of change of total revenue relative to increases in output. That is, the marginal revenue is given by the rate of change of the TR function and so we just calculate the derivative of the TR function.

$$MR = \lim_{Q \to 0} \frac{\Delta TR}{\Delta Q} = \frac{dTR}{dQ}$$

Similarly, the marginal cost (MC) is the rate of change of total cost function (TC) relative to increases in output. That is, the marginal cost is given by the rate of change of the TC function and so we just calculate its derivative.

$$MC = \lim_{Q \to 0} \frac{\Delta TC}{\Delta Q} = \frac{dTC}{dQ}$$

19.3 GDP

GDP is the total value of all the goods and services produced in a country during a specific time period (Fig. 19.4). It is usually calculated on a yearly basis but it may also be calculated quarterly. It provides a snapshot of the economic health of a country, including the size of its economy and its rate of economic growth. GDP data is widely used by economists, investors and governments.

Economists use GDP data to determine the performance of the economy, i.e., whether the economy is growing or experiencing a recession. Investors may use GDP data as a tool to make investment decisions, as poor GDP data

FIGURE 19.4
Gold Bullion Bars. Public Domain.

may indicate a weak economy with lower earnings of corporations and thus lower stock prices. Government will monitor GDP data to ensure that their economic policies are effective, and whether fiscal or monetary intervention is required.

GDP includes all public and private consumption, government expenditure, investment and the balance of trade (exports added and imports subtracted). There are various methods for calculating GDP such as by expenditure (spending), production (output) or income. These methods should yield the same answer.

The *expenditure approach* to calculating GDP involves determining the spending by different groups that participate in the economy. These groups include the government, consumers, private investment and net exports. It uses the following formula:

$$\text{GDP} = \text{C} + \text{G} + \text{I} + \text{NX},$$

where C refers to consumer spending which is the largest component of GDP, so consumer confidence has a major impact on economic growth; G refers to government current and capital spending; I refers to private investment; and, finally, NX refers to the net exports (i.e., total exports – total imports).

The *production approach* to calculating GDP involves estimating the total value of economic output, and deducting the costs of intermediate goods that are consumed in the process such as materials and services. The production approach in a sense looks backwards from a state of completed economic activity.

The *income approach* considers all of the income earned by the factors of production in the economy, which includes wages (salaries), rent, interest and profits. There may be some adjustments to be made to deal with indirect business taxes and depreciation.

GDP per capita measures GDP per person in the population, and it allows a comparison of GDP data between countries. However, care is required in drawing conclusions, as a country with a GDP per capita of $30,000 may appear to be 10 times wealthier than a country with GDP per capita of $3000. However, the cost of goods and services may be substantially different between both countries (i.e., the cost of goods might be significantly cheaper in the second country).

Purchasing power parity (PPP) attempts to solve this problem, and it compares how many goods and services may be purchased for a given unit of money in different countries (i.e., it compares a basket of items in two countries after adjusting for the exchange rate). It is possible (unlikely) that the person with an income of $3000 may be able to purchase more goods than a person with an income of $30,000 due to the differences in prices between both countries.

We distinguish between *Nominal GDP* and *Real GDP*, where nominal GDP uses current market prices and so includes the inflation in prices during the period, whereas for real GDP adjustments are made to deal with inflation in the period (GDP price deflator). This allows a country's GDP to be compared one year to another. Real GDP may also be adjusted for PPP.

There are *limitations* to GDP as a measurement of the health, wealth and happiness of a country. GDP includes market transactions only, so it gives no information on the level of equality and inequality in income distribution in the economy. It gives no information on the well-being or welfare of the country and does not include unpaid voluntary work that has a positive impact on society, or black market transactions that have a negative impact on society. GDP measures the value of all finished products, but it does not take into account what products are being produced (e.g., the country may be active in the arms industry). Finally, GDP takes no account of its impacts on natural resources and its contribution to climate warming, carbon emissions and pollution.

Gross National Income (GNI) is the sum of all income earned by residents of a country, irrespective of whether the economic activity takes place at home or abroad. GNI is calculated as gross domestic income plus its indirect business taxes and depreciation and its net foreign factor income.

GNI is a more modern version of *Gross National Product* (GNP), which uses the production approach in its calculation. GNI is a better metric than GDP for some countries (e.g., the GDP of Ireland is distorted by the activities of multinational companies that seek to optimize their taxation affairs, as well as repatriate their profits to their home country).

19.4 Utility

At the core of consumer decision-making is the concept of utility (the individual benefit from the purchase of a product), and consumers will generally purchase only those products that will provide benefit and pleasure, and customers will seek to maximize their utility in consumption. Further, in general, the more benefit that the consumer feels that the product provides, the more that the consumer is willing to pay for it. Consumers have limited resources, so they need to be rational in their decision-making, and in general consumers will purchase those products that will maximize their utility.

The concept of utility was introduced by Jeremy Bentham in his 1789 book *Principles of Morals and Legislation* [Ben:1789]. Bentham defines utility as the property in any object that tends to produce benefit, pleasure, good, advantage and happiness. He also developed the influential philosophy of utilitarianism (an ethical theory that determines right from wrong based on outcomes, and the utilitarian theory of ethics argues for ethical choices that promote the greatest good).

Consumers generally assign different levels of utility to different goods, and this leads to different levels of demand for different goods. Further, a

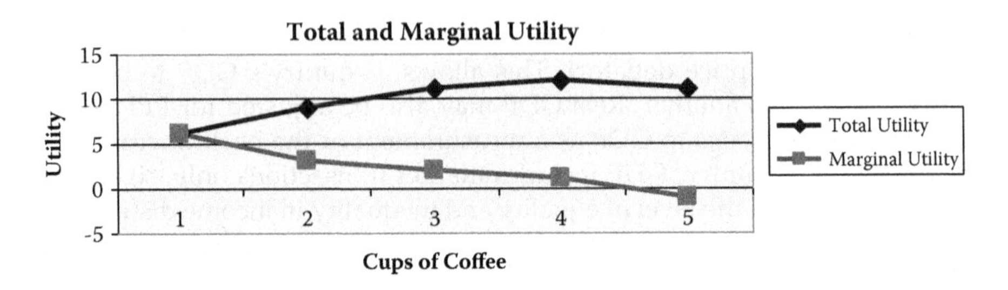

FIGURE 19.5
Total and Marginal Utility.

consumer may buy any number of units of a good, and this leads to the concept of *marginal utility*, which gives the extra utility or satisfaction gained from consuming an additional unit of a good. The total utility is the total satisfaction gained from purchasing the good, and a rational consumer will seek to maximize their utility.

The concept of marginal utility may be illustrated by the purchase and consumption of several cups of coffee (Fig. 19.5). The first cup might give a lot of satisfaction, but the satisfaction of the fifth cup will be considerably less than the first. The total utility will generally increase with each cup, but the marginal utility will generally decrease with each additional cup (it may even become negative).

Scarcity refers to the tension between our almost unlimited desires and the limited resources (time, skilled people, capital, natural resources) to fulfil them. The limitation of resources to unlimited desires means that individuals and economies need to determine how best to allocate scarce resources. That is, they need to decide what goods and services they can produce and which ones they must forego.

The *opportunity cost* refers to the fact that when a consumer or business decides to produce or purchase a particular product it does so at the expense of buying or producing something else. That is, the opportunity cost in consumption is what is foregone to have something else (e.g., we may need to choose between having banoffi pie instead of cheesecake for dessert). Similarly, production requires dedicated resources, so if a business decides to produce a car it is unlikely to have the appropriate resources to produce sewing machines, as these are very different products with very different manufacturing processes and specialized labour.

19.5 Comparative and Absolute Advantages

Absolute advantage and comparative advantage are two important concepts in economics and international trade. They influence how and why companies and economies focus on the production of particular goods.

A country can try to produce all the goods and services that it needs, but this often leads to inefficient utilization of resources. By *specialization*, the country can focus on the production of the things that it does best, rather than dividing up its resources and producing some things that it does well and other things that it produces less efficiently.

Absolute advantage describes the situation where a business or country can produce a higher quality product at a faster rate than another competing business or country. That is, it is superior in the production of the good. *Comparative advantage* differs from the absolute advantage in that it takes opportunity costs into account in its analysis and decision on which goods to produce with limited resources.

Adam Smith argued that countries should specialize in the production of those goods that they can produce most efficiently, and trade with other countries for those goods that they are unable to produce as efficiently [Smi:76]. The term *comparative advantage* was introduced by David Ricardo in the 19th century in his book *On the Principles of Political and Taxation* which was published in 1817 [Ric:1817]. Ricardo explained how countries may benefit from trading even if one of them has an absolute advantage in the production of all goods.

Law of Comparative Advantage

The law of comparative advantage is an important concept in economics, and it is a fundamental principle in the theory of international trade. It states that all countries at all times can mutually benefit from cooperation and voluntary trade, and that all countries are better off if they specialize and produce those goods/services for which they *produce most efficiently and have a comparative advantage*, and trade with other countries for those goods that they are inefficient in producing.

The comparative advantage is from the fact that producing a certain good requires fewer resources and has a lower opportunity cost. An *opportunity cost* is the potential benefit that someone loses out on by selecting one option over another. A country is said to have a comparative advantage in whichever good has the lowest opportunity cost. That is, it has a comparative advantage in whichever good it sacrifices the least to make.

19.6 Elasticity of Demand

The concept of *elasticity* is important in economics, and it indicates how sensitive the demand for a product is to a change in its price. The consumption of some goods (e.g., life-saving pharmaceutical drugs) will hardly fall in response to significant price increases, whereas the demand for other goods (e.g., a particular fruit or chocolate) may change dramatically in response to a change in its price (i.e., consumers may respond to a price increase by purchasing less of the good, or they may choose a different product). In general, people desire things less as they become more expensive, and elasticity indicates how sensitive the product is to price increases. The mathematical definition of price elasticity of demand is:

$$\text{PED} = \%\text{ change in quantity} / \%\text{ change in price}$$
$$\text{PED} = \Delta Q/\Delta P * P/Q$$

$$\text{PED} = \lim_{\Delta P \to 0} \frac{\Delta Q/Q}{\Delta P/P} = \frac{P}{Q} \lim_{\Delta P \to 0} \frac{\Delta Q}{\Delta P} = \frac{P}{Q}\frac{dQ}{dP} = \frac{P}{Q}\frac{1}{dP/dQ}$$

A product where a large change in its price does not affect its demand is said to have *inelastic* demand. A product where a small change in its price results in a large change in its demand is termed *elastic*. Demand is elastic if the absolute value of the elasticity is greater than 1 (i.e., elasticity < -1 or elasticity > 1), where demand is said to be inelastic if $-1 \leq$ elasticity ≤ 1.

The change in the price of a good i can affect the demand of a related good j (e.g., a *substitute good* or *complementary good*). This is given by the formula for cross elasticity of demand (and is positive for substitutes and negative for complements). This is since if the price of good i increases then demand for product j decreases (complementary good), or increases (substitute good).

Example (Elasticity)

What is the elasticity when the price is 12 for the demand function $p = 60 - 3q$?

Solution (Elasticity)

$$\begin{aligned} p &= 60 - 3q \\ \Rightarrow 3q &= 60 - p \\ \Rightarrow 3q &= 60 - 12 \\ \Rightarrow 3q &= 48 \\ \Rightarrow q &= 16 \end{aligned}$$

$$dp/dq = -3$$

$$\begin{aligned} \text{PED} &= 12/16 * 1/-3 \\ &= \tfrac{3}{4} * -1/3 \\ &= -\tfrac{1}{4} \end{aligned}$$

19.7 Role of Money

Money provides a widely accepted way of paying for goods and services, and it is an integral part of the economy. It acts as a *medium of exchange* (it can be exchanged for goods or services), *a unit of account* (it is measured in a specific currency which allows the price of things to be compared) and a *store of value* (it can be saved and used at a later time). Money includes cash, demand deposits and checking accounts, and it plays a key role in facilitating the transfer of goods/services.

The price of money is the opportunity cost of holding the money rather than investing it (the interest rate for bonds provides a return to an investor, so the interest rate for bonds is essentially the price of money). There are various motivations for holding money such as a transaction motive (to pay for goods and services), precautionary motive (money is a relatively safe investment) and the speculative (or asset) motive where having cash available can provide a return (e.g., investors/traders can take advantage of any investment opportunities that arise).

There are various definitions of the money supply in the economy, including the M_1, M_2 and M_3 measures, where $M_1 \leq M_2 \leq M_3$. M_1 consists of all the cash in circulation; M_2 consists of M_1 plus small deposits, short-term deposits and money market funds; M_3 consists of M_2 plus large deposits, long-term deposits and money market funds.

The money supply is greater than the money base by a factor known as the money multiplier, which depends on the banks' reserve ratios and the public's holding of cash relative to deposits.

19.8 Role of Government in the Economy

The role of government in the 19th century was *laissez-faire* (leave alone, i.e., the less that government was involved in the economy the better), and its involvement was limited to the preservation of law and order, the protection of private property, taxation to fund the police and military, while giving business as much freedom as possible.

The role of government today includes intervention in the economy to manage recessions or overheating of the economy when they occur, and government policies (e.g., their spending and taxation policies) can stimulate or depress demand and business activity. The role of government in modern society has increased substantially, with government spending often in the range of 30–40% of GDP. Government revenue is a mixture of taxation on business and employees as well as fees for the services that it provides. The role of government in the economy is summarized in Table 19.1.

TABLE 19.1

Role of Government in Economy

Role	Description
Provision of services	The government provides many services such as health and education, social welfare services such as unemployment and pension payments, law and order, including the police, military and judiciary, fire services, transport and postal services.
Taxation	This involves direct and indirect taxes to raise revenue from the residents of the state. The taxes provide most of the revenue that enables the government to provide its services.
Regulation	The government sets the legal framework for markets to operate, and regulation refer to the rules that the market participants must satisfy in order to enter and participate in a particular market. There may be regulations on mergers and monopolies, the environment and employment.
Redistribution to eliminate poverty/ inequality	An unregulated free market may result in major inequalities and poverty. This may require progressive taxation and income distribution (e.g., higher taxes for high-income earners and financial support for lower income workers and the unemployed). It may include policies for equal access to education, as education plays a key role in poverty reduction.
Manage economic cycle	This involves using fiscal and monetary tools to promote economic growth and manage the economic cycle (e.g., recessions and overheating of economy), and manage key economic parameters such as inflation and unemployment.

19.9 Economic Indicators

An economic indicator is macroeconomic data that provides an indication of the health of the economy. These include *leading indicators* that are used to predict future movements of the economy (e.g., PMI, CCI, share prices); *coincident indicators* that are seen with the occurrence of economic activity (e.g., unemployment levels, retail sales); and *lagging indicators* that are seen only after the economic activity takes place (e.g., GDP, inflation). Table 19.2 summarizes some common economic indicators:

19.10 Competition

Competition is an essential part of the free market economy, and it refers to how a company endeavours to produce and sell a high-quality product at a

TABLE 19.2

Economic Indicators

Indicator	Description
Inflation (CPI)	Inflation is a measure of the rate of increase in the price of a basket of selected goods over a period of time. It is an indication of the level that the prices for goods and services are increasing.
Interest rate	The central bank is responsible for setting interest rates in the economy, and interest rates are a key tool of monetary policy. Central banks reduce rates when they wish to stimulate demand in the economy, and increase rates when they wish to reduce economic activity (to prevent overheating or an economic bubble which could occur in a low-interest economy).
Unemployment	Unemployment is a measure of the health of the economy, and the unemployment rate is the ratio of the number of unemployed in the economy to the size of the labour force. High rates of unemployment indicate economic distress, whereas very low levels of unemployment may indicate potential overheating in the economy.
Purchasing Managers Index (PMI)	The Purchasing Managers Index is an index of the purchasing trends in the manufacturing and service sectors. It is derived from surveys of purchasing managers, and indicates whether market conditions and economic activity are expanding, staying the same or contracting.
Consumer Confidence Index (CCI)	The Consumer Confidence Index is an economic indicator that measures consumer confidence. It indicates how optimistic consumers are in the state of the economy. It is calculated each month from a household survey of consumers' opinions on the economy and their expectations for the future.
Current account deficit/ surplus	The current account deficit is a measure of a country's trade with the rest of the world, where the value of a country's exports is less than a country's imports. A current account surplus is where the value of its exports exceeds that of its imports.
Balance of payments (BOP)	The balance of payments summarizes all transactions (e.g., imports, exports, financial payments) made over a period of time by individuals, companies and governments with others outside the country. It is divided into current account and capital account.

lower price than its competitors. Competition generally results in efficiencies and lower prices for consumers.

Competition in the market place is influenced by several factors such as the features of the product (how unique or different the product is from existing products in the market that are available from competitors); barriers to entry (how easy is it for a firm to enter the market place); the number of sellers (the more sellers of a similar product the greater the competition, and if there are few sellers then competition is low); information availability (where customers are able to determine the prices offered by competitors); and the location of a seller.

Under normal competition, there is a large number of players in the market with minimal barriers to entry. Competition can be healthy leading to goods being produced at the lowest possible cost, and competition among sellers helps to ensure that consumers pay a fair price for the product.

A *monopoly* is a single company that dominates the market, and where there are large barriers to entry into the market. In this situation, a single company sets the price for the product, as there are no competitors offering similar products. There are several examples of monopolies (e.g., a single utility company serving a particular region of a country, or a pharmaceutical company with a patent on a drug, where the royalties from the licensing of the patent acts as a barrier to entry).

An *oligopoly* refers to the situation where a small number of companies dominate a particular market, and the companies cooperate together to reduce competition leading to higher prices for consumers. The Organization of the Petroleum Exporting Countries (OPEC) acts as a *cartel* (a special type of oligopoly), which has formal agreements among its members to fix prices and control production quantities.

19.11 Globalization and Terms of Trade

Globalization is the process by which the world has become increasingly interconnected as a result of a massive increase in international trade and cultural exchange. The process of globalization has been influenced by improvements in transportation of both goods and people; the lowering of barriers and promotion of free trade by organizations such as the WTO; increased communication between people around the world from modern technologies such as the Internet and mobile phone; and the availability of a large pool of low-cost labour (including both low skilled and high skilled workers) from countries such as India and China.

Globalization has led companies to set up branch offices and factories in many countries around the world, and these multinational companies have created inward investment for many countries resulting in job creation and wealth for local economies. It has led to a sharing of ideas and experiences between different cultures, as well as increased awareness of events in distant parts of the world. It is led to increased awareness of global problems such as global warming and climate change as well as the importance of sustainable development.

There are negative impacts associated with globalization in that it operates mainly in the interest of the rich counties that dominate world trade at the expense of developing countries. Often, the driver of opening up a new factory in a country is lower costs, so if cheaper alternatives become available in another country the multinational may leave for that country making local workers redundant. Further, the regulations may be less stringent than in the multinational's home country, and so it is possible that the lack of controls could lead to pollution or damage to the environment or exploitation of local workers. There is also the danger of an erosion of cultural diversity, with local economies and societies becoming more like the western world.

The *terms of trade* (TOT) are used as an economic indicator to monitor a country's economic health in international trade. The ratio is calculated by dividing the price of the country's exports by the price of its imports and multiplying by 100. Another words, changes in import prices and export prices impact the TOT, and an improvement in TOT generally means that export prices have gone up with import prices either steady or going down. Similarly, deterioration in TOT generally means that export prices have dropped with import prices remaining steady or increasing.

A country can purchase more imported goods for every unit of export that it sells when its TOT improves, and so an increase in TOT may be beneficial as a country needs fewer exports to buy a given number of imports. It may also have a beneficial impact on inflation, as it may be indicative of falling import prices relative to export prices.

Similarly, a country must export more goods to import the same number of goods when its TOT deteriorate. This occurs from time to time in some developing countries with an over-reliance on the export of commodities (e.g., oil and copper) and when there is a global fall in commodity prices. This may happen as well with manufactured goods, as globalization has reduced the price of goods.

19.12 Mathematics in Economics

The main mathematics used in economics includes algebra (including matrix algebra), calculus and probability and statistics. The use of mathematics is used to describe economic phenomena, and to draw inferences from their basic assumptions. For example, algebra is used to compute total costs and revenue; calculus is employed to determine the derivative of various curves such as the utility curve; and probability and statistics are used in decision-making and to make forecasts to determine the probability of an event occurring.

Economists create economic models using mathematics based on the assumptions that humans behave rationally (this is not always a valid assumption), and the models are used to make quantitative predictions about economic activity, or for decision-making. All models are simplifications of the underlying reality, and it is essential to test the model to determine its adequacy as a representation of the real world. The models may be tested empirically with quantitative real-world data (statistical data), and validated models are used to make quantitative predictions about economic activity, whereas inadequate models are replaced with models that give more accurate predictions.

Economic models are used in both macroeconomics and microeconomics. The central bank sets monetary policy and therefore needs to have economic models that give accurate predictions on the impacts of changes to key economic parameters. For example, it will need to have a model that gives an accurate prediction on the impact of an increase in the interest rate on inflation and on the rate of growth of GDP.

The mathematics employed in economics may be used for optimization problems such as determining the most efficient way that a business may allocate scarce resources, or the price that a business should charge for its products to maximize profits. The price of a product is generally determined from the intersection of the supply and demand curves, where the price is high when supply is limited and demand high, with the price low when supply is high but demand is low. Statistics is used to process a vast amount of data and may be employed to determine a country's GDP.

Example (Profit Maximization)

A business has the demand schedule $p = 184 - 4q$ and the TC function is given by TC = $q^3 - 21q^2 + 160q + 40$. What output will maximize profit?

Solution (Profit Maximization)

Profit maximization will occur where the marginal revenue is equal to marginal cost (MR = MC). The total revenue is given by the price multiplied by the quantity, so

$$\text{TR} = pq = (184 - 4q)q = 184q - 4q^2$$

The marginal revenue is given by the derivative of the total revenue function and so:

$$\text{MR} = 184 - 8q$$

The marginal cost is given by the derivative of the total cost function and so:

$$\text{MC} = 3q^2 - 42q + 160$$

At maximum profit MC = MR and so:

$$184 - 8q = 3q^2 - 42q + 160$$
$$3q^2 - 34q - 24 = 0$$
$$(3q + 2)(q - 12) = 0$$
$$(3q + 2) = 0 \text{ or } (q - 12) = 0$$
$$q = -2/3 \text{ or } q = 12$$

So the solution is $q = 12$ as a negative quantity cannot be produced.

For more detailed information on economics, see [Man:14].

19.13 Review Questions

1. What is economics?
2. Explain the difference between macroeconomics and microeconomics.
3. What is GDP? Explain how it is calculated.
4. Explain the limitations of GDP.
5. What is utility? Explain the role that it plays in consumption.
6. Explain the difference between absolute and comparative advantages.
7. Explain the difference between the terms elastic and inelastic?
8. What is the role of government in the economy?
9. What is the role of money in the economy?
10. Describe the common economic indicators that are employed.
11. What is globalization? Explain its advantages and disadvantages.
12. Explain the significance of terms of trade in international trade.

19.14 Summary

Economics is the social science that studies the production, distribution and consumption of goods and services. Households and firms are the basic units of the economy, and economics is concerned with the problem of satisfying unlimited desires with limited (or scarce) resources.

Economics includes many economic models, and these are simplified mathematical versions of the economic reality, and they allow understanding and predictions (forecasting) of economic behaviour.

Economics consists of two major fields namely macroeconomics, which is concerned with how the overall economy works, and microeconomics, which is concerned with the behaviour of individuals and businesses and their interactions, and especially how supply and demand interact in individual markets.

Macroeconomics analyses the entire economy including production consumption and savings, as well as issues affecting the economy, e.g., GDP, the rate of economic growth and employment. It looks at the output of the whole economy and how the economy allocates its scarce resources to maximize production.

Microeconomics analyses individual elements in an economy such as households and businesses as well as buyers and sellers. It looks at what happens when individuals make certain choices, where the individuals are buyers, sellers and business owners.

The concept of elasticity indicates how sensitive the demand for a product is to a change in its price. Some goods are inelastic where demand hardly falls in response to significant price increases, whereas the demand for other goods may change dramatically in response to a change in their price. In general, people desire things less as they become more expensive, and elasticity indicates how sensitive the product is to price increases.

20

Software to Support Business Mathematics

The goal of this chapter is to introduce essential software to support business mathematics, and a selection of software such as Excel, Python, Maple, Mathematica, Matlab, Minitab, R, Sage, SPSS and SQL are discussed. The key features of each software are discussed as well as links to more detailed information to allow the readers make a judgement on the most appropriate software to suit their specific needs (Table 20.1).

20.1 Microsoft Excel

Microsoft Excel is a spreadsheet program created by the Microsoft Corporation, and it consists of a rectangular grid of cells in rows and columns that may be used for data manipulation and arithmetic operations. It includes functionality for statistical, engineering and financial applications, and it has graphical functionality to display lines, histograms and charts (Fig. 20.1).

This software program was initially called Multiplan when it was released in 1982, and it was Microsoft's first Office application. It was developed as a competitor to Apple's VisiCalc, and it was initially released on computers running the CP/M operating system.[1] It was renamed to *Excel* when it was released on the Macintosh in 1985, and the first version of Excel for the IBM PC was released in 1987.

It provides support for user-defined macros, and it also allows the user to employ Visual Basic for Applications (VBA) to perform numeric computation and report the results back to the Excel spreadsheet. Lotus 1-2-3 was the leading Spreadsheet tool of the 1980s, but Excel overtook it from the early 1990s.

Excel is used to organize data and perform financial analysis. It is used by both small and large companies, and across all business functions. The main uses of Excel include the following:

- Data entry
- Data management

DOI: 10.1201/9781003308140-20

TABLE 20.1
Software for Business Mathematics

Software	Description
Excel	This is a spreadsheet program created by Microsoft that consists of a grid of cells in rows and columns.
Python	Python is an interpreted high-level programming language that has been applied to many areas including web development, game development, machine learning and artificial intelligence, and data science and visualization.
Maple	Maple is a computer algebra system that can manipulate mathematical expressions and find symbolic solutions to certain kinds of problems in Calculus, Linear Algebra, etc.
Minitab	Minitab is a statistical software package that provides a powerful and comprehensive set of statistics to investigate the data.
R	R is an open-source statistical computing environment that is used for developing statistical software and for data analysis.
Mathematica	Mathematica is a computer algebra program that is used in the scientific, engineering and computer fields.
Matlab	Matlab is a numeric computing environment that supports matrix manipulation, plotting of data and functions, and the implementation of algorithms.
Sage	Sage accounting software is used by accountants and bookkeepers to manage income and expenditure.
SPSS	IBM SPSS Statistics is a statistical software package that allows insights to be quickly extracted from the underlying data.
SQL	Structured Query Language (SQL) is a computer language that tells the relational database what to retrieve and how to display it. It is designed for managing and manipulating data in a relational database management system.
SAS	SAS is a suite of statistical software produced by the SAS Institute for data management, advanced analytics and predictive analysis.

- Accounting
- Financial analysis
- Charts and graphs
- Financial modelling

Excel is used extensively for financial analysis, and many businesses use Excel for budgeting, forecasting and accounting. Spreadsheet software may be used to forecast future performance, as well as calculating revenue and tax due, and completing the payroll. Excel may be used to generate financial reports and charts.

An Excel workbook consists of a collection of worksheets, where each worksheet is a spreadsheet page (i.e., a collection of cells organized in rows and columns). A workbook may contain as many sheets as required, and the columns in a sheet are generally labelled with letters, whereas the rows are

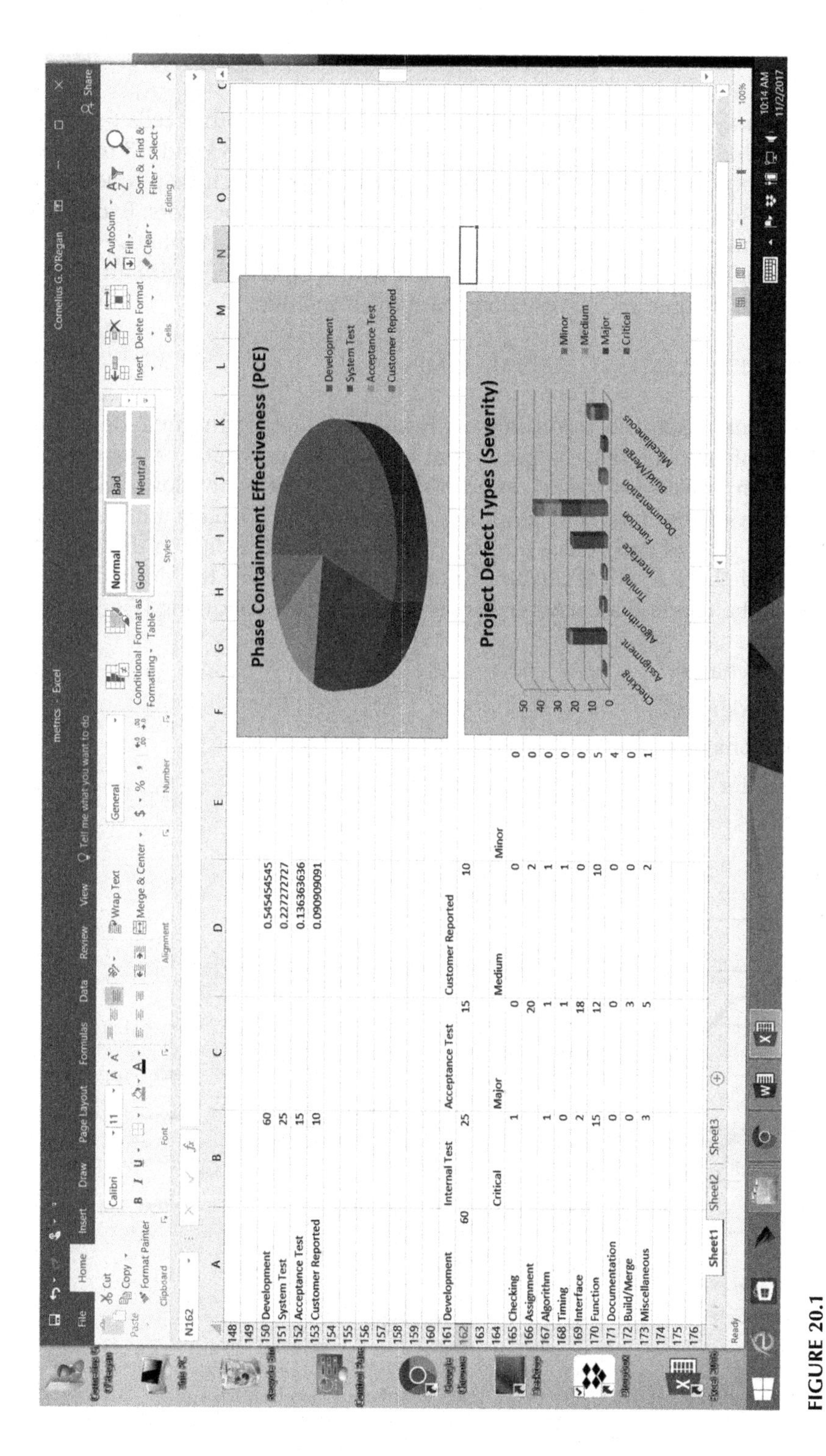

FIGURE 20.1
Excel Spreadsheet Screenshot. Used with Permission of Microsoft.

generally labelled with numbers. Each cell contains one piece of data or information.

A cell may contain a formula that refers to values in other cells (e.g., the effect of the formula = B3 + C3 in cell B1 is to add cells B3 and C3 together and to place the result in the cell B1). A formula may include a function, a reference to other cells, constants, as well as arithmetic operators. Excel uses standard mathematical operators such as +, –, *, /, and it employs ^ for the exponential function. An Excel formula always commences with an equals sign (=), and some of the functions employed include:

AVERAGE, COUNT, SUM, MAX, MIN and IF.

Excel is very useful in recording, analysing and storing numeric data, and various calculations may be performed on the numeric data, or graph tools may be employed for visualization of the data. That is, it allows easy manipulation of the data, and graphing of the data for visualization. A wide variety of graphs and charts may be produced from the data in the spreadsheet, and there is a *chart wizard* to assist with building the desired chart. Some of the charts that may be displayed include the following (Fig. 20.2):

- Bar charts
- Histograms
- Pie charts
- Scatter plots
- Lines

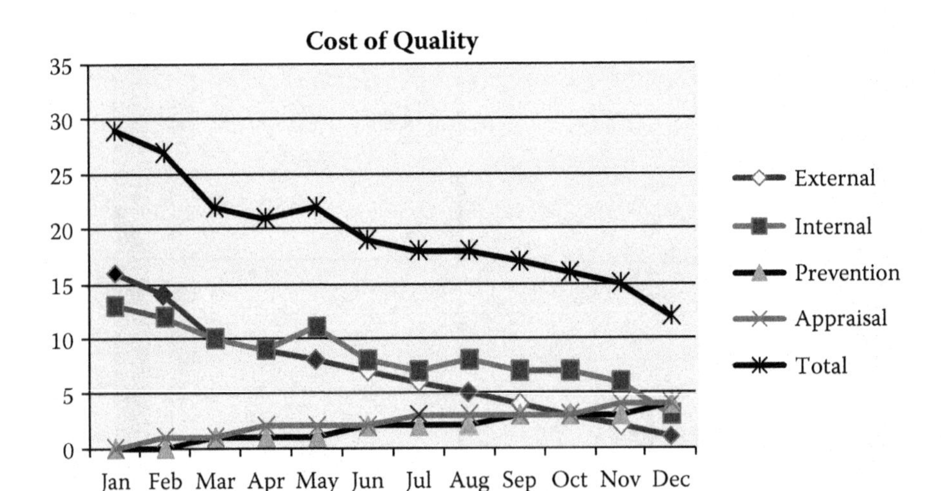

FIGURE 20.2
Excel Screenshot: Cost of Quality Chart. Used with Permission of Microsoft.

Excel may be used for data analysis, and the ability to analyse data is essential in order to make better decisions. This may involve the use of pivot tables, where pivot tables are useful when dealing with large datasets. A pivot table is a data processing technique that is used to explore and summarize large amounts of data, and it may analyse related totals and present summary reports. It allows statistics to be rearranged in order to identify useful information, and it allows a customizable summary table to be created that totals the numbers for any desired field. A pivot table may be employed to aggregate the individual items of a more extensive table (e.g., a database or another spreadsheet) within one or more discrete categories. That is, pivot tables may be used to summarize, sort, reorganize, group, count, total or average data stored in a table. A pivot table allows columns to be transformed into rows and vice versa.

An Excel *macro* is a timesaving tool that allows a sequence of instructions to be executed, and Excel executes the instructions in the macro step by step on the data that it is given. Macros may save a lot of time where the same steps are done on a regular basis, and the macros may record almost anything including numerical operations, formatting raw data, sorting information, moving cells and so on.

Excel allows the user to record the macro using standard commands, without coding it in VBA. The user may start recording the macro by clicking on the button labelled *Record Macro*, and entering a name for the macro and where to save it in the dialog box. Excel will then record everything that the user does, and once the user completes the desired set of actions then the user clicks on *Stop Recording*. Another words, all that is required to record a macro is to click on record, perform some actions and click on stop recording.

Once a macro has been created and saved it may be executed by selecting it and clicking *Run*. Excel then repeats the steps that the user performed during the recording of the macro.

Excel provides several useful statistical functions such as the COUNT function that counts the number of cells in a range containing a number. The COUNTA function counts all cells in a range that are not empty. The COUNTBLANK function counts the number of empty cells in a range of cells. The COUNTIFS function is a useful function that applies one or more conditions to the cells in a given range, and returns only those cells that satisfy the conditions.

The AVERAGE function determines the arithmetic mean of all the cells in a given range. The MEDIAN function determines the middle value of the given range of cells. The MODE function determines the most frequent value in a given range of values. The STDEV function is used to calculate the standard deviation of the entire population given as arguments. The QUARTILES function returns the quartiles of a dataset based on percentile values from 0 to 1. The CORREL function returns the correlation coefficient of two cell ranges, and is an indication of how strong the relationship is between the two variables.

20.2 Python

Python[2] is an interpreted high-level programming language that was designed by Guido van Rossum in the Netherlands in the early 1990s. The design of Python was influenced by the ABC programming language, which was designed as a teaching language, and developed at CWI in Amsterdam. Python has become a very popular programming language, and today the Python Software Foundation (PSF) is responsible for the language and its development. It is an object-oriented language that is based on the C programming language, and the language is versatile with applications in many areas including:

- Rapid web development
- Scientific and numeric computing
- Machine learning applications
- Image processing applications
- Game development
- Machine learning and artificial intelligence
- Gathering data from websites
- Data science and data analytics
- Data visualization
- Business applications such as ERP
- Education on programming

The language has become very popular especially for machine learning and artificial intelligence, as it is reasonably easy to use especially for those who are new to programming. Its syntax is relatively simple and the language is readable and easy to understand. Python applications can run on any operating system for which a Python interpreter exists.

Python includes many libraries such as libraries for web development, libraries for the development of interactive games and libraries for machine learning and artificial intelligence. These allow the computer to learn from past experience through the data collected and stored, or the creation of algorithms that allow the computer to learn by itself.[3] Python includes libraries that enable the development of applications that can multitask and output video and audio media.

Python supports data science and data visualization, and its libraries allow the data to be studied and information to be extracted. The data may be visualized such as in plotting graphs.

Python is able to gather a large amount of data from websites, which enables operations such as price comparison, job listings and so on to be performed. For more information on Python, see https://www.python.org/.

20.3 Maple

Maple is math software that includes a powerful math engine and a user interface that makes it easy to analyse, explore, visualize and solve mathematical problems. It allows problems to be solved easily and quickly in most areas of mathematics. It is a commercial general-purpose computer algebra system that can manipulate mathematical expressions and find symbolic solutions to certain kinds of problems including those that arise in ordinary and partial differential equations. It supports symbolic mathematics, numerical analysis, data processing and visualization. The Canadian software company, Maplesoft™, developed Maple, and its initial release was in the early 1980s (Fig. 20.3).

Maple supports several areas of mathematics including Calculus, Linear Algebra, differential equations, equation solving and symbolic manipulation of expressions. Maple has powerful visualization capabilities including

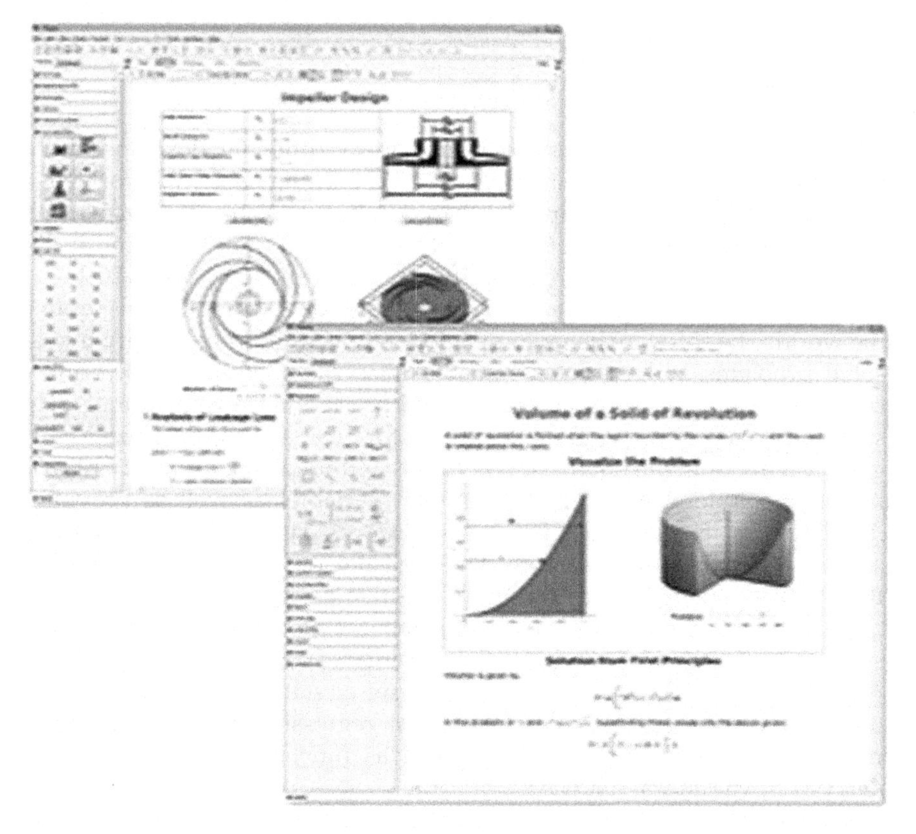

FIGURE 20.3
Maple User Interface. Creative Commons.

support for two-dimensional or three-dimensional plotting as well as animation. Further, Maple includes a high-level programming language that enables users to create their own applications.

Maple makes the use of matrix manipulation tools along with sparse arrays, and it has a wide range of special mathematical libraries. Maple supports 2D image processing and supports several probability distributions. Maple has functionality for code generation in languages such as C, Fortran, Python and Java.

Maple includes the standard arithmetic operators such as +, −, *, / and ^ for exponential; it includes the relational operators <, >, <=, >=, <>, and =; the logical operators and, or, xor, implies and not; and the set operators intersect, union, minus, subset and member. A value may be assigned to a variable with the assign command (:=). Maple includes special constants such as Pi, infinity and the complex number I. For more information on Maple, see https://www.maplesoft.com/.

20.4 Minitab Statistical Software

Minitab is a statistical software package that was originally developed at the University of Pennsylvania in the early 1970s. Minitab, LLC was formed in the early 1980s, and the company is based in Pennsylvania. It is responsible for the Minitab statistical software and its associated products, and it distributes the suite of commercial products around the world (Fig. 20.4).

Minitab statistical software is used by thousands of organizations around the world. The software helps companies and institutions to identify trends, solve problems and discover valuable insights in data by delivering a comprehensive suite of data analysis and process improvement tools. The software is easy to use, and makes it easy for business and organizations to gain insights from the data, discover trends and predict patterns. It assists in identifying hidden relationships between variables as well as providing dazzling visualizations. Minitab has a team of data analytic experts and services to ensure that users get the most out of their analysis, enabling them to make better, faster and more accurate decisions.

Minitab includes a complete set of statistical tools including descriptive statistics, hypothesis testing, confidence intervals and normality tests. It provides a powerful and comprehensive set of statistics to investigate the data. It includes functionality to support regression thereby identifying relationships between variables, as well as functionality to support the analysis of variance (ANOVA).

Minitab supports several statistical tests such as t tests, one and two proportions tests, normality tests, chi-square tests and equivalence test. Minitab's advanced analytics provides modern data analysis, and allows

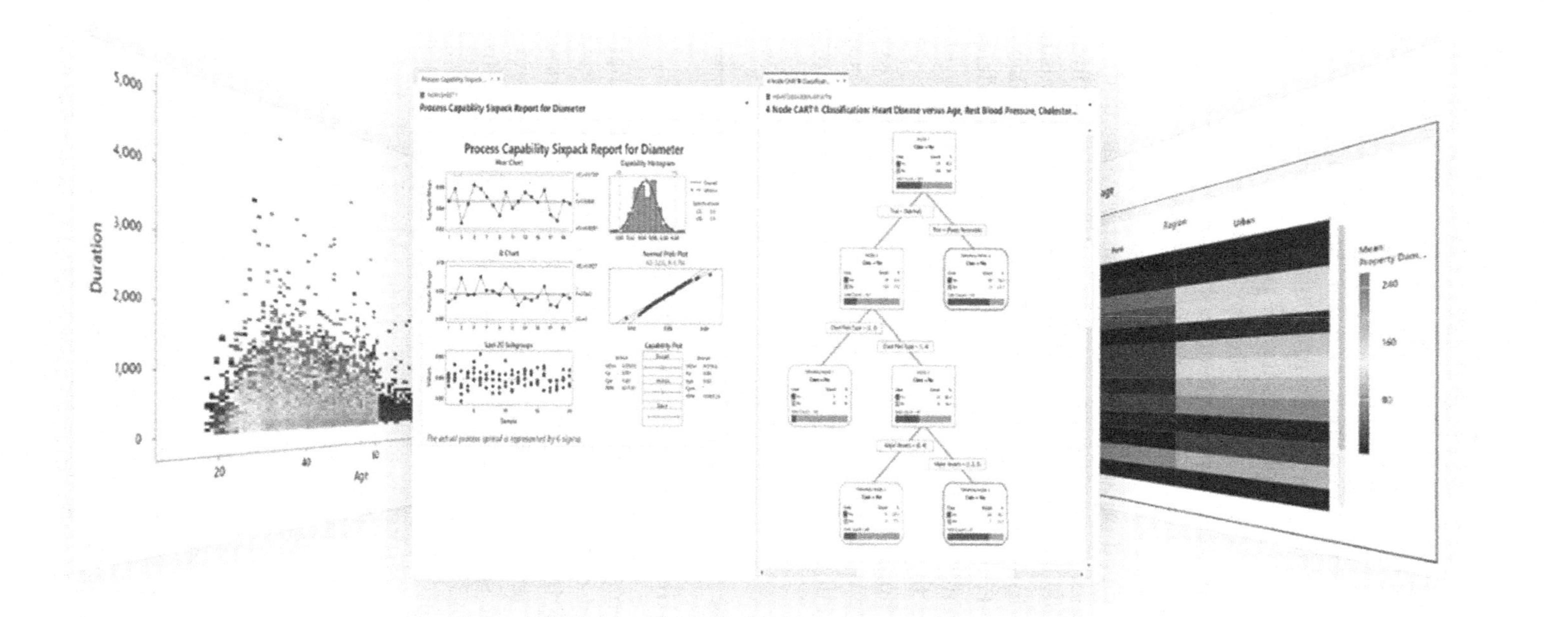

FIGURE 20.4
Minitab Screenshot. Created by and Used with Permission of Minitab, LLC (the Makers of Minitab Statistical Software, www.minitab.com).

the data to be explored further. Minitab's predictive analytics techniques allow predictions and forecasts to be made, and Minitab's powerful visualizations allow the user to decide which graph best displays the data and supports the analysis (Fig. 20.5).

Minitab includes functionality for control charts that allows processes to be monitored over time, thereby ensuring that they are performing between the lower and upper control limits for the process. That is, Minitab may be used for statistical process control thereby ensuring process stability, and for data-driven process improvement to transform the business. Minitab Engage is a tool for managing innovation, and may be used for managing 6-sigma and lean manufacturing deployments.

Minitab has helped organizations over the last 50 years to drive cost containment, enhance quality, boost customer satisfaction and increase effectiveness through its software solutions. Thousands of businesses and institutions worldwide use Minitab® Statistical Software, Minitab Connect™, Real-Time SPC Powered by Minitab®, Salford Predictive Modeler®, Minitab Workspace®, Minitab Engage™ and Quality Trainer® to uncover flaws and opportunities in their processes and address them.

Minitab Solutions Analytics™ provides software and services to enable organizations to make better decisions that drive business excellence. For more detailed information on Minitab, see https://www.minitab.com/.

20.5 R Statistical Software Environment

R is a free open-source statistical computing environment that is used by statisticians and data scientists for developing statistical software and for data analysis. R includes various libraries that implement various statistical and graphical techniques such as statistical tests, linear and non-linear modelling, time series analysis and many more. It allows the user to clean, analyse, plot and communicate all of their data all in one place (Fig. 20.6).

R is an interpreted language and the user generally accesses it through a command line interpreter, and it has thousands of packages to assist the user. R is a very popular statistical software tool and it is widely used in academia and industry. It can produce high-quality graphs and the advantages of R include:

- Free open-source software
- Large community
- Integrates with other languages (e.g., C and C++)

R was created in the 1990s by Ross Ihaka and Robert Gentleman at the University of Auckland in New Zealand, and it was based on the S statistical

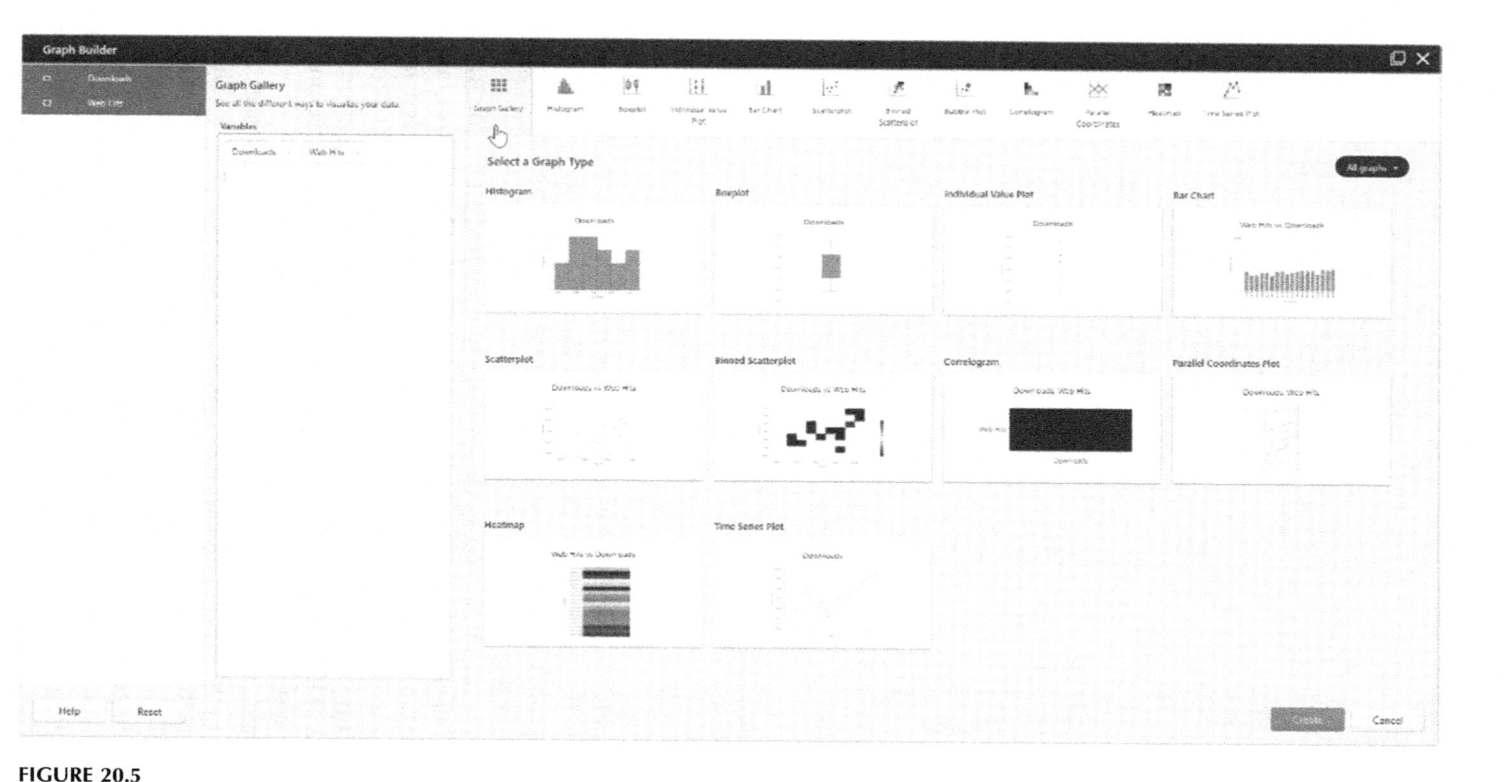

FIGURE 20.5
Minitab Graphs. Created by and Used with Permission of Minitab, LLC (the Makers of Minitab Statistical Software, www.minitab.com).

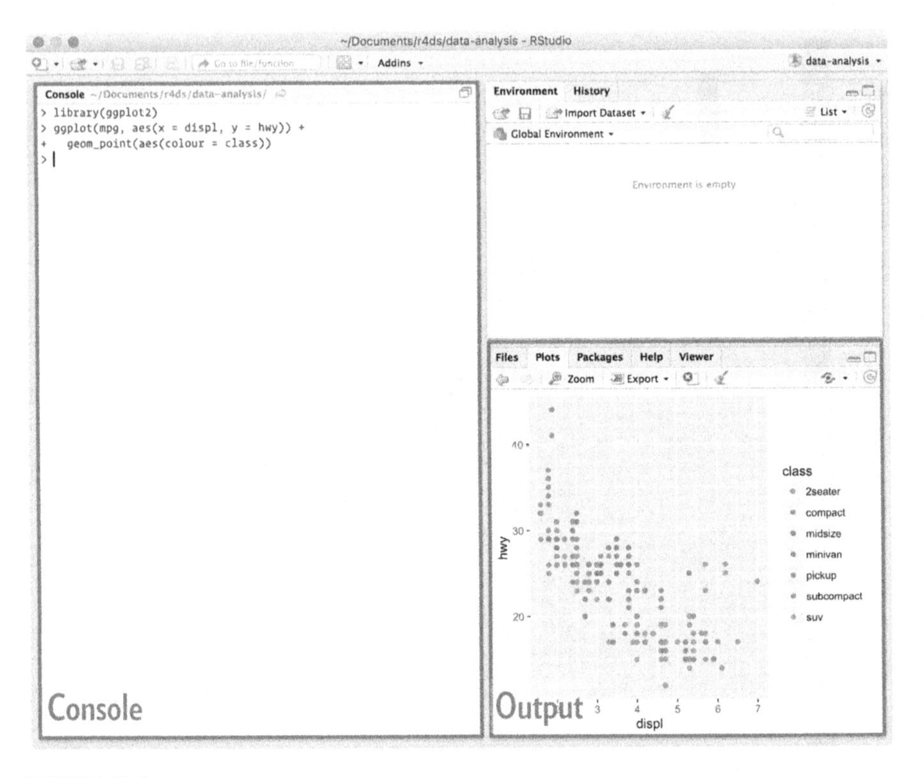

FIGURE 20.6
RStudio.

programming language that was developed by John Chambers and others at Bell Labs in the 1970s. The R Development Core Team is now responsible for its development, and R programming plays a key role in statistics, machine learning and data analysis.

RStudio is an integrated development environment (IDE) for R, and its functional user interface provides an easier way of using R. Programs may be written in R using the RStudio IDE. R may be downloaded from the Comprehensive R Archive Network (CRAN) https://cloud.r-project.org, and RStudio may be downloaded from http://www.rstudio.com/download. After installing RStudio, there will be two key regions in the interface, and R code is typed in the console panel (Fig. 20.4).

R packages may then be installed where an R package consists of functions, data and documentation that extend the capabilities of R. The use of packages is the key to the successful use of R in data science, as it has a large number of packages available and it is easy to install and use them. A core set of packages is included with the basic installation, and other packages may be installed as required. For example, the package tidyverse may installed with a single line of code that is typed in the console:

```
install.packages("tidyverse")
library("tidyverse")
```

The package must then be loaded with the library command before the functions, objects and help files may be used. The statistical features of R include:

- Basic statistics including measures of central tendency
- Static graphics
- Probability distributions (e.g., Binomial and Normal)
- Data analytics (tools for data analysis)

For more information on R and Rstudio, see https://www.r-project.org/.

20.6 Mathematica

Mathematica is a powerful tool for problem solving, and it includes symbolic calculations with nice graphical output. It is a way for doing mathematics with a computer, and this powerful computer algebra program is used in the scientific, engineering and computer fields. Symbolic mathematics involves the use of computers to manipulate equations and expressions in symbolic form, as distinct from manipulating the numerical quantities represented by the symbols (Fig. 20.7).

Mathematica was developed by Stephan Wolfram of Wolfram Research in the late 1980s, and it supports many areas of mathematics including basic arithmetic, algebra, geometry and trigonometry, calculus, complex analysis, vector analysis, matrices and linear algebra, and probability and statistics.

It has a large number of predefined functions for mathematics and other disciplines, and includes functionality for the visualization of data and functions. This includes good graphical capabilities that are useful in plotting functions and data in two or three dimensions. For example, the Mathematica command *RevolutionPlot3D* constructs the surface formed by revolving an expression around an axis, and Fig. 20.8 is generated from the command:

RevolutionPlot3D[$x^4 - x^2$, {x, −1, 1}]

Mathematica may be used to solve simple arithmetic problems, as well as solving complex problems in differential equations. It has approximately 5000 built-in functions covering the vast majority of areas in technical

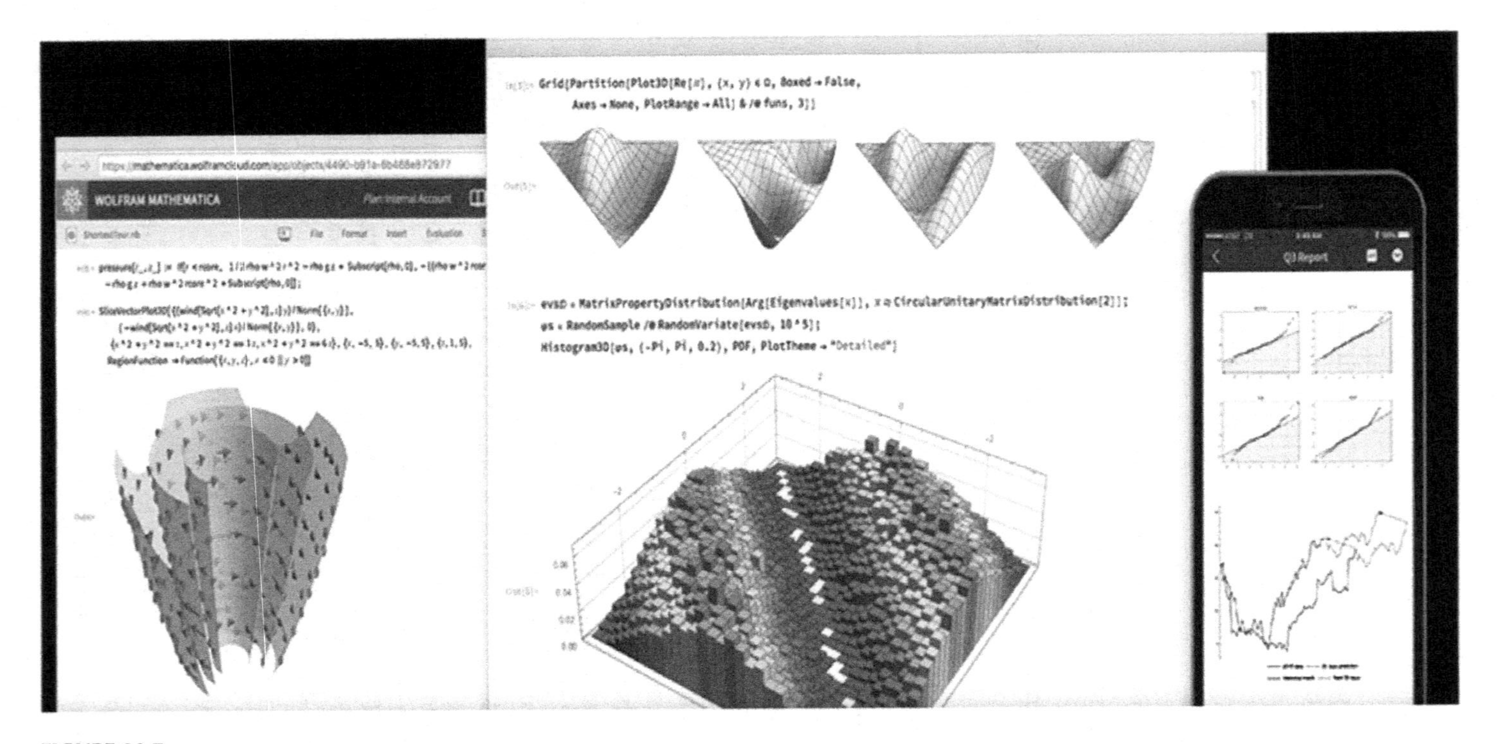

FIGURE 20.7
Mathematica in Operation. Provided Courtesy of Wolfram Research, Inc., the Makers of Mathematica, www.wolfram.com.

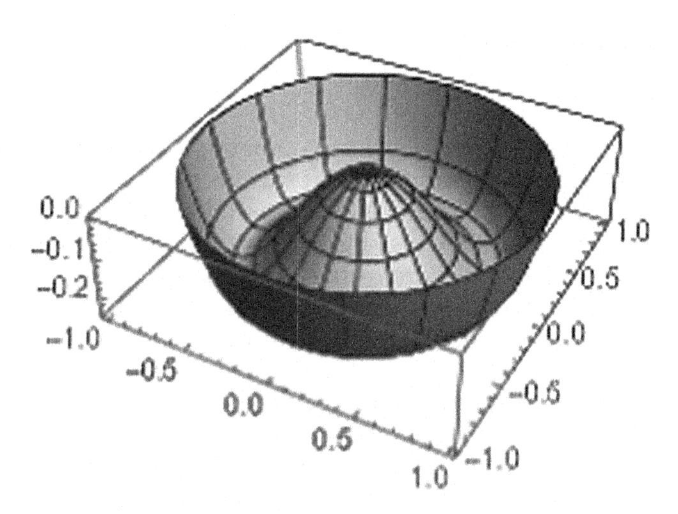

FIGURE 20.8
Surface Generated with RevolutionPlot3D. Provided Courtesy of Wolfram Research, Inc., the makers of Mathematica, www.wolfram.com.

computing, and that work together in the integrated system. The Wolfram programming language was developed by Wolfram Research and is the programming language of Mathematica.

This is a multi-paradigm language and emphasizes functional programming and symbolic computation. The language can perform integration, differentiation, matrix manipulation and solve differential equations using a set of rules. Mathematica has been applied to many areas of computing including:

- Machine learning
- Neural networks
- Image processing
- Data science
- Geometry
- Visualization

Mathematica has also been applied to many other areas, and it produces documents as well as code. Its visualizations are aesthetically pleasing and powerful, and Mathematica also produces publication-quality documents, and it has thousands of examples in its documentation centre. It has built in algorithms across many areas that aim to be of industrial strength. Finally, Mathematica is integrated with the cloud, and this allows sharing as well as cloud computing. There is more detailed information on Mathematica at Wolfram Research, Inc. (https://www.wolfram.com/mathematica/).

20.7 Matlab

Matlab is a commercial high-level programming language that is used to perform mathematical computing, and this numeric computing environment was developed by MathsWork. It is used by engineers and scientists to organize, explore and analyse the data, and the Matlab language may be employed to develop programs based on algorithms from a variety of domains. Matlab allows the user to create customized visualizations and to automatically generate the Matlab code to reproduce them with new data.

Matlab manages array and matrix problems, and it may be used to solve complex algebraic equations as well as analysing data and plotting graphs. It allows the user to create customized visualizations, as well as to use the built in charts. Matlab has many applications including:

- Machine learning
- Deep learning
- Robotics
- Computer vision
- Image processing
- Control systems

At the heart of Matlab is a high-level programming language that allows engineers and scientists to express matrix and array mathematics directly. Matlab has a large library of toolboxes from everything from wireless communication, to control systems, to signal and image processing, to robotics and to AI. It is easy to use and learn, and it allows ideas to be explored with the results and visualizations seen quickly.

It includes pre-built apps and allows the user to create their own apps. It includes App Designer, which allows a non-specialist to create professional apps by laying out the visual components of the GUI as well as programming the app behaviour. Matlab may be extended with thousands of packages and tools, and its capabilities include:

- Data analysis
- Graphics
- Algorithm development
- App building

There is more detailed information on Matlab at Mathworks (https://www.mathworks.com/products/matlab.html).

20.8 Sage Accounting Software

The Sage Corporation is a British corporation located in the north of England, and it is one of the leading providers of enterprise resource planning (ERP) software in the world. Sage has many software products including Sage accounting software, which is an accounting package that supports standard financial statements such as balance sheet, profit and loss account, and cash statements. Sage accounting software has become the industrial standard for accounting software, and it includes the following features:

- Dashboards for visuals of company performance
- Core accounting
- Manages income and expenditure
- Payroll
- Invoicing
- Budgeting and forecasting
- Manage cash flow
- Calculation and submission of VAT returns
- Tax management
- Preparation of financial statements
- Profit analysis
- Accounts receivable and payable
- Inventory tracking

Sage accounting is now a cloud-based accounting software platform that can be accessed by a web browser, and so it is accessible on a variety of devices in different locations. Sage accounting employs a single subscription plan with additional charges for certain features. For more information on Sage, see https://www.sage.com/.

20.9 SPSS Statistics

Statistical Package for Social Sciences (SPSS) was developed by Norman Nie, Dale Bent and Hadai Hull in Chicago in the late 1960s, and it was a widely used program for statistical analysis in social science. The company was incorporated as SPSS Inc. in the mid-1970s, and the company also sold software for market research, survey research and statistical analysis. SPSS Inc. was acquired by IBM in 2010, and today it is called IBM® SPSS®.

IBM SPSS Statistics is a statistical software platform that includes a set of features that allows insights to be quickly extracted from the underlying data. It allows an organization to analyse its data to discover information to improve decision-making. SPSS plays an important role in making sense of complex patterns and associations, and thereby assists users in drawing the right conclusions and making accurate predictions.

SPSS addresses the entire analytics lifecycle from data preparation and management to analysis and reporting. It may be used to improve forecasts and plans by the input of missing values with expected values through regression. SPSS has automated methods to identify anomalies as well as functionality to communicate the results effectively through tables and visualization.

It provides functionality to predict the value of a target variable based on the values of predictor variables, and it also may be used to model linear and non-linear relationships. SPSS provides support for hypothesis testing, linear and non-linear regression, ANOVA, Bayesian statistics, visualizations with graphs and charts, forecasting and many more. For more detailed information on SPSS, see https://www.ibm.com/products/spss-statistics.

20.10 SQL and Relational Databases

The concept of a relational database was first described in a paper "A Relational Model of Data for Large Shared Data Banks" by Codd [Cod:70]. A relational database is a database that conforms to the relational model, and it may be defined as a set of relations (or tables).

Codd was a British mathematician, computer scientist and IBM researcher, who initially worked on the SSEC (Selective Sequence Electronic Computer) project in New York, and then on the IBM 701 and 702 computers. He was the creator of STEM (statistical database expert manager).

He developed the *relational database model* in the late 1960s, and he published an internal IBM paper on the Relational Model in 1969. Today, this is the standard way that information is organized and retrieved from computers, and relational databases are at the heart of systems from hospitals' patient records to airline flight and schedule information.

IBM was promoting its IMS hierarchical database in the 1970s, and it showed little interest for Codd's new relational database model. It made business sense for IBM to preserve revenue for the IMS/DB model, rather than embarking on a new technology. However, IBM agreed to implement Codd's ideas on the relational model for the *System R research project* in the 1970s, and this project demonstrated the power of the model, as well as demonstrating good transaction processing performance. The project introduced a data query language that was initially called SEQUEL (later renamed to SQL), and this language was designed to retrieve and manipulate data in the IBM database. Codd continued to develop and extend his relational model, and he received the ACM Turing Award in 1981 in recognition of its development.

A binary relation R (A, B) where A and B are sets is a subset of the Cartesian product $(A \times B)$ of A and B. The domain of the relation is A, and the codomain of the relation is B. The notation aRb signifies that there is a relation between a and b and that $(a, b) \in R$. An n-ary relation R $(A_1, A_2, \ldots, A_n)$ is a subset of the Cartesian product of the n sets, i.e., a subset of $(A_1 \times A_2 \times \ldots \times A_n)$. However, an n-ary relation may also be regarded as a binary relation R (A, B) with $A = A_1 \times A_2 \times \ldots \times A_{n-1}$ and $B = A_n$.

The data in the relational model are represented as a mathematical n-ary relation. That is, a relation is defined as a set of n-tuples, and a table provides a visual representation of the relation, with the data organized in rows and columns. The data stored in each column of the table are of the same data type.

The basic relational building block is the domain or data type (often called just type). Each row of the table represents one n-tuple (one tuple) of the relation, and the number of tuples in the relation is the cardinality of the relation. Consider the PART relation taken from [Dat:81], where this relation consists of a heading and the body. There are five data types representing part numbers, part names, part colours, part weights and locations in which the parts are stored. The body consists of a set of n-tuples. The PART relation is of cardinality six (Fig. 20.9).

Strictly speaking, there is no ordering defined among the tuples of a relation, since a relation is a set, and a set may be unordered. However, in practice, relations are often considered to have an ordering.

It is often the case that within a given relation that there is one attribute with values that is unique within the relation, and can thus be used to identify the tuples of the relation. For example, the attribute P# of the PART relation has this property since each PART tuple contains a distinct P# value, which may be used to distinguish that tuple from all other tuples in

P#	PName	Colour	Weight	City
P1	Nut	Red	12	London
P2	Bolt	Green	17	Paris
P3	Screw	Blue	17	Rome
P4	Screw	Red	14	London
P5	Cam	Blue	12	Paris
P6	Cog	Red	19	London

FIGURE 20.9
PART Relation.

the relation. P# is termed the *primary key* for the PART relation, and a candidate key that is not the primary key is termed the *alternate key.*

An index is a way of providing quicker access to the data in a relational database, as it allows the tuple in a relation to be looked up directly (using the index) rather than checking all tuples in the relation.

The consistency of a relational database is enforced by a set of constraints that provide restrictions on the kinds of data that may be stored in the relations. The constraints are declared as part of the logical schema and are enforced by the database management system. They are used to implement the business rules for the database.

20.10.1 Structured Query Language (SQL)

Codd proposed the Alpha language as the database language for his relational model. However, IBM's implementation of his relational model in the System-R project introduced a data query language that was initially called SEQUEL (later renamed to SQL). This language did not adhere to Codd's relational model but it became the most popular and widely used database language. It was designed to retrieve and manipulate data in the IBM database, and its operations include *insert, delete, update, query,* schema creation and modification and data access control.

SQL is a computer language that tells the relational database what to retrieve and how to display it. It was designed and developed at IBM by Donald Chamberlin and Raymond Boyce, and it became an ISO standard in 1987.

The most common operation in SQL is the query command, which is performed with the SELECT statement. The SELECT statement retrieves data from one or more tables, and the query specifies one or more columns to be included in the result. Consider the example of a query that returns a list of expensive books (defined as books that cost more than $100.00).

```
SELECT*4
        FROM Book
        WHERE Price > 100.00
        ORDER by title.
```

The *Data Manipulation Language* (DML) is the subset of SQL used to add, update and delete data. It includes the INSERT, UPDATE and DELETE commands. The *Data Definition Language* (DDL) manages table and index structure, and includes the CREATE, ALTER, RENAME and DROP statements.

There are extensions to standard SQL that add programming language functionality. A stored procedure is executable code that is associated with the database. It is usually written in an imperative programming language, and it is used to perform common operations on the database. Oracle is recognized as a world leader in relational database technology, and an Oracle database consists of a collection of data managed by an Oracle

Database Management System. Oracle is the main standard for database technology.

20.11 SAS

The Statistical Analysis System (SAS) arose from a project conducted by several universities in the United States to computerize the vast amount of agricultural data being collected. The consortium was led by North Carolina State University (NCSU) as it had access to a large powerful mainframe S/360 computer, and the first version of the software was released in the early 1970s. It included functionality for multiple regression and analysis of variance and the resulting program was called SAS.

The SAS Institute was formed in the mid-1970s as a private company by several employees from NCSU including James Goodnight and Anthony Barr. SAS was developed further in the 1980s and 1990s with the addition of new statistical procedures.

SAS today is a suite of statistical software produced by the SAS Institute for data management, advanced analytics, multivariate analysis, and predictive analysis. SAS can mine, alter, manage and retrieve data from a variety of sources and perform statistical analysis on it. For more detailed information on SAS, see https://www.sas.com/.

20.12 Review Questions

1. Why is software used to support business mathematics?
2. What are the main features of Excel?
3. What are the main features of Minitab?
4. What are the advantages of an open-source package such as R?
5. What are the main features of Mathematica?

20.13 Summary

The goal of this chapter was to discuss a selection of software available to support business mathematics, including Excel, Python, Maple, Mathematica, Matlab, Minitab, R, Sage, SPSS, SQL and SAS.

Excel is a spreadsheet program that consists of a grid of cells in rows and columns. Python is an interpreted high-level programming language with applications in web development, artificial intelligence and data science. Maple is a computer algebra system that can find symbolic solutions to certain kinds of problems. Minitab is a statistical software package with a comprehensive set of statistics to investigate the data. R is an open-source statistical computing environment that is used for developing statistical software and for data analysis. Mathematica is a computer algebra program that is used in the scientific, engineering and computer fields.

Matlab is a computing environment that supports matrix manipulation, plotting of data and functions and the implementation of algorithms. Sage is used by accountants to manage income and expenditure. IBM SPSS Statistics is a statistical software package that allows insights to be quickly extracted from the underlying data. SQL is a database language that tells the relational database what to retrieve and how to display it. SAS is a suite of statistical software for data management, advanced analytics and predictive analysis.

Notes

1 The CP/M operating system was developed by Gary Kildall at Digital Research, and Kildall did a lot of early work on operating systems for early microprocessors.
2 Python is named after the famous Monty Python's Flying Circus comedy series on British TV, which starred John Cleese, Michael Palin, Graham Chapman and others.
3 There is a need for care with machine learning algorithms as some algorithms have been biased. These include a predictive policing algorithm that attempted to identify people and locations at risk of crime that led to racial discrimination of minorities, and Amazon's AI machine learning hiring algorithm that discriminated against females.
4 The asterisk (*) indicates that all columns of the Book table should be included in the result.

21

Epilogue

We embarked on a long journey in this book and set ourselves the objective of providing a concise introduction to the mathematics used in the business field. The book was influenced by the author's industrial experience, as well as a course on business mathematics taught by the author at Algonquin College in Kuwait in 2017.

The early chapters focused on the mathematical foundations with chapters on number theory, algebra, set theory, and sequences and series. The first chapter introduced number theory, and we discussed prime number theory including the greatest common divisor and least common multiple of two numbers. We discussed the various numbers systems such as the natural numbers, the integers and rational and real numbers. We showed how fractions are added or multiplied together, and we discussed ratios, proportions and percentages.

Chapter 2 introduced algebra, which uses letters to stand for variables or constants in mathematical expressions. Algebra is the study of such mathematical symbols and the rules for manipulating them, and it is a powerful tool for problem solving in science, engineering and business. We discussed simple and simultaneous equations and methods to solve them, including the method of elimination, the method of substitution and graphical techniques. We showed how quadratic equations may be solved by factorization, completing the square, using the quadratic formula or graphical techniques. We presented the laws of logarithms and indices, and discussed exponentials and natural logarithms.

Chapter 3 discussed sets, relations and functions. A set is a collection of well-defined objects and it may be finite or infinite. A relation $R(A, B)$ between two sets A and B indicates a relationship between members of the two sets, and it is a subset of the Cartesian product of the two sets. A function is a special type of relation such that for each element in A there is at most one element in the co-domain B. Functions may be partial or total, and injective, surjective or bijective.

Chapter 4 gave an introduction to sequences and series, and we discussed arithmetic sequences and series (where two successive terms differ by a constant d), and geometric sequences and series (where two successive terms differ by a ratio r to the previous element). We discussed the counting principle and the pigeonhole principle and permutations and combinations.

Chapter 5 discussed simple interest and applications, where simple interest is earned on the principal only, and the amount earned is determined from

DOI: 10.1201/9781003308140-21

the principal invested, the rate of interest and the period that the principal is invested for. We discussed future and present values, where the future value is the amount that the principal will grow to at a given rate of simple interest over the period of time, whereas the present value of an amount to be received at a given date in the future is the principal that will grow to that amount at a given rate of simple interest over that period of time. We discussed short-term promissory notes that are a written promise by one party to pay a certain sum of money (with or without interest) on a particular date to another party. A promissory note may be interest-bearing (where the rate of interest is stated in the note), or non-interest-bearing (where there is no rate of interest specified) and these are termed treasury bills.

Chapter 6 discussed compound interest, and while simple interest is calculated on the principal only, compound interest is more complicated in that interest is also applied to the interest earned in previous compounding periods. Compound interest is generally employed for long-term investments and loans, and we showed how present and future values may be determined. The concept of the time value of money is that the earlier that cash is received the greater its value to the recipient. Similarly, the later that a cash payment is made, the lower its cost to the payer and the lower its value to the payee.

Chapter 7 discussed banking and financial services, and we discussed the roles of various types of banks including the central bank, commercial banks and retail banks. We discussed annuities in the banking sector, and we showed how a mortgage or loan is paid back with an annuity. We discussed the foreign exchange market including spot and forward trading, and discussed corporate bonds, which are used by businesses as an alternative source of finance from bank loans or shareholder funding.

Chapter 8 discussed trade discounts that are a reduction in the list price of a manufactured product, and the discount is generally stated as a percentage of the list price. Trade discounts are used by manufacturers, distributors and wholesalers as pricing tools for their products, and in communicating changes of prices to their customers. A cash discount may be given to encourage early or prompt payment of an invoice, and the rate of discount and the discount period for when the discount may be applied are specified on the invoice.

Chapter 9 discussed statistics including sample spaces; sampling; the abuse of statistics; frequency distributions and various charts such as bar charts, histograms, pie charts and trend graphs. Various statistical measures such as the average of a sample including the arithmetic mean mode and median were discussed, as well as the variance and standard deviation of a sample. We discussed correlation and regression, and hypothesis testing.

Chapter 10 discussed probability theory including a discussion on discrete and continuous random variables; probability distributions such as the Binomial and Poisson distributions. We discussed the normal and unit normal distributions, and discussed confidence intervals. We discussed

Bayesianism and how this helps in updating probabilities in the light of new information.

Chapter 11 discussed the fundamentals of the insurance field including the basic mathematics underlying motor and health insurance. The concept of a life annuity is discussed as well as the basic mathematics underlying pensions. We discussed the role of the actuary in the insurance field.

Chapter 12 discussed data science and data analytics, where data science is a multi-disciplinary field that extracts knowledge from data sets that consist of structured and unstructured data, and large data sets may be analysed to extract useful information.

Chapter 13 is concerned with metrics and problem solving, and this includes a discussion of the balanced scorecard which assists in identifying appropriate metrics for an organization. The Goal, Question, Metrics (GQM) approach was discussed, and this allows appropriate metrics related to the organization goals to be defined. A selection of sample metrics for an organization was presented, and problem-solving tools such as fishbone diagrams, Pareto charts and trend charts were discussed.

Chapter 14 discussed matrices including 2×2 and general $n \times m$ matrices. Various operations such as the addition and multiplication of matrices are considered, and the determinant and inverse of a matrix was discussed. The application of matrices to solving a set of linear equations using Gaussian elimination was considered, and we showed how the inverse of a matrix is determined.

Chapter 15 discussed operations research, and we discussed linear programming, cost volume profit analysis (CVPA) and game theory. Linear programming is a mathematical model for determining the best possible outcome of a particular problem, where the problem is subject to various constraints that are expressed as a set of linear equations and linear inequalities. CVPA is used to analyse the relationship between the costs, volume, revenue and profitability of the products produced. Game theory is the study of mathematical models of strategic interaction among rational decision makers.

Chapter 16 discussed basic financial statements including the balance sheet, the profit and loss account and cash management. The balance sheet is a snapshot of the net worth of a company and it is a summary of everything that is owned by the company less everything that it owes. The profit and loss account is the earning statement of the company during a period.

Chapter 17 gave a short introduction to calculus, including an overview of limits, continuity, differentiation and integration.

Chapter 18 presented various applications of differentiation and integration, and gave a brief introduction to Fourier series, Laplace transforms and differential equations.

Chapter 19 provided a short introduction to economics, and included a short discussion of macroeconomics and microeconomics. The important economic concepts of gross domestic product, inflation and employment were discussed, as well as the theoretical concepts of utility and elasticity.

Chapter 20 discusses a selection of tools to support business mathematics and includes a discussion of Excel, Python, Maple, Mathematica, Matlab, Minitab, R, Sage, SPSS, SQL and SAS.

Chapter 21 is the concluding chapter and it summarizes the journey that we have travelled in this book.

References

Adam Smith. 1776. *An Enquiry into the Nature and Causes of the Wealth of Nations*. London.
Annamaria Olivieri. 2011. *Introduction to Insurance Mathematics*. Springer Verlag.
David Ricardo. 1817. *On the Principles of Political Economy and Taxation*. London.
Donald Gross and John Shortle. 2008. *Fundamentals of Queueing Theory*. 4th Edition. Wiley.
Edmond Halley. 1693. An estimate of the degrees of the mortality of mankind, drawn from curious tables of the births and funerals at the City of Breslaw; with an attempt to ascertain the price of annuities upon lives. *Philosophical Transactions of the Royal Society of London*.
E.F. Codd. 1970. A relational model of data for large shared data banks. *Communications of the ACM*, 13(6), 377–387.
F.M. Dekking, C. Kraaikamp, H.P. Lopuhaa and L.E. Meester. 2010. *A Modern Introduction to Probability and Statistics*. Springer Texts in Statistics.
Gargi Keeni et al. 2000. The evolution of quality processes at Tata Consultancy Services. *IEEE Software*, 17(4), 79–88.
Gerard O' Regan. 2014. *Introduction to Software Quality*. Springer.
Gerard O' Regan. 2019. *Concise Guide to Software Testing*. Springer.
Gerard O' Regan. 2020. *Mathematics in Computing*. 2nd Edition. Springer.
Gerard O' Regan. 2021. *Guide to Discrete Mathematics*. 2nd Edition. Springer.
Hamdy A. Taha. 2016. *Operations Research: An Introduction*. 10th Edition. Pearson.
Harvey M. Deitel. 1990. *Operating Systems*. 2nd Edition. Addison Wesley.
Jeremy Bentham. 1789. *An Introduction to the Principles of Morals and Legislation*.
John Bird. 2005. *Basic Engineering Mathematics*. 4th Edition. Elsevier.
John Maynard Keynes. 1936. *The General Theory of Employment, Interest and Money*. Macmillan, London.
John von Neumann. 1928. On the theory of Parlor Games. Zur Theorie der Gesellschaftsspiele. *Mathematische Annalen*, 100, 295–300.
Judy Feldman Anderson and Robert L. Brown. 2005. *Risk and Insurance*. Education and Examination Committee of the Society of Actuaries.
Kaora Ishikawa. 1985. *How to Operate QC Quality Circles Activities*. QC Circle Headquarters. Union of Japanese Scientists and Engineers, Tokyo.
Kaplan and Norton. 1996. The Balanced Scorecard. *Translating Strategy into Action*. Harvard Business School Press.

Michael Brassard and Diane Ritter. 1994. *The Memory Jogger. A Pocket Guide of Tools for Continuous Improvement and Effective Planning*. Goal I QPC, Methuen, MA.
N. Gregory Mankiw. 2014. *Principles of Economics*. 7th Edition. Cengage Learning.
Norman Fenton. 1995. *Software Metrics: A Rigorous Approach*. Thompson Computer Press.
O. Schreier and E. Sperner. 2013. *Introduction to Modern Algebra and Matrix Theory*. 2nd Edition. Dover Publications.
Philip Crosby. 1979. *Quality Is Free. The Art of Making Quality Certain*. McGraw Hill.
Peter Rose and Sylvia Hudgins. 2012. *Bank Management and Financial Services*. 9th Edition. McGraw Hill.
Remy Charlip. 1993. *Fortunately. Atheneum Books for Young Readers*. 1st Aladdin Books Edition.
Richard L. Burden and J. Douglas Faires. 1989. *Numerical Analysis*. 4th Edition. PWS Kent.
Sheldon M. Ross. 2014. *Introduction to Probability and Statistics for Engineers And Scientists*. 5th Edition. Elsevier Publications, New York.
Sir Thomas Heath (Trans.). 1956. *Euclid. The Thirteen Books of the Elements*. Vol. 1. Dover Publications. (First published in 1925)
Song Y. Yan. 1998. *Number Theory for Computing*. 2nd Edition. Springer.
Terry Gasking. 1991. *How to Master Finance. A No-Nonsense Guide to Understanding Business Accounts*. Business Books Ltd.
Thomas Finney. 1988. *Calculus and Analytic Geometry*. 7th Edition. Addison Wesley.
Victor Basili and H. Rombach. 1988. The TAME project. Towards improvement-oriented Software environments. *IEEE Transactions on Software Engineering*, 14(6).
William Boyce and Richard DiPrima. 1992. *Elementary Differential Equations and Boundary Value Problems*. 5th Edition. Wiley.

Glossary

ACM	Association for Computing Machinery
AI	Artificial Intelligence
ANOVA	Analysis of Variance
ATM	Automated Teller Machine
BI	Business Intelligence
BOP	Balance of Payments
BP	Breakeven Point
BSC	Balanced Scorecard
CCI	Consumer Confidence Index
CIA	Central Intelligence Agency
COPQ	Cost of Poor Quality
CPI	Consumer Price Index
CP/M	Control Program for Microprocessors
CRAN	Comprehensive R Archive Network
CVPA	Cost Volume Profit Analysis
CWI	Centrum voor Wiskunde en Informatica
DDL	Data Definition Language
DML	Data Manipulation Language
DP	Dynamic Programming
DPIA	Direct Privacy Impact Assessment
ECB	European Central Bank
ERP	Enterprise Resource Planning
ESI	European Software Institute
EVA	Economic Value Added
FC	Fixed Cost
FIP	Fair Information Processing principles
FV	Future Value
FX	Foreign Exchange
GCD	Greatest Common Divisor
GDP	Gross Domestic Product
GDPR	General Data Protection Regulation
GNI	Gross National Income

GNP	Gross National Product
GQM	Goal Question Metric
GUI	Graphical User Interface
HR	Human Resources
IBAN	International Bank Account Number
IBM	International Business Machines
IDE	Integrated Development Environment
IoT	Internet of Things
IRR	Internal Rate of Return
IT	Information Technology
KLOC	Thousands Lines of Code
KPI	Key Performance Indicator
LCM	Least Common Multiple
LP	Linear Programming
MC	Marginal Cost
MD	Markdown
MR	Marginal Revenue
MTBF	Mean Time Between Failure
MTTF	Mean Time to Failure
MTTR	Mean Time to Repair
MV	Maturity Value
NCSU	North Carolina State University
NPF	Net Price Factor
NPV	Net Present Value
NSA	National Security Agency
OPEC	Organisation of Petroleum Exporting Countries
OR	Operations Research
OTC	Over the Counter
PCE	Phase Containment Effectiveness
PED	Price Elasticity of Demand
PIN	Personal Identification Number
P&L	Profit and Loss
PMI	Purchasing Managers Index
PPP	Purchasing Power Parity
PSF	Python Software Foundation
PV	Present Value
RSA	Rivest, Shamir and Adleman
SaaS	Software as a Service
SAS	Statistical Analysis System
SP	Selling Price
SPC	Statistical Process Control
SPSS	Statistical Package Social Science
SQL	Structured Query Language
SWIFT	Society Worldwide Inter-bank Financial Telecommunication
T-Bill	Treasury Bill

TC	Total Cost
TOT	Terms of Trade
TVC	Total Variable Cost
TR	Total Revenue
VAT	Value Added Tax
VBA	Visual Basic for Applications
VC	Variable Cost
WTO	World Trade Organisation
ZB	Zettabyte

Index

Printed in the United States
by Baker & Taylor Publisher Services